Lecture Notes in Control and Information Sciences

Edited by M. Thoma and A. Wyner

Vol. 81: Stochastic Optimization
Proceedings of the International Conference, Kiew, 1984
Edited by I. Arkin, A. Shiraev, R. Wets
X, 754 pages, 1986.

Vol. 82: Analysis and Algorithms of Optimization Problems
Edited by K. Malanowski, K. Mizukami
VIII, 240 pages, 1986.

Vol. 83: Analysis and Optimization of Systems
Proceedings of the Seventh International Conference of Analysis and Optimization of Systems
Antiba, June 26-27, 1986
Edited by A. Bensoussan, J. L. Lions
XVI, 901 pages, 1986.

Vol. 84: System Modelling and Optimization
Proceedings of the 12th IFIP Conference
Budapest, Hungary, September 2–6, 1985
Edited by A. Prékopa, J. Szelezsán, B. Strazicky
XII, 1046 pages, 1986.

Vol. 85: Stochastic Processes in Underwater Acoustics
Edited by Charles R. Baker
V, 205 pages, 1986.

Vol. 86: Time Series and Linear Systems
Edited by Sergio Bittanti
XVII, 243 pages, 1986.

Vol. 87: Recent Advances in System Modelling and Optimization
Proceedings of the IFIP-WG 7/1 Working Conference
Santiago, Chile, August 27-31, 1984
Edited by L. Contesse, R. Correa, A. Weintraub
IV, 199 pages, 1987.

Vol. 88: Bruce A. Francis
A Course in H_∞ Control Theory
XI, 156 pages, 1987.

Vol. 88: Bruce A. Francis
A Course in H_∞ Control Theory
X, 150 pages, 1987.
Corrected - 1st printing 1987

Vol. 89: G. K. H. Pang/A. G. J. McFarlane
An Expert System Approach to Computer-Aided Design of Multivariable Systems
XII, 223 pages, 1987.

Vol. 90: Singular Perturbations and Asymptotic Analysis in Control Systems
Edited by P. Kokotovic, A. Bensoussan, G. Blankenship
VI, 419 pages, 1987.

Vol. 91 Stochastic Modelling and Filtering
Proceedings of the IFIP-WG 7/1 Working Conference
Rome, Italy, Decembre 10-14, 1984
Edited by A. Germani
IV, 209 pages, 1987.

Vol. 92: L. T. Grujić, A. A. Martynyuk, M. Ribbens-Pavella
Large-Scale Systems Stability Under Structural and Singular Perturbations
XV, 366 pages, 1987.

Vol. 93: K. Malanowski
Stability of Solutions to Convex Problems of Optimization
IX, 137 pages, 1987.

Vol. 94: H. Krishna
Computational Complexity of Bilinear Forms
Algebraic Coding Theory and Applications to Digital Communication Systems
XVIII, 166 pages, 1987.

Vol. 95: Optimal Control
Proceedings of the Conference on Optimal Control and Variational Calculus
Oberwolfach, West-Germany, June 15-21, 1986
Edited by R. Bulirsch, A. Miele, J. Stoer and K. H. Well
XII, 321 pages, 1987.

Vol. 96: H. J. Engelbert/W. Schmidt
Stochastic Differential Systems
Proceedings of the IFIP-WG 7/1 Working Conference
Eisenach, GDR, April 6-13, 1986
XII, 381 pages, 1987.

For information about Vols. 1–80 please contact your bookseller or Springer-Verlag.

Lecture Notes in Control and Information Sciences
Edited by M. Thoma and A. Wyner

156

R. P. Hämäläinen,
H. K. Ehtamo (Eds.)

Differential Games – Developments in Modelling and Computation

Proceedings of the Fourth International Symposium on Differential Games and Applications
August 9-10, 1990, Helsinki University of Technology, Finland

Springer-Verlag Berlin Heidelberg GmbH

Editors
Raimo Pertti Hämäläinen
Harri Kalevi Ehtamo

Systems Analysis Laboratory
Helsinki University of Technology
Otakaari 1M, 02150 Espoo
Finland

ISBN 978-3-540-53787-8 ISBN 978-3-540-47105-9 (eBook)
DOI 10.1007/978-3-540-47105-9

Originally published by Springer-Verlag Berlin Heidelberg New York in 1991.

61/3020-543210 Printed on acid-free paper.

PREFACE

The Fourth International Symposium on Differential Games and Applications was organized at the Systems Analysis Laboratory, Helsinki University of Technology, with the participation of the International Federation of Automatic Control (IFAC), Mathematics of Control Committee, and the Institute of Electrical and Electronics Engineers (IEEE), Control Systems Society, on August 9 and 10, 1990.

The program of the meeting had two main tracks one concentrating on classical zero-sum differential games, on computational questions as well as on engineering applications. The other track dealt with differential games in economics and management problems. The present volume includes papers presented in the first area. The second track is covered in a related volume in this Lecture Notes Series entitled "Dynamic Games in Economic Analysis".

In the present volume the papers have been grouped into four sections according to their topic. The first thirteen papers deal with zero-sum differential games of pursuit-evasion type and their applications to aerospace problems. Next there is a section on search and discrete games followed by seven articles on robust controller design and five papers on numerical and hierarchical methods.

Pursuit-evasion games are models of problems where one player, the pursuer, tries to drive the state of the system into a certain domain, target set, against the will of the other player, the evader. However, if both players are symmetric in the sense that they can change roles during the manoeuver, the classical one-target pursuit-evasion approach is not adequate. These kind of pursuit-evasion problems call for formulation and analysis of so called two-target, or combat games. Grimm and Well describe current developments in this area. They give a short summary of classical pursuit-evasion game approaches applied to air combat, and review the research work concerning the qualitative solution of two-target games. Two new approaches to define various winning strategies for two-target games are also discussed. Järmark presents a realistic short-range combat game of two aggressive aircrafts under radar information about the positions of the aircrafts. Missile/target pursuit-evasion problems, with one or several missiles and one target, are treated by Guelman, Levchenkov *et al.*, Breitner *et al.*, Atir, Nikolskii, and Grigorenko. Various dynamic models containing realistic approximations for thrust and drag are considered and both analytical and numerical methods to find the optimal barrier strategies and trajectories are developed. Several types of different singular surfaces are frequently encountered when solving differential games. Once a singular surface is reached the game solution requires the determination of the type of singularity and the continuation of the backwards integration accordingly. A method for constructing singular paths and manifolds in optimal control and differential games is discussed by Melikyan. Bardi and Sartori consider a minimum time problem where the system is affected by an unpredictable disturbance and the controller follows a policy of total risk aversion. The problem is closely related to robust controller design. The main result concerns the case where unknown parameters cause a small perturbation in a deterministic system.

The rapid developments in the field of artificial intelligence during the past decade have produced new approaches and techniques in problem solving. The objective of Shinar is to outline a new way to analyze complex dynamic conflicts based on combining artificial intelligence techniques with the concepts of differential game theory. Meyer and Trigeiro discuss a software implementation aimed at the modelling of an air-to-air combat game of two aircrafts. When the manipulator arms operate in a region of space they can be considered as two players avoiding collision. Ardema and Skowronski formulate the control of a two armed robot system as a game of collision avoidance; i.e., as a game problem of reaching one target set, in the manipulator's state space, while avoiding another, anti target set.

The second section includes four articles dealing with search games and discrete games. Approximation of pursuit-evasion type differential games by discrete-time games is considered by Bardi and Soravia. It is shown that the upper and lower value functions of the discrete-time game converge towards the upper and lower value functions of the corresponding differential game as the length of the time interval goes to zero. Auger considers discrete, two-player search games between an infiltrator and a guard. A possible strategy for finding a solution to the resulting infinite game is to obtain the solutions to the corresponding finite games and to take the limit as time tends to infinity. Laporte *et al.* use lattice methods of solid state physics to solve certain type of discrete pursuit games. The implementation of these methods is considered in the context of naval warfare and submarine search. This approach takes into account ill-defined data and provide interesting information about the possible strategies. A deterministic graphical game is a two-person zero-sum game played on a directed graph with n nodes. Baston and Bostock give an algorithm for solving discounted deterministic graphical games when the discount factor is greater than unity. Such games are called supercounted games.

The next section consists of seven papers devoted to robust controller design and stabilization. H-infinity optimal control has received increasing attention in the past few years. It has been recognized that the L_2-minimization solution to the linear-quadratic game problem is inadequate to guarantee robust stability properties of the closed-loop system. There is a close link between H-infinity minimization and differential game approaches as is pointed out by Başar. In his article the versatility of linear-quadratic differential game theory in the design of optimal disturbance attenuating controllers for linear continuous-time systems subject to unknown disturbances is discussed. The disturbance attenuation problem is solved under different information patterns and also for uncertain nonzero initial state. A new type of H-infinity control problem with a mixed attenuation of disturbances and initial uncertainties for time-varying systems is formulated and solved by Uchida and Fujita. Kojima *et al.* study robust stabilization of time-delay systems by using Riccati type operator equations. Ichikawa derives disturbance attenuating controllers for infinite horizon continuous time tracking problems under the overtaking optimality concept. Engwerda considers the design of robust optimal controllers for tracking asymptotically admissible target paths in dynamic economic models. Botkin *et al.* and Kaitala *et al.* discuss the design of stabilizing controllers for aircraft landing and take-off. Botkin *et al.* consider the aircraft landing problem in the presence of wind disturbance as a minmax differential game problem and develop numerical methods for linearized state dynamics. Kaitala *et al.* design a deterministic controller to stabilize the relative path inclination of the aircraft during its take-off. This performance of the controller is then simulated under different windshear models.

The last five papers deal with numerical methods for dynamical and hierarchical games. Başar and Srikant analyze the iterative computation of Nash policies for two groups of players in a setting where the players' interactions within each group are strong but the interaction between the two groups is weak. It is shown that in the iterative scheme the higher order solutions are Nash equilibria of appropriate quadratic games, even though the original problem may have non-quadratic cost functions. Breton and L'Écuyer propose a finite element computational approach for the numerical solution of zero-sum stochastic games with continuous or very large state and action spaces. Tolwinski presents a new approach to the numerical solution of zero-sum finite-state Markov games. The method is related to the learning algorithms considered in the neural network literature and it is based on the computation of the Bellman-Isaacs value function. Haurie and Zaccour give a qualitative analysis of a stochastic cooperative game model of power exchange between interconnected power utilities. Different approaches to the design of acceptable pricing schemes are proposed and discussed. This is an interesting example of the application of dynamic game theory to real world energy management problem. Finally, Loulou *et al.* simulate energy trade game consisting of linear programs.

The present collection of articles shows that there is a strong new boom of interest in the area of dynamic games. The papers included in this volume give a relatively wide coverage of the new problems in the area. Thus it is likely to serve as a useful source of up to date problem formulations and references for established researchers as well as for new researchers who are entering this field.

Perhaps the most important result of the Symposium was the foundation of the International Society of Dynamic Games (ISDG). The founding members of the society were Tamer Başar, Pierre Bernhard, Alain Haurie, Raimo P. Hämäläinen, Geert Jan Olsder, Josef Shinar, Boleslaw Tolwinski, and Klaus H. Well. The first president of the Society is professor Tamer Başar, and the Society's Headquarters are hosted by professor Raimo P. Hämäläinen at the Systems Analysis Laboratory, Helsinki University of Technology. The society is open to all interested. Its main goal is to improve the exchange of information among the researchers and take responsibility of continuing the biannual Symposium Series on Dynamic Games and Applications.

August 1990,

Raimo P. Hämäläinen   Harri Ehtamo

LIST OF CONTRIBUTING AUTHORS

ARDEMA, M.D.
Santa Clara University, Department of Mechanical Engineering
Santa Clara, CA 95053 — USA

ATIR, H.
RAFAEL, M.O.D.
P.O.Box 2250-(35), Haifa 31 021 — Israel

AUGER, J.M.
University of Southampton, Faculty of Mathematical Studies
Southampton SO9 5NH — UK

BARDI, M.
Università di Padova, Dipartimento di Matematica P. e. A.
Via Belzoni 7, 35131 Padova — Italy

BAŞAR, T.
University of Illinois, Coordinated Science Laboratory
1101 W. Springfield Avenue, Urbana, IL 61801 — USA

BASTON, V.J.
University of Southampton, Department of Mathematics
Southampton SO9 5NH — UK

BERNHARD, P.
INRIA Sophia Antipolis
2004 Route des Lucioles, BP 109, 06561 Valbonne Cedex — France

BOSTOCK, F.A.
University of Southampton, Faculty of Mathematical Studies
Southampton SO9 5NH — UK

BOTKIN, N.D.
USSR Academy of Sciences, Institute of Mathematics and Mechanics
S. Kovalevskaja str. 16, 620066 Sverdlovsk — USSR

BREITNER, M.
Technical University of Munich
Weidenstrasse 14, 8080 Fürstenfeldbruck — Germany

BRETON, M.
GERAD H.E.C.
5255 Decelles, Montréal H3T-1V6 Canada

ENGWERDA, J.
Tilburg University, Department of Econometrics
P.O. Box 90153, 5000 LE Tilburg The Netherlands

FUJITA, M.
Kanazawa University, Department of Electrical and Computer Engineering
Kodatsuno 2-40-20, Kanazawa 920 Japan

GRIGORENKO, N.L.
Moscow State University, Dept. of Comput. Mathematics and Cybernetics
Leninskye Gory, 119899 Moscow USSR

GRIMM, W.
DFVLR Institute for Flight System Dynamics
Oberpfaffenhofen, D-8031 Wessling Germany

GUELMAN, M.
Rafael M.O.D.
P. O. Box 2250, Haifa Israel

HAURIE, A.
University of Geneva, Faculty of SES-COMIN
2 rue de Candolle, 1211 Geneva 4 Switzerland

ICHIKAVA, A.
Shizuoka University, Department of Electrical Engineering
3-5-1 Johoku Hamamatsu 432 Japan

JÄRMARK, B.
SAAB-SCANIA AB
S-58188 Linköping Sweden

KAITALA, V.
Helsinki University of Technology, Systems Analysis Laboratory
Otakaari 1 M, SF-02150 Espoo Finland

KOJIMA, A.
Waseda University, Department of Electrical Engineering
Okubo 3-4-1, Shinjuku, Tokyo 169 Japan

LAPORTE, V.
Thomson Sintra ASM
1 Avenue Aristide Briand, 94117 Arcueil Cedex France

L'ÉCUYER, P.
Université Laval, Dept. d'informatique
Québec G1K 7P4 Canada

LEITMANN, G.
University of California, Department of Mechanical Engineering
6131 Etcheverry Hall, Berkeley, CA 94720 USA

LESSARD, E.
GERAD H.E.C.
5255 Decelles, Montréal H3T 1V6 Canada

LEVCHENKOV, A.Y.
USSR Academy of Sciences, Institute for Problems in Mechanics
Prospect Vernadskogo 101, 117526 Moscow USSR

LOULOU, R.
GERAD H.E.C.
5255 Decelles, Montréal H3T 1V6 Canada

MELIKYAN, A.A.
USSR Academy of Sciences, Institute for Problems in Mechanics
Prospect Vernadskogo 101, 117526 Moscow USSR

MEYER, S.K.
STR Corporation
10805 Parkridge Boulevard, Reston, Virginia 22091 USA

NICOLAS, J.-M.
Thomson Sintra ASM
1 Avenue Aristide Briand, 94117 Arcueil Cedex France

NIKOLSKII, M.S.
USSR Academy of Sciences, Steklov Institute of Mathematics
Vavilova 42, Moscow 117966, GSP-1 USSR

PANDEY, S.
University of California, Department of Mechanical Engineering
6131 Etcheverry Hall, Berkeley, CA 94720 USA

PASHKOV, A.G.
USSR Academy of Sciences, Institute for Problems in Mechanics
Prospect Vernadskogo 101, 117526 Moscow USSR

PATSKO, V.S.
USSR Academy of Sciences, Institute of Mathematics and Mechanics
S. Kovalevskaja str. 16, 620066 Sverdlovsk USSR

PESCH, H.J.
Technical University of Munich
Weidenstrasse 14, 8080 Fürstendfeldbruck Germany

SARTORI, C.
Università di Padova, Dipartimento di Matematica P. e. A.
Via Belzoni 7, 35131 Padova Italy

SAVARD, G.
College Militaire Royal de Saint Jean
Richelain, Québec J0J 1R0 Canada

SHIMEMURA, E.
Waseda University, Department of Electrical Engineering
Okubo 3-4-1, Shinjuku, Tokyo 169 Japan

SHINAR, J.
Technion, Israel Inst. of Technology, Faculty of Aerospace Engineering
32000 Haifa Israel

SKOWRONSKI, J.M.
University of Southern California, Department of Mechanical Engineering
University Park, Los Angeles, CA 90089-1453 USA

SORAVIA, P.
Università di Padova, Dipartimento Matematica P. e. A.
Via Belzoni 7, 35131 Padova Italy

SRIKANT, R.
University of Illinois, Coordinated Science Laboratory
1101 W. Springfield Avenue, Urbana, IL 61801 USA

TEREKHOV, S.D.
USSR Academy of Sciences, Institute for Problems in Mechanics
Prospect Vernadskogo 101, 117526 Moscow USSR

TOLWINSKI, B.
Colorado School of Mines, Department of Mathematics
Golden, CO 80401 USA

TRIGEIRO, W.W.
STR Corporation
10805 Parkridge Boulevard, Reston, Virginia 22091 USA

UCHIDA, K.
Waseda University, Department of Electrical Engineering
Okubo 3-4-1, Shinjuku, Tokyo 169 Japan

WELL, K.H.
DFVLR Institute for Flight System Dynamics
Oberpfaffenhofen, 8031 Wessling Germany

ZACCOUR, G.
GERAD H.E.C.
5255 Decelles, Montréal H3T 1V6 Canada

ZARKH, M.A.
USSR Academy of Sciences, Institute of Mathematics and Mechanics
S. Kovalevskaja str. 16, 620066 Sverdlovsk USSR

TABLE OF CONTENTS

MODELLING AIR COMBAT AS DIFFERENTIAL GAME
RECENT APPROACHES AND FUTURE REQUIREMENTS

W. Grimm, K.H. Well
Institute for Flight Systems Dynamics
German Aerospace Research Establishment DLR
D–8031 Oberpfaffenhofen, FRG

Abstract. The paper deals with differential game models of one–versus–one air combat. After a brief retrospection to classical pursuit–evasion game approaches recently introduced two–target game formulations are addressed. The possible complexity of the qualitative solution is pointed out. It results from the fact that two players need not be able to achieve the same outcome from the same initial position. The paper concentrates on two recently published quantitative approaches to define strategies for the winning player. The "combat game" is an extension of the classical game of degree. "Pareto–optimal security strategies" have their origin in n–person game theory. Finally the design of "reprisal strategies" to exploit nonoptimal behaviour of the opponent is outlined.

Keywords. pursuit–evasion game, two–target game, air combat, firing envelope, winning strategy, reprisal strategy.

A. Introduction: Intention of the Paper

The classical book on differential games by Isaacs [15] contains the first comprehensive presentation of the mathematical modeling of conflicts between two parties. The pursuit–evasion game is suited for all applications, where one player (the "pursuer") tries to drive the state of the system into a certain domain ("target set") against the will of the other player (the "evader"). If both opponents are equivalent in the sense that they have comparable weaponry, the role assignment ("pursuer" and "evader") is inadequate. In this case the "two–target game" (Blaquiere et al. [4]) is the appropriate model: Each player has his own target set, where he wants to move the state of the system before entering the other target set. Analogous to the identification of the capture and evasion zone in the classical "game of kind", one can look for the "qualitative solution". In the presence of two target sets four different outcomes are possible: If one of the target sets is reached first the respective player "wins". If both target sets are reached at the same time the outcome is called a "mutual kill". The game terminates in a "draw" if none of the target sets is reached in a certain time. The qualitative solution is the determination of the best guaranteed outcome for each player in every point of the game space. The qualitative solution does not completely describe the strategies how to achieve this outcome. This is the objective of the quantitative solution. In the classical pursuit–evasion game optimal strategies are defined by the "game of degree". Something similar is needed for the two–target game.

The main intention of this paper is to describe the present research in this direction. The paper begins with a short summary of classical pursuit–evasion game approaches applied to air combat.

Arriving at the two–target game the research work concerning the qualitative solution is briefly reviewed. The central point of the contents are two recently developed quantitative approaches for the two–target game. Open questions of theoretical and practical nature are pointed out. To limit the material under consideration, the scope is confined to deterministic differential game formulations of the one–versus–one medium range air combat problem. Since the research work of the authors concentrates on numerical methods, the chance of numerical solvability is addressed, too. The progress required in this direction is outlined. In particular, real–time algorithms are needed, which can be implemented on onboard computers after being tested on flight simulators.

B. Air Combat Modelled as a Pursuit–Evasion Game

B.1. The scenario under consideration. In the sequel "air combat" means the encounter of two fighter aircraft, where each one is equipped with one missile. It is assumed that "firing envelopes" separate effective and ineffective launch positions in the joint state space. Firing envelopes can be determined if both missile and target stick to given guidance laws after launch. A reasonable evasion strategy of the target needs positional information of the missile – information, which is not always available with present sensor technology. Another stringent assumption is that missiles are not fired outside the firing envelopes. In reality pilots often try to secure their survival by prematurely launching a missile.

B.2. Complete and simplified game formulations. As "complete model" for the game dynamics we consider the point mass model for both aircraft: The state vector consists of the position and the velocity vectors of both vehicles; controls are power setting, angle of attack and bank angle on each side. Realistic functions for drag and thrust are assumed. The firing envelope of the pursuing aircraft takes the role of the terminal manifold. Open–loop results for a few specified initial conditions can be obtained numerically (see Järmark [16], Moritz et al. [25]). In particular, the firing envelope is evaluated with missile/target–simulations.

Simplified game setups allow a global solution. The following ways to simplify the equations of motion are common:

a) restriction to the horizontal plane, either with constant speed (Breakwell, Merz [5]) or variable speed (Järmark et al. [19], Rajan et al. [30], Shinar [31]).

b) restriction to the vertical plane (Guelman et al. [13]).

c) energy approach: In Kelley's "differential turning model" the state of an aircraft consists of specific energy and heading (see Kelley, Lefton [23]).

Simultaneously, the firing envelope is substituted by simpler functions. E.g. game termination is given by a prescribed value of the Euklidean distance (Shinar [31], Rajan et al. [30]). Merz and Hague [24] take account for the fact that medium–range missiles are only effective in a tail–chase situation. Accordingly, game is over as soon as one aircraft is just ahead of the other one and relative heading is sufficiently small. Additionally, Kelley and Lefton [23] required a certain energy advantage of the pursuer.

B.3. Numerical methods to compute open–loop solutions. Similar as in optimal control there are "direct" and "indirect" solution methods. "Direct" means to affect a control history directly by varying a finite set of defining parameters. Moritz et al. [25] adapted the direct approach to the solution of zero sum games. "Indirect" means to solve the multipoint boundary value problem (MPBVP) constituted by the necessary conditions of optimality (Isaacs' "main equation", transversality conditions etc.). "Differential Dynamic Programming" (DDP, Järmark [18]) and the multiple shooting technique implemented in the MPBVP–solver BNDSCO (Bulirsch [6], Oberle, Grimm [26]) are indirect methods. Differential game solutions obtained with BNDSCO are published by Shinar et al. [33]. Indirect methods require an a priori identification of singular surfaces – an unsolved problem to date.

B.4. Techniques to develop real time algorithms. The basic idea is to approximate the optimal closed–loop control by periodical updates of the open–loop control for the current state. However, this approach becomes invalid if some of the notorious "singular surfaces" occur. Singular perturbation technique is a tool to derive simple guidance laws (Shinar [31]). More sophisticated methods stake on the rapid development of modern computers, concerning both storage and computation speed. Large memory is required if the global solution is stored as an "extremal trajectory map" (Rajan, Ardema [29]). The only task of the guidance algorithm is to search for the trajectory fitting to the current state. High computation speed is needed for the real time application of DDP (Järmark [17]).

C. The Two–Target Game – Definition and Qualitative Solution

C.1. The transition single–target/two–target game. It was realized early that the role assignment of pursuit–evasion games is not adequate for two comparably equipped opponents in air combat. Although the two–target game was theoretically prepared rather early by Blaquiere et al. [4], the idea was not taken up immediately. Instead, there are attempts to give an equivalent position to each player in the framework of single target games. Merz and Hague [24] exchanged the roles of pursuer and evader to determine the capture zone of each player. Järmark [16] and Moritz et al. [25] define the difference of the missile/target distances as the performance index in a zero sum game. The distance is measured at the moment when the closing speed between the aircraft and the attacking missile takes a prescribed value. Thus, two terminal conditions are imposed, one for each player – a game setup which does not belong to the categories defined in standard literature (Isaacs [15], Blaquiere et al. [4]).

C.2. Definition of the two–target game. In a two–target game two players 1 and 2 are represented by their controls $u \in U \subset \mathbb{R}^{m_1}$, $v \in V \subset \mathbb{R}^{m_2}$, where U and V are compact sets. The state vector $x \in \mathbb{R}^n$ is the solution of the ODE–system

$$\dot{x} = f(x, u, v)\ . \tag{1}$$

Player i tries to drive x into his own target set which is mathematically expressed as

$$T_i := \{ x \in \mathbb{R}^n \mid d_i(x) \leq 0 \}\ . \tag{2}$$

Game termination is defined as the earliest instant when $x \in T_1 \cup T_2$. The game is terminated by the time limit T if $x \notin T_1 \cup T_2$ for $t \leq T$:

$$t_f = \min \{ t \mid x(t) \in T_1 \cup T_2 \text{ or } t = T \} \tag{3}$$

The following outcomes are possible:

1) Player 1 wins the game, i.e. $x(t_f) \in T_1 \setminus T_2$.
2) The game terminates in a "draw", i.e. none of the target sets is reached within $[0, T]$: $x(t_f) \notin T_1 \cup T_2 \quad \Rightarrow \quad t_f = T$.
3) The game ends in a "mutual kill", i.e. both target sets are reached at the same time $t \le T$: $x(t_f) \in T_1 \cap T_2$.
4) Player 2 wins the game, i.e. $x(t_f) \in T_2 \setminus T_1$.

Player 1's preference ordering is 1–2–3–4. Player 2's preference ordering is just the reversal. Note that player 1 prefers a draw over a mutual kill whereas player 2 does the opposite. A different definition of the preference orderings would lead to a different game solution as recognized by Kelley, Lefton [22] and Prasad, Ghose [27].

The following assumption serves to ensure the existence of the trajectory and some quantities defined later on. It replaces the impractical notion of an "invariant game space".

Assumption: The initial state of the game is taken from a compact domain $D \subset \mathbb{R}^n$. It is assumed that for each $x_0 \in D$ and piecewise continuous admissible control functions u(t), v(t), $0 \le t \le T$, (1) has a unique solution. Moreover, all the solutions are assumed to be within a compact domain $F \subset \mathbb{R}^n$ with $D \subset F$. d_1 and d_2 are C^1–functions on F. f is a continuous function on $F \times U \times V$.

In air combat x consists of the position and velocity vectors of both aircraft (= players 1 and 2). u and v comprise their aerodynamic and thrust controls. The firing envelopes take the roles of the T_i. The resulting two–target game is a more realistic model for air combat than a pursuit–evasion game with preassigned roles (section B.2). But it also suffers from the assumptions stated in B.1.. Moreover, a pilot captured in the firing envelope of his opponent may be able to achieve a mutual kill by launching his missile before he is actually hit. This possibility is also ignored in the game formulation above.

C.3. Definition of the qualitative solution. Let $\gamma_1 : F \times \mathbb{R} \to U$ and $\gamma_2 : F \times \mathbb{R} \to V$ denote feedback strategies of player 1 and 2, respectively. The triple $(x_o, \gamma_1, \gamma_2)$ uniquely determines the outcome of the game. Let s denote the mapping which maps $(x_o, \gamma_1, \gamma_2)$ into the set of the numbers $\{1, 2, 3, 4\}$ which indicate the outcome of the game. According to his preference ordering player 1 wants to minimize the number of the outcome. The worst case on this attempt is given by

$$b_1(x_o) := \min_{\gamma_1} \max_{\gamma_2} s(x_o, \gamma_1, \gamma_2) . \tag{4}$$

$b_1(x_o)$ is called the "best secured outcome" for player 1. The analogon for player 2 is

$$b_2(x_o) := \max_{\gamma_2} \min_{\gamma_1} s(x_o, \gamma_1, \gamma_2) . \tag{5}$$

In general $b_1(x_o) \geq b_2(x_o)$ holds. Following this relation ten different combinations (b_1, b_2) are possible:

b_1	possible values of b_2
1	1
2	1, 2
3	1, 2, 3
4	1, 2, 3, 4

The qualitative solution consists in the determination of (b_1, b_2) for each $x_o \in D$. The sets

$$W_1 := \{ x_o \in D \mid b_1(x_o) = b_2(x_o) = 1 \} , \qquad (6)$$
$$W_2 := \{ x_o \in D \mid b_1(x_o) = b_2(x_o) = 4 \}$$

are the "winning zones" of players 1 and 2, respectively. There is complete uncertainty about the actual outcome on both sides, if the initial state is taken from

$$U := \{ x_o \in D \mid b_1(x_o) = 4, b_2(x_o) = 1 \} . \qquad (7)$$

In U each player can be captured by playing any strategy, but only if his opponent adapts his reaction to this strategy. In other words: In W_i (U) player i can win without (only with) a priori knowlegde of his opponent's strategy.

C.4. Examples for qualitative game solutions. To obtain the winning zones one starts with the solution of two classical pursuit–evasion games with exchanged roles: the games with player i as pursuer, player $j \neq i$ as evader and ∂T_i as terminal manifold. The further procedure was formalized by Shinar and Davidovitz [32] and illustrated by an air combat application (Davidovitz, Shinar [7]) with T_i modeled as the "no–escape firing envelope" of aircraft (=player) i. Other examples are two–target game versions of the "game of two cars" and the "homicidal chauffeur" (Getz, Pachter [9]). A completely different approach to identify at least parts of W_i is to check certain sufficient conditions (Getz, Leitmann [8], Skowronski, Stonier [35]).

D. The Two–Target Game – Quantitative Approaches

The qualitative solution only determines the winning strategy on the boundary surface of a winning zone. The intention of the following quantitative approaches is to define a winning strategy in the interior of a winning zone.

D.1. The combat game. In the classical pursuit–evasion game optimal strategies in the pursuer's capture zone are defined by the "game of degree". The idea is extended to the winning zone W_1 of a two–target game to determine an optimal (feedback) strategy $\gamma_1 : F \times \mathbb{R} \longrightarrow U$ of the winning player (No. 1).

Let γ_1 be the saddle point of the cost functional

$$J(\gamma_1, \gamma_2) = \varphi(x_f, t_f) + \int_0^{t_f} L(x, u, v, t)\, dt \tag{8}$$

where $u(t) = \gamma_1(x(t), t)$, $v(t) = \gamma_2(x(t), t)$ and game termination is defined by $x \in T_1$. Player 1 is the minimizer and player 2 the maximizer of J. Unlike the ordinary game of degree admissible strategies γ_1 are restricted to those which prevent x from moving into T_2:

$$d_2(x(t)) \geq 0 \quad \forall\ t \in [0, t_f] \tag{9}$$

This "event–constrained" differential game is called "1–game" in the combat game concept introduced by Ardema et al. [2]. Similarly, the "2–game" is played in W_2. An example for a combat game solution is given by Ardema et al. [1]. Heymann et al. [14] proposed an interior penalty method to approximate the solution of an event–constrained differential game.

Along a subarc where (9) is active, i.e. where (9) holds with "=", differentiation yields

$$\frac{\partial d_2}{\partial x}(x) \cdot f(x, u, v) = 0 . \tag{10}$$

As can be seen from (10) the "event constraint" couples both players but only player 1 is responsible for it. Player 2 can choose any value for v. Player 1 must react such that (10) is satisfied. The exact determination of u and v in this case is yet unknown. One (trivial) condition is that player 2 must not be able to pull the joint state into his target set:

$$\max_u \min_v \frac{\partial d_2}{\partial x}(x) \cdot f(x, u, v) \geq 0 \tag{11}$$

In other words: The trajectory must not cross the "usable part" on player 2's target set.

As long as (9) does not become active an ordinary game of degree is solved. The usual procedure is to solve the boundary value problem for the open–loop solution. Then, however, one should be cautious about the solution. First, it can happen that a solution starts from outside the capture zone of player 1 (the capture zone with respect to T_1 in the sense of the classical "game of kind"; W_1 is a subset of the capture zone). This pathology is depicted as case 1 in Fig. 1. There is an example in the book of Basar and Olsder [3], who call the phenomenon "termination dilemma". It is due to the fact that the boundary value problem is only made up of necessary (not sufficient) conditions for the game solution. Cases 2 and 3 in Fig. 1 are two–target game specific. As long as x_0 is chosen within the capture zone of player 1 there will be a solution terminating on T_1. Even if it does not touch T_2 it is possible that it starts within or crosses W_2.

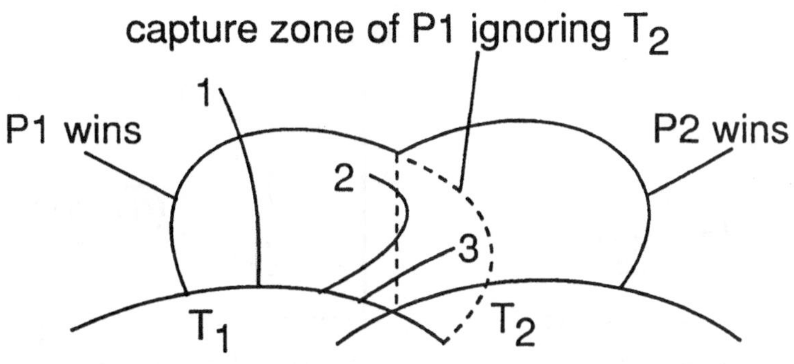

Fig. 1: Parasitic solutions of the combat game.

To exclude cases 2 and 3 player 2's target set T_2 should enter the game via the performance index. Choosing

$$J = -d_2(x(t_f)) \quad \text{or (better but more complicated)} \tag{12a}$$

$$J = \max\{-d_2(x(t)) \mid 0 \le t \le t_f\} \tag{12b}$$

and omitting the constraint (9) one obtains a solution terminating on T_1 as long as x_o is taken from the capture zone of player 1. $J < 0$ is equivalent with $x_o \in W_1$. If $J \ge 0$ the original two-target game would terminate on T_2 prior to $t = t_f$.

D.2. Pareto–Optimal security strategies are based on the definition of the "security levels"

$$\bar{d}_1(x_o, \gamma_1) := \sup_{\gamma_2} d_1(x_f), \tag{13a}$$

$$\underline{d}_2(x_o, \gamma_1) := \inf_{\gamma_2} d_2(x_f), \tag{13b}$$

which express the chance of capture and survival, respectively. Note that both $\bar{d}_1$ and $\underline{d}_2$ exist because of the assumption in section C.2.. Similarly, security levels $\underline{d}_1$ and $\bar{d}_2$ can be defined for player 2. In Fig. 2 the security levels can be found graphically by considering the set Γ of those pairs $(d_1(x_f), d_2(x_f))$, which result on applying all possible strategies γ_2.

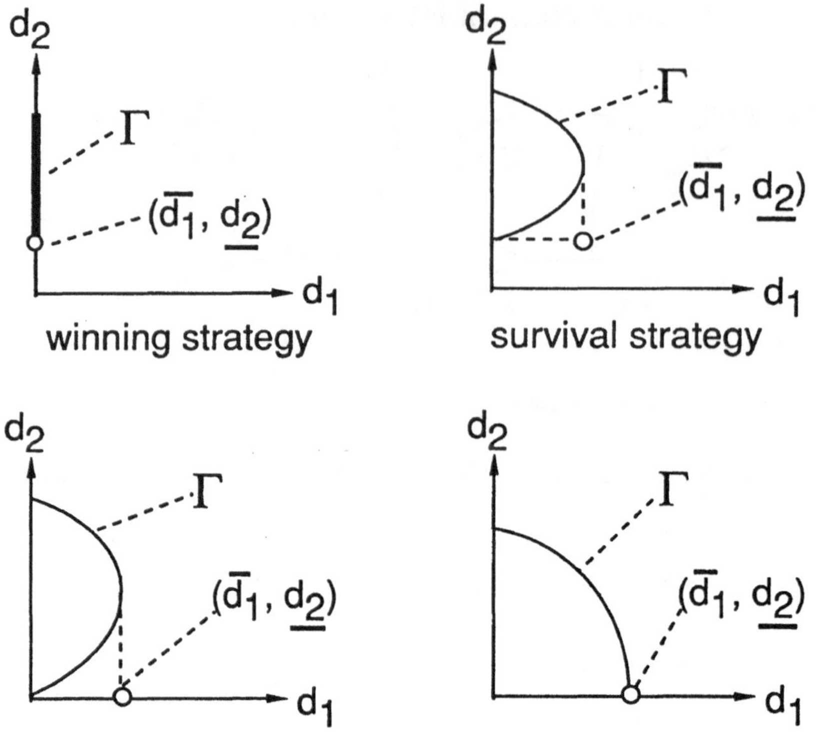

Fig. 2: The security levels for different strategies.

The first example is a winning strategy characterized by $\bar{d}_1(x_0, \gamma_1) = 0$ and $\underline{d}_2(x_0, \gamma_1) > 0$. In this case one can conclude $x_0 \in W_1$, $b_1(x_0) = 1$. In the second example player 1 can enforce a draw because of $\underline{d}_2(x_0, \gamma_1) > 0$ ("survival strategy") $\Rightarrow b_1(x_0) \leq 2$. In the third example the worst possible outcome on applying γ_1 is a mutual kill $\Rightarrow b_1(x_0) \leq 3$. In the fourth example capture by the opponent can occur ($b_1(x_0) \leq 4$). Note that the last two types cannot be distinguished by the security levels. In both cases we have $\bar{d}_1(x_0, \gamma_1) > 0$ and $\underline{d}_2(x_0, \gamma_1) = 0$.

"Pareto optimal security strategies" (POSS; see Ghose, Prasad [11]) are defined as the set of minimal elements of

$$\bar{J}(x_0, \gamma_1) := (\bar{d}_1(x_0, \gamma_1), -\underline{d}_2(x_0, \gamma_1)) . \tag{14}$$

"Minimal" refers to the following ordinal relation on the elements of $\mathbb{R}^2$:

$$(a_1, b_1) \leq (a_2, b_2) :\Leftrightarrow a_1 \leq b_1 \text{ and } a_2 \leq b_2 \tag{15}$$

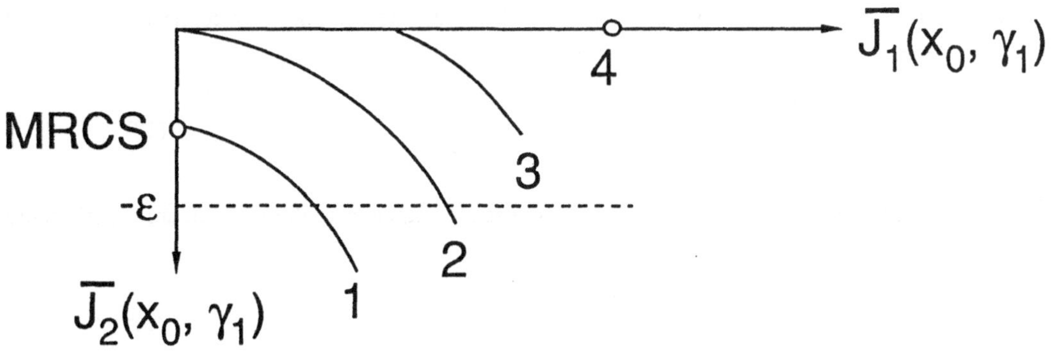

Fig. 3: $\bar{J}$–image of the POSS–set of player 1 for different initial conditions x_0.

Fig. 3 shows the $\bar{J}$–image of the POSS–set for different initial states x_0. Obviously, $x_0 \in W_1$ iff there is a POSS with $\bar{d}_1(x_0, \gamma_1) = 0$ and $\underline{d}_2(x_0, \gamma_1) > 0$. This situation is depicted in case 1. The winning POSS is the leftmost point of the POSS–set. The corresponding strategy γ_1 is not just any winning strategy but that with the largest "safety margin" $\underline{d}_2$. Therefore, γ_1 is called the "minimum risk capture strategy" (MRCS; see Ghose, Prasad [10]). γ_1 is the solution of the "strategy optimization problem"

$$\max_{\gamma_1} \underline{d}_2(x_o, \gamma_1) \quad \text{subject to} \quad \bar{d}_1(x_o, \gamma_1) = 0 . \tag{16}$$

In cases 2 and 3 there are POSS's with $\underline{d}_2(x_0, \gamma_1) > 0$, i.e. player 1 can achieve a draw $\Rightarrow$ $b_1(x_0) = 2$. In case 4, where the POSS–set only consists of one element, one only can conclude $b_1(x_0) \in \{3, 4\}$. The two cases cannot be distinguished by the POSS–set.

As long as a certain distance ε to the adversary target set is guaranteed the "closest approach survival strategy" (CASS) is defined:

$$\min_{\gamma_1} \bar{d}_1(x_o, \gamma_1) \quad \text{subject to} \quad \underline{d}_2(x_o, \gamma_1) \geq \varepsilon \tag{17}$$

In Fig. 3 ε is chosen such that in case 1 the CASS is not identical with the MRCS. In case 2 (17) has a solution, whereas in case 3 the admissible strategy set is empty.

The combat game (8), (9) looks for a saddle point in the strategy space. The strategy optimization problems (16), (17) are minmax problems (insert the definition of the security levels to see the minmax character). This is the main difference between this concept and the combat game. If the MRCS accidentally has saddle point property, then it coincides with the combat game solution with performance index (12a).

The "strategy optimization problems" (16), (17) are exclusively designed for closed–loop controls. Numerical examples for open–loop controls u(t) for player 1 are given by Grimm, Prasad, Well [12]. In this case the same control history must satisfy the optimality criterion and the restriction $\underline{d}_2 \geq \varepsilon$ (in the case of the CASS). Applying the optimal feedback strategy $\gamma_1(x, t)$ there are different control histories $u_1(t)$ and $u_2(t)$ in the optimality criterion and the constraint $\underline{d}_2 \geq \varepsilon$. The problem divides into two independent subproblems: the search for u_1 and u_2. The only connection is given by the initial point:

$$\gamma_1(x_o, t_o) = u_1(t_o) = u_2(t_o) \tag{18}$$

To couple u_1 and u_2 one might restrict the set of admissible strategies to those which are Lipschitz–continuous in the following sense:

$$\| \gamma_1(x_1, t) - \gamma_1(x_2, t) \| \leq C \| x_1 - x_2 \| \tag{19}$$

Varying C in (19) between 0 and ∞ can be regarded as a transition from open–loop to closed–loop controls.

D.3. Reprisal strategies. Generally, the initial state of an air combat scenario is such that none of the pilots is in a winning position in the sense defined above. The goal of the prelaunch maneuver is to reach one's winning zone first. For instance, player 1 could update his CASS (if it exists) for the current state thus retaining the chance of survival. If player 2 plays nonoptimally player 1 might reach his winning zone first. In this case nonoptimal behaviour of player 2 would indirectly be fed back via the state vector. "Reprisal strategies" directly include the current control action of the enemy. Examples are given by Kelley et al. [20, 21], who also introduced the name. A simple idea is illustrated in Fig. 4.

$$T_1: d_1(x, t) \leq 0$$
$$T_2: d_2(x, t) \leq 0$$

$$\text{minimize } \tan\alpha = \frac{\dot{d}_2}{\dot{d}_1}$$

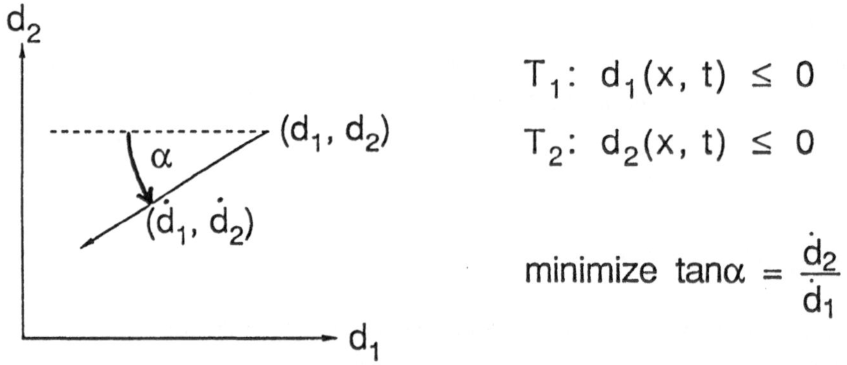

Fig. 4: A simple reprisal strategy: the "ratio criterion".

As long as none of the target sets is reached the pair $(d_1(x(t)), d_2(x(t)))$ moves in the first quadrant of the (d_1, d_2)–plane. Since player 1 wants to drive d_1 to 0 while keeping $d_2 > 0$ it is reasonable for him to minimize the angle α at each time. This is equivalent to minimizing the ratio $\dot{d}_1/\dot{d}_2$ ("ratio criterion"). Usually both u and v explicitly appear in the ratio. One could consider the worst case by solving a minmax–problem, but this would not be a true reprisal strategy. A reprisal strategy results if the actual value of v is put in. In practice one must replace v by an estimate obtained from the observation of the opponent's motion. The idea was applied to air combat by Prasad et al. [28].

A more daring concept needs a prediction of the opponent's complete strategy γ_2. If γ_2 is fixed the "worst case" numbers $\bar{d}_1$ and $\underline{d}_2$ in $\bar{J}$ can be replaced by the numbers actually achieved by the particular γ_2. One can define an analogon to the POSS: "Pareto optimal response strategies" (PORS) associated with the particular γ_2. If player 2 really plays the assumed γ_2 the PORS take the greatest possible advantage. Otherwise safety is not guaranteed at all and x may quickly move into W_2.

Altogether the search for prelaunch strategies seems to be a dilemma between maintaining safety and improving one's chance of success. The risk increases with the extent of assumptions made about the opponent's future actions.

E. Future Requirements

- Both quantitative approaches, the "combat game" and the POSS, need further theoretical development. The combat game leads to state constrained differential games, which received little attention so far. In particular, state constraints which couple both players have not yet been examined. The result depends on the decision, which player is responsible for satisfying the constraint. Another question is in how far the existence of an open–loop game solution indicates a winning position.

  The strategy optimization problems need to be prepared such that they appear as minmax problems for open–loop controls. A hint is given in Equ. (19).
  Necessary conditions for optimality in both concepts are required.
- In the near future the above concepts cannot be applied to realistically modelled air combat problems. The mathematical and numerical tools are missing. Instead, the concepts should be simplified without losing the essential features of air combat with missiles. For instance, the "security levels" could be replaced by estimates based on "representative" offensive and defensive strategies (see Grimm et al. [12]).
- Via parametrization of control functions optimal control problems can be reduced to finite dimensional optimization problems. For this problem class efficient algorithms exist. Similarly, a differential game can be converted into a finite dimensional saddle point problem. However, algorithms for this problem type are missing yet.
- The design of "reprisal strategies" (section D.3.) should be continued. Unlike rigorous theoretical concepts one can start with simple heuristic ideas and gradually increase complexity. Useful results should be available in near future.
- To get an idea about size and location of winning zones in air combat, some winning positions should be determined in an off–line study. Points at issue like the alternative between low speed/high maneuverability and high speed/low maneuverability could be clarified.
- AI is not necessarily a competing concept on analyzing air combat. AI is useful to solve problems of fuzzy nature. For precisely defined quantitative problems DG–concepts should be preferred. Combinations of both like the prototype presented by Shinar et al. [34] are promising.

F. Conclusions

The classical pursuit–evasion game is not suited to model a conflict between two equivalent opponents in air combat. The two–target game with the firing envelopes as target sets is a more realistic mathematical formulation. Yet some practical aspects are missing like the option to fire a missile outside its firing envelope.

Concerning the qualitative solution the procedure to obtain the winning zones is well developed theoretically. Its application is restricted to simple, low dimensional problems.

There are different concepts to define a quantitative solution of a two–target game. The "event-–constrained" differential game is essentially a classical game of degree augmented by a constraint to avoid the target set of the other player. "Pareto–optimal security strategies" are another quantitative approach coming from n–person game theory. Both approaches are still in the definition phase. Theory is not yet complete.

Parameter optimization is recommended to solve the saddle point or minmax problems arising in the quantitative concepts above. However, efficient algorithms for finite dimensional saddle point problems are not yet developed.

As long as an aircraft has not reached a favourable position application of game theory makes no sense. In this phase heuristic reprisal strategies are needed, which take advantage of strategic errors of the opponent.

Altogether differential games seem to be the appropriate framework for a limited section of military aircraft guidance. But it takes many more years of research until the first results can be implemented.

Acknowledgement

The authors are indebted to Prof. U.R. Prasad (Indian Institute of Science, Bangalore, India) for his valuable comments on section D of this paper.

Literature

[1] Ardema, M.D., Heymann, M., Rajan, N., "Analysis of a Combat Problem: The Turret Game", Journal of Optimization Theory and Applications, Vol. 54, No. 1, July 1987.

[2] Ardema, M.D., Heymann, M., Rajan, N., "Combat Games", Journal of Optimization Theory and Applications, Vol. 46, No. 4, pp. 391 – 398, 1985.

[3] Basar, T., Olsder, G.J., "Dynamic Noncooperative Game Theory", Academic Press, 1982.

[4] Blaquiere, A., Gerard, F., Leitmann, G., "Quantitative and Qualitative Games", Academic Press, 1969.

[5] Breakwell, J.V., Merz, A.W., "Minimum Required Capture Radius in a Coplanar Model of the Aerial Combat Problem", AIAA Journal, Vol. 15, No. 8, August 1977, pp. 1089 – 1094.

[6] Bulirsch, R., "Die Mehrzielmethode zur numerischen Lösung von nichtlinearen Randwertproblemen und Aufgaben der optimalen Steuerung", Report of the Carl–Cranz–Gesellschaft, Heidelberg, 1971.
Reprint: Munich University of Tehnology, Mathematical Institute, Munich, 1985.

[7] Davidovitz, A., Shinar, J., "Two–Target Game Model of an Air Combat with Fire–and–Forget All–Aspect Missiles", Journal of Optimization Theory and Applications, Vol. 63, No. 2, Nov. 1989, pp. 133 – 166.

[8] Getz, W.M., Leitmann, G., "Qualitative Differential Games with Two Targets", Journal of Mathematical Analysis and Applications, Vol. 68, pp. 421 – 430, 1979.

[9] Getz, W.M., Pachter, M., "Two–Target Pursuit–Evasion Differential Games in the Plane", Journal of Optimization Theory and Applications, Vol. 34, pp. 383–403, 1981.

[10] Ghose, D., Prasad, U.R., "Analysis of Security Strategies for a Two–Traget Game", Proceedings of the AIAA Guidance, Navigation & Control Conference, Boston, Ma., August 14 – 16, 1989.

[11] Ghose, D., Prasad, U.R., "Solution Concepts in Two–Person Multicriteria Games", Journal of Optimization Theory and Applications, Vol. 63, No. 2, pp. 167 – 189, Nov. 1989.

[12] Grimm, W., Prasad, U.R., Well, K.H., "Open–Loop Guidance for Prelaunch Maneuvering in Medium Range Air Combat", Proc. of the 27th IEEE Conference on Decision and Control, Austin, Texas, pp. 1442 – 1447, Dec. 1988.

[13] Guelman, M., Shinar, J., Green, A., "Qualitative Study of a Planar Pursuit Evasion Game in the Atmosphere", Proc. of the AIAA Guidance, Navigation and Control Conference, Minneapolis, Minnesota, August 15 – 17, 1988.

[14] Heymann, M., Rajan, N., Ardema, M.D., "On Optimal Strategies in Event–Constrained Differential Games", Proc. of the 24th IEEE Conference on Decision and Control, Fort Lauderdale, Florida, pp. 1115 – 1118, Dec. 1985.

[15] Isaacs, R., "Differential Games", Wiley, New York, 1965.

[16] Järmark, B.S.A., "A Missile Duel between Two Aircraft", Journal of Guidance and Control, Vol. 8, No. 4, pp. 508 – 513.

[17] Järmark, B.S.A., "Closed–Loop Controls for Pursuit–Evasion Problems Between Two Aircraft", Presented at the Second International Symposium on Differential Game Applications, Williamsburg, Virginia, USA, August 21 – 22, 1986.

[18] Järmark, B.S.A., "Differential Dynamic Programming in Differential Games", Control and Dynamic Systems, Vol. 17, edited by C.T. Leondes, Academic Press, New York, 1981.

[19] Järmark, B.S.A., Merz, A.W., Breakwell, J.V., "The Variable–Speed Tail–Chase Aerial Combat Problem", Journal of Guidance and Control, Vol. 4, May – June 1981, pp. 323 – 328.

[20] Kelley, H.J., "A Threat–Reciprocity Concept for Pursuit/Evasion", In: Differential Games and Control Theory II, Roxin, E.O., Liu, P.–T., Sternberg, R.L., Eds., Lecture Notes in Pure and Applied Mathematics, Vol. 30. New York: Marcel Dekker, 1977.

[21] Kelley, H.J., Cliff, E.M., Lefton, L., "Reprisal Strategies in Pursuit Games", Journal of Guidance and Control, Vol. 3, pp. 257 – 260.

[22] Kelley, H.J., Lefton, L., "A Preference–Ordered Discrete–Gaming Approach to Air–Combat Analysis", IEEE Transactions on Automatic Control, Vol. AC–23, No. 4, August 1978.

[23] Kelley, H.J., Lefton, L., "Calculation of Differential–Turning Barrier Surfaces", Journal of Spacecraft and Rockets, Vol. 14, No. 2, Feb. 1977, pp. 87 – 95.

[24] Merz, A.W., Hague, D.S., "Coplanar Tail–Chase Aerial Combat as a Differential Game", AIAA Journal, Vol. 15, No. 10, October 1977.

[25] Moritz, K., Polis, R., Well, K.H., "Pursuit–Evasion in Medium Range Air Combat Scenarios", Comp. Math. Appl., Vol. 13, No. 1/3, pp. 167 – 180, 1987.

[26] Oberle, H.J., Grimm, W., "BNDSCO – A Program for the Numerical Solution of Optimal Control Problems", Internal Report No. 515–89/22, Institute for Flight Systems Dynamics, DLR, Oberpfaffenhofen, FRG, 1989.

[27] Prasad, U.R., Ghose, D., "Bicriterion Differential Games with Qualitative Outcomes", Private communication, 1990.

[28] Prasad, U.R., Grimm, W., Berger, E.G., "A Feedback Guidance for Prelaunch Maneuvering in Medium–Range Air Comabt with Missiles", Proc. III. International Symposium on Differential Games and Applications, Antibes, France, June 16 – 17, 1988. Published in Springer Lecture Notes in Control and Information Sciences, Vol. 119, eds. T.S. Basar/P. Bernhard, pp. 86 – 96.

[29] Rajan, N., Ardema, M.D., "Computation of Optimal Feedback Strategies for Interception in a Horizontal Plane", Journal of Guidance and Control, Vol. 7, No. 5, Sept.–Oct. 1984, pp. 627 – 629.

[30] Rajan, N., Prasad, U.R., Rao, N.J., "Pursuit–Evasion of Two Aircraft in a Horizontal Plane", Journal of Guidance and Control, Vol. 3, No. 3, May–June 1980, pp. 261 – 267.

[31] Shinar, J., "Validation of Zero–Order Feedback Strategies for Medium–Range Air–to–Air Interception in a Horizontal Plane", NASA Technical Memorandum 84237, NASA Ames Research Center, Moffett Field, California 94035, USA, April 1982.

[32] Shinar, J., Davidovitz, A., "Unified Approach for Two–Target Game Analysis", Proc. of the 10th IFAC World Congress on Automatic Control, Vol. VIII, Munich, FRG, 1987, pp. 72 – 77.

[33] Shinar, J., Negrin, M., Well, K.H., Berger, E.G., "Comparison between the Exact and an Approximate Feedback Solution for Medium Range Interception Problems", Proc. of the Joint Automatic Control Conference, Charlottesville, Va., June 17 – 19, 1981.

[34] Shinar. J., Siegel, A.W., Gold, Y.I., "A Medium–Range Air Combat Game Solution by a Pilot Advisory System", Proceedings of the AIAA Guidance, Navigation and Control Conference, Boston, Ma., August 14 – 16, 1989.

[35] Skowronski, J.M., Stonier, R.J., "The Barrier in a Pursuit–Evasion Game with Two Targets", Comput. Math. Applic., Vol. 13, No. 1–3, pp. 37–45, 1987.

Air Combat Aiming with a Game in the Lag of the Radar Data.

Bernt Järmark
Aircraft division, SAAB-SCANIA AB Linköping, Sweden

Abstract: A particular incompleteness in radar information is utilized and manipulated in a head-on gunnery duel between two fighter aircraft. It might be possible for one particular aircraft to put out a decoy of him self and by this deceive the other one. One aircraft assumes to have an extra degree of freedom to control its aiming and trajectory, whereas the other one does not have this facility. His trajectory is determined by his aiming. Due to the circumstances there are several game formulations. In a couple of this formulations there is a great advantage of having an aircraft with this facility. Still, it might be room for discussions, what sort of game will occur in real life.

1. Introduction

In order to accomplish an air-to-air gunnery in a head on encounter, between two fighter aircraft, the aiming and shooting have to be accomplished by the onboard systems. Both want to have a gunning as long time as possible at the same time as they do not want to collide. The process is of a very short duration, four to six seconds. The calculations of the aiming control and break away decision are based on information from the radar measurement of the hostile aircraft including the filter calculation. Besides the fact that the radar signal is exposed to noise the information can also be wrong to some extent due to lags in the Kalman filter used. This incompleteness can be used in a couple of ways. It might be possible to jam aiming for the opponent or he can be deluded to break aiming earlier than he has to. The game definition is far from clear. The study in this paper is based on the filter limitations, which is a new approach in this class of studies.

In an earlier study, [1] a passive target (flying on a straight line) is assumed. Perfect information and no lagging components besides the differential equations are also assumed. A three dimensional pointmass model is used and the process is optimized. An extension is found in [2] where the target is also maneuvering. A short period dynamic model in the vertical plane is assumed. The restriction due to the radar tracking is indicated in [3], where a very simple filter model of the opponent´s states is simulated. The models used in this paper are simplified in purpose to demonstrate the game mechanism, as there is a possibility to calculate optimal controls. The models for the aircraft are pointmasses, moving in a plane with constant velocity and controlled by the crossacceleration. One aircraft has decoupling facilities, using a small pitch deflection added to the ordinary angle of attack to achieve aiming besides to some extent controlling the trajectory. Behind the measured data of the opponents position, velocity and acceleration there is a simple Kalman filter assuming cartesian coordinates of the target.

In the game considered we will assume both aircraft to be aggressive. The decoupled aircraft is supposed to aim as long and good as possible at the same time doing a maneuver within the limits of his decoupling abilities with the purpose to give the opponent misleading information of his future position. The trajectory of the aircraft with no decoupled facility is determined by his aiming. Then his pass distance can only be produced by breaking aiming and making an avoidance maneuver. His alternative is to give up aiming and concentrate on jamming and collision avoidance.

The purpose of this paper is first to give an illustration of a typical aerial target tracking implementation of the Kalman filter. Then maximizing the difference of the real final position at the pass and what the opponent predicts the final position to be. An important issue is to put attention to the game mechanism when both aircraft are aggressive, including decision of the turn direction.

2. Kalman Filter

With the purpose to have a simple optimization performed we assume simple models of the aircraft, also the game is taking place in the vertical plane (no gravity). For each aircraft we have,

$$\dot{z} = v_z \tag{1}$$

$$\dot{v}_z = Acc \tag{2}$$

A simple Kalman filter is then,

$$\dot{z}_e = v_{ze} + K_1 \cdot (z_r - z_e) \tag{3}$$

$$\dot{v}_{ze} = Acc_e + K_2 \cdot (z_r - z_e) \tag{4}$$

$$\dot{Acc}_e = -B \cdot Acc_e + K_3 \cdot (z_r - z_e) \tag{5}$$

$$z_r = z + w, \quad w = N(0, \sigma) \tag{6}$$

where z is the position, v_z is the velocity in z-direction (both initially set to zero), Acc is the crossacceleration, subindex e refer to estimated variables and w is measurement (white) noise, emanating from e.g. glinting and signal noise ratio. The Kalman gains are K_i; i=1, 2 or 3. See reference 4 for details of calculating the Kalman filter. The measurement noise is assumed to be stationary. In practice this is not the case, since, at closer distances the glint is dominating and further out the signal noise ratio will dominates the measurement noise, which constitutes a nonstationary process.

As the maneuver, Acc, is not known by the measuring aircraft, a model of the maneuver must be used in the filter. If we assume the target is switching from maximum turn in one direction to maximum turn in the other direction, $\pm Acc_{max}$, it will possess a Poisson process. An approximation of this is a Marcov model, Eqs.(1,2,7), where the input is a white noise, v. This is easier to handle than the Poisson process.

$$\dot{Acc} = B\cdot(-Acc + v), \qquad v = N(0, Acc_{max}) \tag{7}$$

There is a theoretic background in determining the fictitious parameter B from the standard deviations σ and Acc_{max}. The filter used in this paper as well as in most aircraft implemented filters are theoretically simple and the value of B is chosen empirically. The Kalman gains, K_i, are calculated with assumed noises and from a Riccati equation [4]. In practice this is backed up by simulations. The filter is calculated for a moderate noise environment. The response is as in figure 1, where a step in target acceleration at a certain time is applied. We can notice the lag in the estimates of both the acceleration and the velocity, while the position is less affected. Initially, position and acceleration are zero and velocity is set to 20 meter/second in this demonstration.

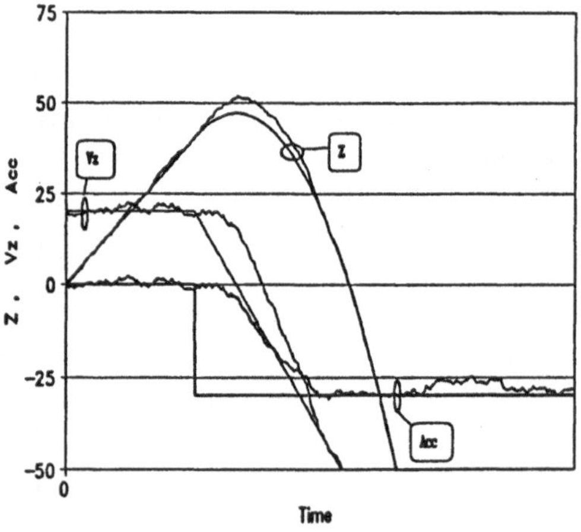

Figure 1. Stepresponse of the filter.

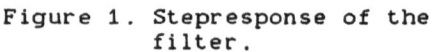

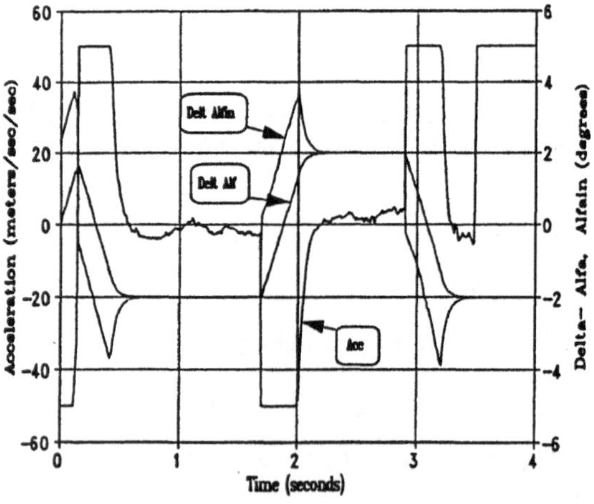

Figure 2. Control histories and $\delta\alpha$ vs time, $\delta\alpha_{max}=2^\circ$.

3. Game Formulations

Depending on the assumption we do on the practical restrictions the game might be formulated in several ways. There are delays due to bullet's traveltime and radar data of the target (the opponent). Also, there are assumptions of the combatant's behavior etc. A deterministic case where both break at the same time in order to contribute with a half pass distance each is studied in [2]. Concerning the kill probability there is a model in [2],

$$p_k = Const \cdot (R_o - R_{break})^3 / R_o^2 \tag{8}$$

The place where the gun is activated is R_o and R_{break} is the relative distance at break away. Obviously, a late break will strongly increase the kill probability. Thus, it is an advantage to be able to aim long as possible and delude the opponent to break aiming earlier. Additionally, taken the realistic measurements into account a plausible game formulations could be;

> Both produce at least a half pass distance relative to the predicted position of the opponent at pass and assume that the opponent does the same. Both aim as long they can.

3.1 Optimal Control Problems

With the purpose to illustrate the possibility to take advantage of the lags in the filter the game problem will be to deceive the opponent in prediction of the pass position. The optimal control problem is,

$$\text{Maximize } \{ k_z \cdot z(t_f) - z_p(t_f;\ \delta t) \} \tag{9}$$

where k_z is a weighting factor, $k_z = 1$ gives maximal misleading. A smart maneuver by the aircraft with decoupling facility can make him aiming and producing a predicted position, $z_p(t_f;\ \delta t)$ like a decoy in the other ones system, which deviates considerably from the real position at t_f. This forces the other aircraft to make his avoidance maneuver much earlier then he otherwise should have to. Each player will predict the opponents position, $z_p(t_f;\ \delta t)$, at an estimated pass time equal to the final time, t_f, determined by the closing velocity and present distance. The break time, t_b, is then

$$t_b = t_f - \delta t \tag{10}$$

The prediction is based on the measured variables from Eqs.(3-5) and the time it takes for the aircraft respectively to accomplish the avoidance maneuver, δt. The predicted position is then given by,

$$z_p(t_f;\ \delta t) = z_e(t_b) + v_{ze}(t_b) \cdot \delta t + \tfrac{1}{2} Acc_e(t_b) \cdot \delta t \cdot \delta t \tag{11}$$

The aircraft which has decoupling facilities uses a small pitch deflection, $\delta\alpha$, added to the angle of attack in order to achieve aiming besides to some extent controlling the trajectory with the crossacceleration, Acc. The aiming is achieved by striving for closing the aiming triangle, see Appendix A. Introduce an additional equation to Eqs.(1,2) for this aircraft,

$$\delta\dot{\alpha} = k_{\delta\alpha}\cdot(\delta\alpha_{in} - \delta\alpha), \quad |\delta\alpha| \le \delta\alpha_{max} \tag{12}$$

The gain $k_{\delta\alpha}$ is in this study set to .2. A typical value of $\delta\alpha_{max}$ is two degrees. The two controls are Acc in Eq.(2) and $\delta\alpha_{in}$, where the latter one controls the aiming as it can not affect the trajectory and Acc is mainly used for trajectory shaping subject to the working range of $\delta\alpha$. We have a state constraint on $\delta\alpha$, as is shown in Eq.(12), and a control constraint on the acceleration, $\pm Acc_{max}$, set to 50 meter/seconds².

Obviously aiming can not be satisfied arbitrarily close to the target if we want to avoid collision. A break has to be made, the time for this corresponds to t_b above. This time is in the first examples considered as fixed (3.5 seconds). At least beyond this point maximum acceleration should be applied. The optimal control derived in Appendix B execute an acceleration large as possible including switching subject to the aimingcondition (A3). Typical optimal control histories are depicted in figure 2, with $\delta\alpha_{max} = 2^\circ$ and a passive target. Notice that the noise comes through particularly apparent on the acceleration. Also, the $\delta\alpha$ is plotted and demonstrates the effectiveness of the hyperplane technique applied to this case.

4. Game Examples

Consider the two aircraft, one has one control the Acc_1 and the other one has two controls the Acc_2 and the $\delta\alpha_{in}$. Refer to the aircraft as AC1 and AC2 respectively. The most difficult part is what to base the break on and particularly, which direction should it take. A couple of conditions will be shown. These will then be the platform for a discussion of the break problem. The advantage of having a more sophisticated aircraft depends strongly on these conditions. An other question is if both combatants know each others filters i.e. do we have perfect information or not.

The prediction times used by each aircraft is a function of the opponent's aiming time i.e. the break time.

$$\delta t_1 = t_f - t_{b2} \tag{13}$$

$$\delta t_2 = t_f - t_{b1} \tag{14}$$

where the subindex 1 and 2 indicate AC1 and AC2 respectively. The game will then be to match the break times such that each aircraft can produce at least a half of the given pass distance (50 meters each), which must be a fair assumption as both care to the same degree for noncollision. The optimization is first of all to maximize the prediction error for AC1 then adjust the break times till the pass condition is satisfied for both. Obviously, a smaller t_b is needed to achieve a larger side step distance, z_f ($z(t_f)$ in the object function (9)). On the other hand the opponent then gets a larger δt and this gives an even harder demand on the first aircraft, as he needs to turn away more as the prediction error grows (assuming that the turn is in a proper direction) etc.

4.1 Break direction à priori determined

Both aircraft are supposed to make an up and a down break turn respectively and produce at least 50 meters, z_{fj}, away from the predicted final position of the hostile aircraft, z_{pi} ($z_{pi}(t_f;\ \delta t_i)$ in the object function (9)), where j=1 when i=2 and vice versa. In this case AC2 after a while reaches his limitations in both Acc_2 and $\delta\alpha_{in}$, which means he is not able to aim the full time out. When this happens his break is initiated. Hence, AC2 will obtain more than 50 meters in his contribution to the pass distance. The flight paths are shown in figures 3, 4 note that there are different scales in z- and x-direction, which exaggerates the picture of the maneuvers. Clearly, AC2 deceives AC1's prediction and forces him to make a large avoidance as z_{p2} is much negative. The individual contributions to the pass distance, $z_{fj} - z_{pi}$, are for AC1 -50 meters while AC2 produces around 92 meters or 165 meters due to limitations in aiming, $|\delta\alpha| \leq 2^\circ$ or 4° respectively. The breaktimes are t_{b1} = 2.035 seconds and t_{b2} = 2.9775 seconds using $|\delta\alpha| \leq 2^\circ$ and t_{b1} = 1.585 seconds and t_{b2} = 2.655 seconds using $|\delta\alpha| \leq 4^\circ$.

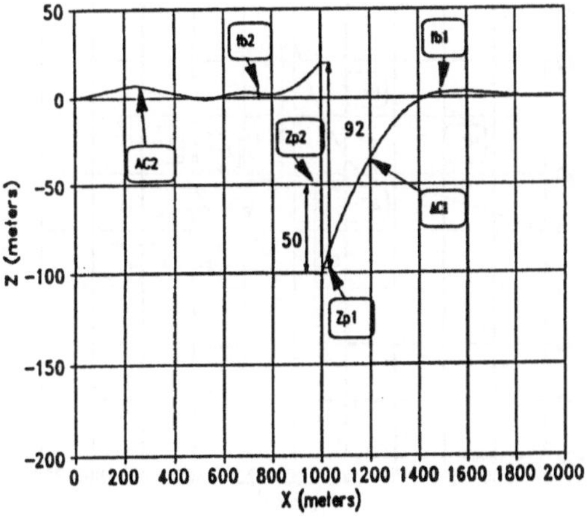

Figure 3. Flight paths, $\delta\alpha_{max}=2^\circ$.

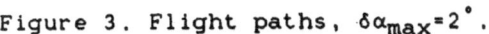

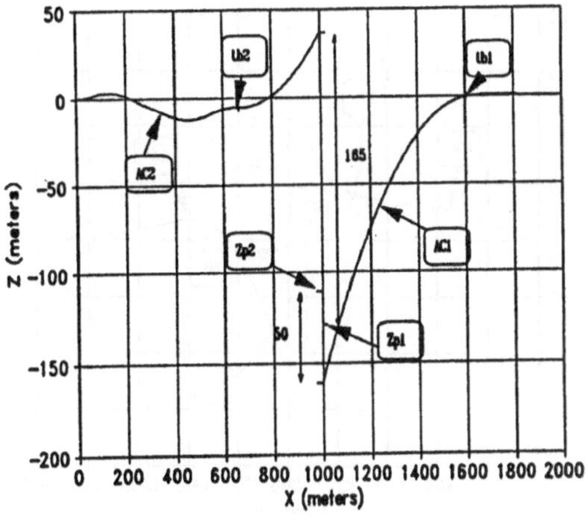

Figure 4. Flight paths, $\delta\alpha_{max}=4^\circ$.

Using this strategy in determining the break condition there is a considerable advantage of having an aircraft with decoupling facility. Assume the kill probability, Eq.(8), is a good measure of killing the opponent. The exchange-ratio of kill probability will then be 3.14 and 4.70 using $|\delta\alpha| \leq 2^\circ$ and 4° respectively in advantage to AC2.

4.2 Break direction based on prediction of the opponent's position

If there are no regulations expressing the turn directions the decision must be based on available information in each aircraft. It is likely, they will turn away from the predicted position of the hostile aircraft in that direction, in which is simplest to obtain a large side step. We use k_z in the object function (9) to control the prediction of AC2's final position and place it so that AC1 turns in a desirable direction, e.g. a down turn and then use t_{b1} to obtain 50 meters from this point. Then AC2 knows from the beginning that he can plan for an up turn. In figure 5 z_{p2} has been forced by k_z to be 0.2 meters t_{b1} = 2.434 seconds and t_{b2} = 2.915 seconds resulting in z_{f1} = -49.9 meters, z_{p1} = -2 meters and z_{f2} = 49.8. In this case the x-axis is supposed to be a sort of reference line. That is, the predicted positions must be on the opposite side of the line, then the aircraft are supposed to turn to the opposite side of the predicted position.

However, due to aiming AC1 has reached 7-8 meters above this line and would not be willing turning down on the information of z_{p2}. A better condition would be to use the line of boresight and place the z_{pi} on opposite side of that line before either one breaks. Thereafter, the nonbreaking aircraft will turn in the opposite direction since he would get a stronger indication by his prediction of what the opponent is doing. As AC2 has control over AC1's prediction of AC2, then AC2 has control over the turn directions. An additional example where z_{p2} is

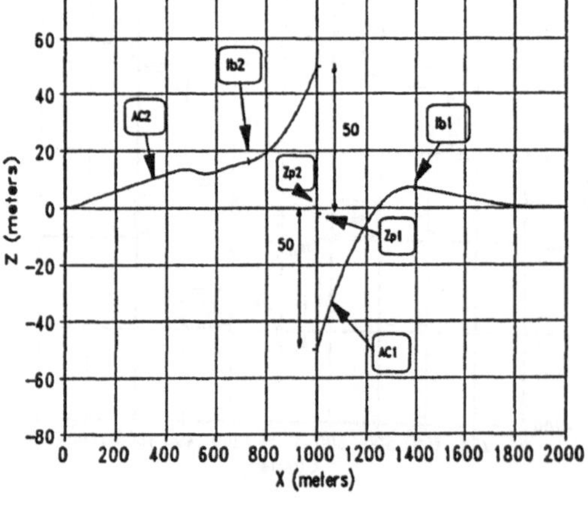

Figure 5. Flight paths, $\delta\alpha_{max}=2^\circ$.

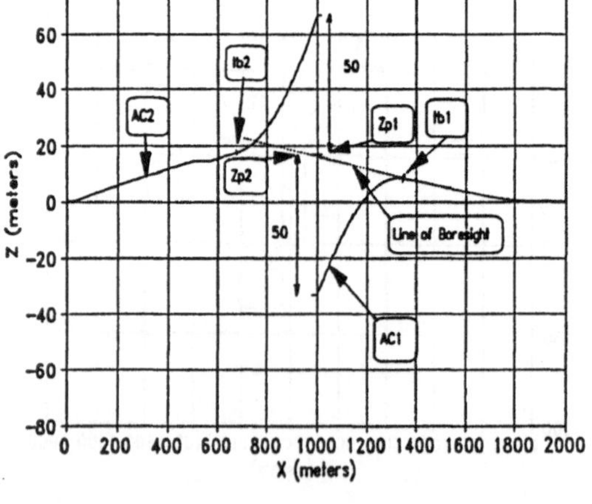

Figure 6. Flight paths, $\delta\alpha_{max}=2^\circ$.

above the line of bore sight, as is shown in figure 6. The obtained break times are; t_{b1} = 2.62 seconds and t_{b2} = 2.71 seconds.

Unfortunately, there is not much to gain by the decoupled aircraft in respect to tricking the break for the opponent. The ratio of kill probability turns out to be in the two examples; 1.72 and 1.11 respectively. It might even be a disadvantage in this respect, if there are even harder demands of where to put the z_{p2}.

This forms actually a slightly different optimal control problem. Anyway we use the formulation already used i.e. the object function (9), which would not significantly affect the results.

4.3 Aiming errors

As a guideline, an acceptable aiming error at 1000 meters must be within ±0.3° in order to obtain a good hit. As an illustration of the aiming errors we pick the examples corresponding to figures 3 and 5. Also, the measurement noise in Eq.(6) is not applied, since we like to isolate the effect on the errors emanating from the maneuvers. The estimated errors as well as the real errors are shown for both aircraft in figures 7, 8, respectively. The controls of the aiming are based on the estimated errors as before. The results show the following: The real errors are larger than the estimated ones but still acceptable for AC2 till just after AC1 breaks, while AC1 has difficulties to keep the error within the ±0.3°. The maneuvers by AC2, which original purpose were to deceive AC1 in prediction of AC2's pass position, also affect AC1's aiming significantly.

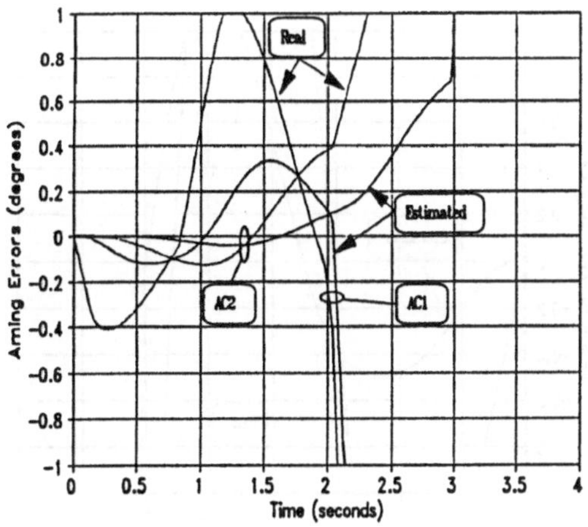

Figure 7. Estimated and real aiming errors for AC1 and AC2 vs time.

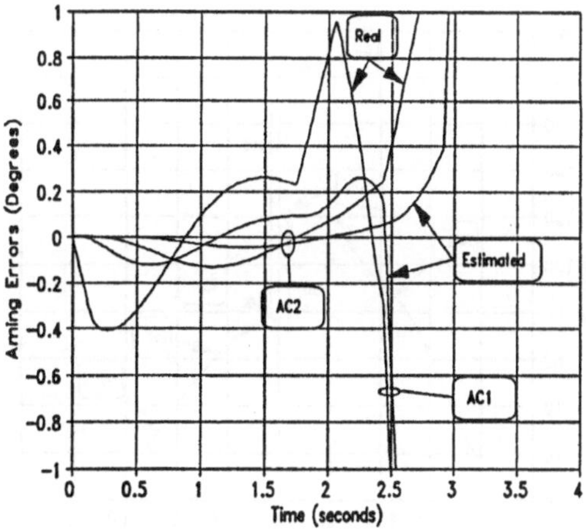

Figure 8. Estimated and real aiming errors for AC1 and AC2 vs time.

4.3.1 AC1 is aiming and AC2 jams AC1's aiming

If the aiming is based on the radar signal it can be extremely jammed by maneuvering. The bullets traveltime ranges from about 1.7 seconds at two kilometers distance till about 0.3 seconds just before break. A sort of prediction is actually made in the aiming triangle, Eq.(A1), where a nonmaneuvering target is assumed. If the target makes a maneuver there is an additional term consisting of target acceleration [3] to be included in Eq.(A1), which makes the aiming even more difficult to perform. For a pure deception in aiming an other optimal control problem should be considered,

$$\text{Max}\{ \int |\epsilon(t)| \cdot dt \} \tag{15}$$

The object function (15) contains prediction quantities from the radar, which also the original optimal control problem (object function (9)) does. Thus we accept to use the object function (9) as an approximation. A test example was run, with t_{b1} = 3 seconds corresponding to a mean value of the bullets traveltime. The flight paths are shown in figure 9, where AC1 is aiming and AC2 is maximizing the prediction error. We used AC1 as the aiming aircraft and AC2 as the jamming one. This would not make any difference since the Acc_2 is a good control to achieve aiming as the aiming itself is not improved by a $\delta\alpha$ in the particular model used. A down avoidance turn is performed by AC1 at t_{b1} and k_z is used to make 50 meters at predicted pass for AC1. As a side effect AC2 also disturbs aiming for AC1, see figure 10. An acceptable error must be within $|\epsilon| \leq 0.3^\circ$, which is widely passed. The tactic of AC2 is very safe as the pass distance is large, over 120 meters.

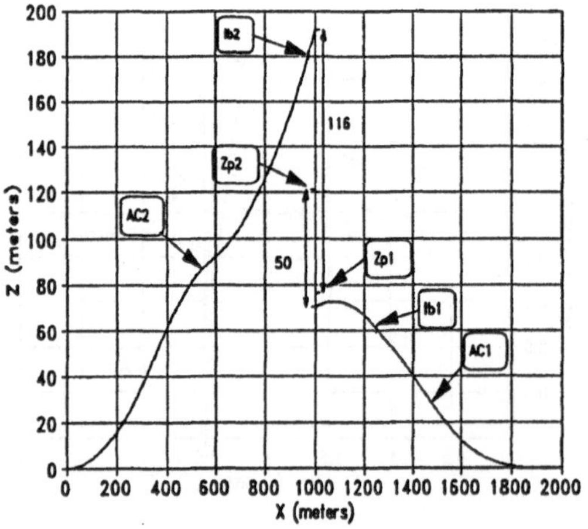

Figure 9. Flight paths with jammed aiming.

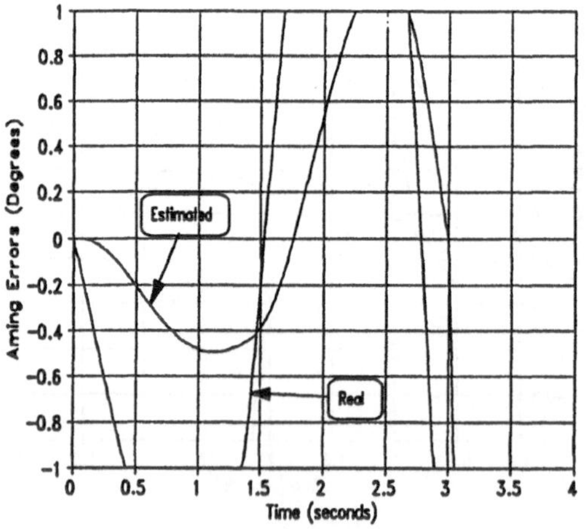

Figure 10. Estimated and real aiming errors for AC1 with jammed aiming.

5. DISCUSSION

The results obtained are based on the assumptions made in this paper. Simplified models of the aircraft (pointmasses) and motion in a plane. Including more short period dynamic might somewhat smooth out the effect of switches in the acceleration. The game illustration will still be left. The game ingredients are more clearly visualized when using the simple models.

The filter used is designed for other purposes than optimizing the predictor used, Eq.(11). A better predictor than Eq.(11) adapted to the filter used might be possible to find. However, the effects of the delayed information would still be a significant contributor to the game problem discussed.

One definite advantage with the decoupling facility is the irregular flight path such aircraft can perform causing a much larger aiming error in the opponent's measuring equipment than his more regular flight path returns to the decoupled aircraft. It takes a while till the system can detect that it does not aim perfectly. Then it is to late to switch strategy. The optimization used here is not designed for this particular deception aiming problem. Thus, if optimizing with respect to deceiving the aiming it might be possible to gain more. A passive target would not raise any aiming problem, while an optimally deceived radar can easily drive the aiming error out off what is acceptable for a hit.

The best to do for an aircraft without decoupling facility is to give up aiming and concentrate on jamming the opponent's aiming. This means the aircraft with decoupling facility can force other aircraft not to shoot in the head-on-encounter. Thus he has removed a threat.

5.1 The break problem

The break problem is not trivial. It is reasonable to refer the size of the side step to the predicted pass position of the hostile aircraft. Which side of it i.e. the turn direction is more questionable. If we base the turn direction on the à priori determined direction or on the prediction of the opponent there is a large difference in the outcome, cf. sections 4.1 and 4.2. There might be a motivation for using a sort of à priori determined direction. In the onboard autonomy system there must then be a calculation of what the opponent might do. This will include a simulation of the opponent's Kalman filter i.e. both knows each others prediction of position at pass.

Assume both the combatants know both z_{p1} and z_{p2}. Assume further that we have $z_{p1}<z_{p2}$ and $z_{p1}<\theta$, which AC2 has control over. The number θ is such that it is below the line of boresight from AC2. The AC2 will for sure turn up on this condition, this is also obvious for AC1 and additionally indicated by the condition $z_{p1} < z_{p2}$. If $z_{p2} - z_{p1} < 50$ then AC1 must assume that AC2 is above z_{p2} in order to produce 50 meters above z_{p1}, he can not be sure about how much. If he turns up he might get into an area where AC2 will be. To be safe he has to turn down. This is satisfied in the examples behind figures 3, 4.

APPENDIX: A. AIMING CONDITION

The two aircraft are closing with constant velocity (250 meters/seconds each) in an near collision course. Then angles and deviations can be considered small. The geometry in figure A1 is exaggerated. The gun may point within $\pm\delta\alpha_{max}$ out from the velocity vector. An aiming error, ϵ, is introduced, which has to be close to zero on a given tolerance level for a good hit. This error can be derived from figure A1 using the sinus theorem.

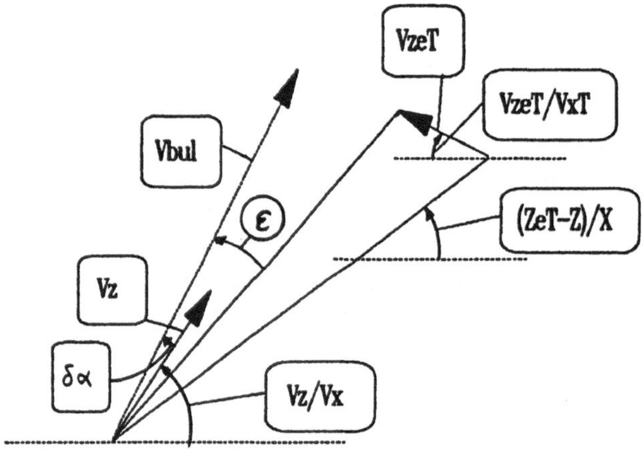

Figure A1. Aiming Geometry.

$$\epsilon = v_z/v_x + \delta\alpha - (1 + K_{bul})\cdot(z_{eT} - z)/x - K_{bul}\cdot v_{zeT}/v_{xT} \qquad (A1)$$

In order to close the aiming triangle we have to drive ϵ to zero. As there is no direct control in Eq.(A1) we have to make use of time derivative of ϵ and use the technique in reference 2, Eq.(7), yielding the aiming condition

$$0 = \epsilon + \{Acc/v_x + k_{\delta\alpha}\cdot(\delta\alpha_{in} - \delta\alpha) - (1 + K_{bul})\cdot[(v_{zeT} - v_z)/x + (v_x + v_{xT})\cdot(z_{eT} - z)/x^2] - K_{bul}\cdot Acc_{eT}/v_{xT}\}\cdot Dt \qquad (A2)$$

where Dt is the time increment it is supposed to take to drive an error to zero, [2]. It is convenient to rewrite Eq.(A2) with the controls explicit and summarize the remaining terms in an auxiliary function Eps(•).

$$Acc/v_x + k_{\delta\alpha}\cdot\delta\alpha_{in} + Eps(\epsilon/Dt,\ \delta\alpha,\ z_{eT},\ z,\ v_{zeT},\ v_z,\ v_{xT},\ v_x,\ Acc_{eT},\ x,\ k_{\delta\alpha},\ K_{bul}) = 0 \qquad (A3)$$

APPENDIX: B. OPTIMAL CONTROL CONDITIONS

Substitute Eq.(11) into the object function (9), the object function to be maximized turns out to be

$$V = k_z\cdot z(t_f) - z_e(t_b) - v_{ze}(t_b)\cdot\delta t - \tfrac{1}{2}Acc_e(t_b)\cdot\delta t\cdot\delta t \qquad (B1)$$

That is an interior time control problem. Forming the Hamiltonian, first for the no aiming case,

$$H = V_z\cdot v_z + Vv_z\cdot Acc + V_{ze}\cdot[v_{ze} + K_1\cdot(z + w - z_e)] + Vv_{ze}\cdot[Acc_e + K_2\cdot(z + w - z_e)] + VAcc_e\cdot[-B\cdot Acc_e + K_3\cdot(z + w - z_e)] \qquad (B2)$$

The adjoint variables start by a capital V. The control is Acc and the noise, w, is considered as zero. Maximizing the Hamiltonian gives the optimal control to be

$$Acc = Acc_{max}\cdot Sign\{V_{vz}\} \qquad (B3)$$

The differential equations for the adjoints including the transversality conditions will then be

$$\dot{V}_z = -V_{ze}\cdot K_1 - V_{vze}\cdot K_2 - V_{Acce}\cdot K_3\ , \qquad V_z(t_f) = k_z \qquad (B4)$$

$$\dot{V}_{vz} = -V_z, \qquad V_{vz}(t_f) = 0 \qquad (B5)$$

$$\dot{V}_{ze} = V_{ze}\cdot K_1 + V_{vze}\cdot K_2 + V_{Acce}\cdot K_3, \qquad V_{ze}(t_b) = -1 \qquad (B6)$$

$$\dot{V}_{vze} = -V_{ze}, \qquad V_{vze}(t_b) = -\delta t \qquad (B7)$$

$$\dot{V}_{Acce} = -V_{vze} + V_{Acce}\cdot B, \qquad V_{Acce}(t_b) = -\tfrac{1}{2}\delta t\cdot\delta t \qquad (B8)$$

We also have

$$V_{ze}(t) = V_{vze}(t) = V_{Acce}(t) = 0 \qquad \text{for all } t > t_b \qquad (B9)$$

Introducing aiming, the state constraint in Eq.(12) has to be included in the optimal control problem. This is done by the Hyperplane technique [5] transferring it to a state-control constraint,

$$\delta\alpha - A_o\cdot(\delta\alpha + \delta\alpha_{max}) \le \delta\alpha_{in} \le \delta\alpha - A_o\cdot(\delta\alpha - \delta\alpha_{max}) \qquad (B10)$$

The parameter A_o is chosen as large as practically useful. Too large might raise numerical problem, too low will satisfy the constraint pessimistic and not use the full working area of it. A value of 4 gave a distinct approach, with a smooth arc, of $\delta\alpha$ to it´s maximal value, see figure 2. A larger A_o will give a sharper arc.

When $t \geq t_b$ we have the same conditions given by Eqs.(B3-B9). Next step is to consider if the acceleration is on it's limit or not. The $\delta\alpha_{in}$ is on it's limit when the acceleration is off it's limit and vice versa. Consider acceleration as the control variable maximizing the Hamiltonian and let the other control, $\delta\alpha_{in}$, satisfy Eq.(A3). Form the new Hamiltonian with the interesting terms written down,

$$H = V_{vz}\cdot Acc + V_{\delta\alpha}\cdot(Eps - Acc/v_x - k_{\delta\alpha}\cdot\delta\alpha) + \text{remaining terms from Eq.(B2)} \quad \text{(B11)}$$

The optimal acceleration will then be

$$Acc = Acc_{max}\cdot Sign\{V_{vz} - V_{\delta\alpha}/v_x\} \quad \text{(B12)}$$

The adjoint equations to be modified are those for V_z and V_{vz}, also an extra adjoint variable for $\delta\alpha$ has to be introduced, $V_{\delta\alpha}$. Two cases have to be considered with respect to saturation or not of the acceleration.

The case when $|Acc| = Acc_{max}$;

$$\dot{V}_z = -V_{ze}\cdot K_1 - V_{vze}\cdot K_2 - V_{Acce}\cdot K_3 + V_{\delta\alpha}\cdot(1 + K_{bul})\cdot[1/Dt + (v_x + v_{xT})/x]/x, \quad V_z(t_f) = k_z \quad \text{(B13)}$$

$$\dot{V}_{vz} = -V_z + V_{\delta\alpha}\cdot[1/Dt/v_x + (1 + K_{bul})/x], \quad V_{vz}(t_f) = 0 \quad \text{(B14)}$$

$$\dot{V}_{\delta\alpha} = V_{\delta\alpha}/Dt, \quad V_{\delta\alpha}(t) = 0 \text{ if } t \geq t_b \quad \text{(B15)}$$

where v_x is the velocity of the aircraft (in x-direction), x is the distance to the target (initially 2000 meters), subindex T stands for target and K_{bul} is the ratio of target velocity and the mean velocity of a fired bullet, v_{bul}. Numerically, K_{bul} is in the order of 0.2. See Appendix A for the time increment Dt.

The case when $|Acc| < Acc_{max}$;

$$\dot{V}_z = -V_{ze}\cdot K_1 - V_{vze}\cdot K_2 - V_{Acce}\cdot K_3 + V_{vz}\cdot v_x\cdot(1 + K_{bul})\cdot[1/Dt + (v_x + v_{xT})/x]/x, \quad V_z(t_f) = k_z \quad \text{(B16)}$$

$$\dot{V}_{vz} = -V_z + V_{vz}\cdot[1/Dt + v_x\cdot(1 + K_{bul})/x], \quad V_{vz}(t_f) = 0 \quad \text{(B17)}$$

$$\dot{V}_{\delta\alpha} = k_{\delta\alpha}\cdot(V_{\delta\alpha} - V_{vz}\cdot v_x)\cdot A_o + V_{vz}\cdot v_x/Dt, \quad V_{\delta\alpha}(t) = 0 \text{ if } t \geq t_b \quad \text{(B18)}$$

B.1 The computational procedure

Assume δt is given. As the velocities in the x-direction of the aircraft can be considered constant and perfectly measurable as well as the present distance, the pass time t_f is known. Then we also know t_b, Eq.(10). The adjoint equations, the applicable ones of Eqs.(B4-B8) and Eqs(B13-B18), can then be integrated backwards in time. Then we pick up the timevalues when the control in Eq.(B3) or Eq(B12) changes sign and store them. These are called switchingtimes. The state equations Eq.(1-5) can now be integrated forward in time using the stored switching times to create the control associated with Eq.(B3) or Eq.(B12). The predicted position from Eq.(11) and the final position can then be determined. Besides storing the switching times during backward integration the timeintervals while the acceleration is saturated must be stored during forward integration and then be used to determine, which of the sets of adjoints above is applicable. This makes the procedure iterative. When initiating the backward integration a guess of time intervals must be done.

References

1) Forsling, G. and Järmark, B. A. S., **Optimal Fuselage Aiming of a Modern Fighter Aircraft**, Theory and Applications of Nonlinear Control Systems, (Editors C. Byrnes and A. Lindquist), Elsvier Science Publishers B.V. (North Holland), 1986.

2) Järmark, B. A. S. and Speyer, J. L., "Optimal Fuselage Aiming with Decoupled Short Period Dynamics", AIAA, GNC Conf. Minneapolis August 1988.

3) Järmark B., "Peksiktning med Frikopplad Kortperiod Dynamik", Report TKU-MI-88:48, Saab-Scania AB, Linköping, August, 1988.

4) Järmark, B., "Gaming in Fuselage Aiming", To be presented at the 17th ICAS-90 Congress, 9-14 September, 1990, Stockholm.

5) Mårtensson, K., "New Approaches to the Numerical Solution of Optimal Control Problems", ISBN 91-44-08851-5, Studentlitteratur, 1972, Lund, Sweden.

CONTROL STRATEGIES IN A PLANAR PURSUIT EVASION GAME WITH ENERGY CONSTRAINTS

M. Guelman
RAFAEL, M.O.D., P.O. Box 2250
Haifa, Israel

ABSTRACT

In this paper a coplanar pursuit evasion game in the atmosphere opposing a pursuer employing thrust amplitude and angle of attack as control variables, to a maneuvering evader flying with constant velocity is considered. The aerodynamic forces acting on the pursuer are explicitly modelled as functions of the angle of attack. The pursuer and evader strategies in the boundary of the capture set are determined for the case of a circular target set. Singular arcs both for the pursuer and evader are shown to exist.

1. INTRODUCTION

Realistic pursuit evasion games, opposing a missile to a maneuvering aircraft in the atmosphere characterize themselves by the fact that the pursuer possesses a limited amount of energy (kinetic, propulsive) able to be expended while the evader has in practical terms, relative to the pursuer, an unlimited amount of energy. The missile is designed to be faster and more maneuverable than the aircraft. This kinematical advantage, is however, temporary. The missile rocket motor accelerating it to a high velocity is of short duration. In the coasting phase, the high but finite kinetic energy of the missile is rapidly dissipated by the work done against the aerodynamic drag. On the other hand, the evader can maintain its velocity for periods of time greater, by orders of magnitude, than those of the pursuer.

In the missile versus aircraft pursuit evasion game considered in this paper the energetic-kinematic dissymetry leads to the existence of a finite capture set. While the pursuer advantages the evader kinematically it can assure its capture, but with a limited amount of energy at its disposal, the pursuer will lose at a certain instant of time its kinematic advantage and the pursuer will not be able to reach the target from every initial state.

The solution of the appropriate game of kind, yields the optimal barrier strategies of the players.

Analysis of planar pursuit evasion with variable speeds was performed in [1], [2] employing a real space coordinate system. The optimal strategies for both players are similar, to turn towards the final line of sight direction with an asymptotically decaying rate. The case of a coasting pursuer [3] was previously studied using two different aerodynamic models, for small and large angles of attack, and the capture region of this pursuer when opposed to a constant velocity maneuvering evader was investigated. In [4] a suboptimal guidance law for the coasting pursuer was derived based on the game strategies.

In this work the pursuer employs both propulsive and aerodynamic forces to control its motion and the optimal strategies on the barrier will be considered for analysis.

2. PROBLEM DEFINITION

The geometry of the planar pursuit evasion game in the atmosphere is depicted in Fig. 1.A vehicle P, called the pursuer, possessing a velocity V_P is pursuing in the same plane, a second vehicle, the evader E, assumed to be flying with a constant velocity V_E.

The pursuer possesses a rocket motor aligned with its longitudinal axis able to generate a thrust T. The pursuer is able to control its angle of attack α and the thrust amplitude T.

The pursuit evasion game equations are defined by

$$\dot{R} = V_E\cos(\gamma_E-\Theta) - V_P\cos(\gamma_P-\Theta) \tag{1}$$

$$\dot{\Theta} = [V_E\sin(\gamma_E-\Theta) - V_P\sin(\gamma_P-\Theta)]/R \tag{2}$$

$$\gamma_P = (M_f c_1/M)V_P\sin2\alpha + (T/MV_P)\sin\alpha \tag{3}$$

$$\dot{V}_P = -(M_f c_1/M)\ (k-\cos2\alpha)V_P^2 + (T/M)\cos\alpha \tag{4}$$

$$\dot{M} = -T/gI_{sp} \tag{5}$$

$$\gamma_E = \Gamma_E u_E, |u_E| \leq 1 \tag{6}$$

where (1)-(2) are the kinematic equations, (3)-(5) the pursuer equations and (6) the evader equation. c_1 and k are two pursuer parameters defined by the aerodynamic coefficients and flight conditions [3], M_f is the final pursuer mass, g is the gravitation constant and I_{sp} is the specific impulse. Γ_E is the evader maximum turning rate, a constant for a constant speed vehicle in horizontal flight.

The pursuer angle of attack α and the rocket motor thrust T

$$0 \leq T \leq T_m \tag{7}$$

are the pursuer controls and u_E is the evader control. The system states are the P-E range R, LOS direction Θ, pursuer mass M, velocity direction and amplitude γ_P and V_P, and evader flight direction γ_E.

Let us define new normalized variables, $r = R/R_{ref}$, $v = V_P/V_E$, $m = M/M_f$ and $t' = tV_E/R_{ref}$. R_{ref} is the minimum admissible turning radius of the pursuer in the final coasting phase defined by $R_{ref} = 1/c_1$. With these new variables, and with a dot denoting now the derivative with respect to normalized time t' the pursuit game equations are defined as follows,

$$\dot{r} = \cos(\gamma_E-\Theta) - v\cos(\gamma_P-\Theta) \tag{8}$$

$$\dot{\Theta} = [\sin(\gamma_E-\Theta) - v\sin(\gamma_P-\Theta)]/r \tag{9}$$

$$\gamma_E = \sigma u_E \tag{10}$$

$$\gamma_P = (v/m)\sin 2\alpha + (T'/vm)\sin\alpha \tag{11}$$

$$\dot{v} = -(v^2/m)\ (k-\cos 2\alpha) + (T'/m)\cos\alpha \tag{12}$$

$$\dot{m} = -T'/c \tag{13}$$

where now σ is the ratio of pursuer's minimum turning radius to that of the evader (in general $\sigma<1$), T' is the normalized thrust defined by $T' = T/(M_f c_1 V_E^2)$ and $c = gI_{sp}/V_E$.

The game terminates with capture when the pursuer approaches the evader to the normalized distance $r = r_f$ provided $m \geq 1$, i.e., the game target set is defined as a closed circular cylinder of radius r_f above the hyperplane $m = 1$,

$$T = \{x \in E^6 : r \leq r_f, \; m \geq 1\},$$

where $x = (r,\Theta,\gamma_E,v,\gamma_P,m)^T$ is the state vector and no additional conditions are imposed on $\Theta,\gamma_E,v,\gamma_P$.

3. BARRIER STRATEGIES

The pursuer playability domain [5], for the game previously defined is a bounded region in space. The boundary of this domain is the so called barrier, an hypersurface formed by trajectories. To the trajectories belonging to the boundary correspond well defined strategies, α^*, T^*, u_E^*, that can be determined employing the basic theorem of qualitative games [5]:

$$\min_{\alpha,T} H(x,p,\alpha,T,u_E^*) = \max_{u_E} H(x,p,\alpha^*,T^*,u_E) = H(x,p,\alpha^*,T^*,u_E^*) = 0 \quad (14)$$

where H is the Hamiltonian of the game,

$$H = p_r[\cos(\gamma_E-\Theta)-v\cos(\gamma_P-\Theta)] + p_\Theta[\sin(\gamma_E-\Theta)-v\sin(\gamma_P-\Theta)]/r + p_{\gamma E}\sigma u_E +$$

$$p_{\gamma P}[(v/m)\sin2\alpha+(T'/mv)\sin\alpha] + p_v[-(v^2/m)(k-\cos2\alpha)+(T'/m)\cos\alpha] - p_m T'/c \quad (15)$$

and $p = (p_r,p_\Theta,p_{\gamma E},p_v,p_{\gamma P},p_m)^T$ is the vector of adjoint variables.

From the necessary conditions, it follows that the evader strategy on the barrier maximizes H and is given by,

$$u_E^* = \text{sign}(p_{\gamma E}), \quad p_{\gamma E} \neq 0 \quad (16)$$

In order to find the pursuer strategies α^* and T^* let us rewrite the Hamiltonian as follows,

$$H = a_o(\alpha,p,x) + a_1(\alpha,p,x)T' + a_2(p,x) \tag{17}$$

where

$$a_o = (v/m)[p_{\gamma P}\sin 2\alpha + p_v v\cos 2\alpha] \tag{18}$$

$$a_1 = (1/mv)[p_{\gamma P}\sin\alpha + p_v v\cos\alpha] - p_m/c \tag{19}$$

$$a_2 = p_r[\cos(\gamma_E-\Theta)-v\cos(\gamma_P-\Theta)]+p_\Theta[\sin(\gamma_E-\Theta)-v\sin(\gamma_P-\Theta)]/r+p_{\gamma E}\sigma u_E - kp_v v^2/m \tag{20}$$

Equation (17) defines a ruled surface in the H, α, T' space schematically depicted in Fig. 2. This surface is generated by a straight line parallel to the H, T' plane with slope a_1 and intersecting the H, α plane at,

$$H_o = a_o(\alpha,p,x) + a_2(p,x) \tag{21}$$

In order to minimize H as a function of α and T' we shall consider the two cross sections $H = H_o$ and $H = H_T$ $(T'=T_M)$.

H_o is minimized by α_o defined by,

$$\sin 2\alpha_o = -p_{\gamma P}/d \tag{22}$$

$$\cos 2\alpha_o = -p_v v/d \tag{23}$$

where $d = [(p_v v)^2+(p_{\gamma P})^2]^{1/2}$

H_T is minimized by α_T defined by,

$$\frac{v\sin 2\alpha_T + (T'/v)\sin\alpha_T}{(v^2+(T'/v)^2+2T'\cos\alpha_T)^{1/2}} = -p_{\gamma P}/d \tag{24}$$

$$\frac{v\cos 2\alpha_T + (T'/v)\cos\alpha_T}{(v^2+(T'/v)^2+2T'\cos\alpha_T)^{1/2}} = -p_v v/d \tag{25}$$

In Fig. 3 are depicted both α_o and α_T. From the graphic construction it can be readily seen, that excepted for $p_{\gamma P}=0$,

$$|\alpha_T| > |\alpha_o| \tag{26}$$

Furthermore, for same p and x, it follows from (26) that

$$a_1(\alpha_T,p,x) < a_1(\alpha_o,p,x) \tag{27}$$

For $p_{\gamma P} \neq 0$, there are three different cases for H_o and H_T, as schematically depicted in Fig. 4. From the analysis of the three possible different cases for H_o and H_T, it follows that for given p and x the values of α^* and T^* that minimize H, are defined as follows,

1) $a_1\ (\alpha_T,p,x) \geq 0$

$\alpha^* = \alpha_o,\ T^* = 0$

2) $a_1(\alpha_o,p,x) \leq 0$

$\alpha^* = \alpha_T,\ T^*, = T_M$

3) $a_1(\alpha_o,p,x) > 0,\ a_1(\alpha_T,p,x) < 0$

a) $H_o(\alpha_o) < H_T(\alpha_T)$

$\alpha^* = \alpha_o,\ T^* = 0$

b) $H_o(\alpha_o) > H_T(\alpha_T)$

$\alpha^* = \alpha_T,\ T^* = T_M$

For any $P_{\gamma P}$ different from zero, $\alpha_o \neq \alpha_T$ and there are <u>no singular arcs</u>, T^* is either 0 or T_M. For $P_{\gamma P} = 0$, $\alpha_o = \alpha_T = 0$. In this case there exists the possibility of a singular arc ($a_1 \equiv 0$).

4. SINGULAR ARCS

Further analysis of the problem, requires a solution of the adjoint equations. Fortunately, the adjoint equations can be analytically integrated in terms of the system state variables and their final values:

$$p_r = \cos(\Theta-\Theta_f) \tag{28}$$

$$p_\Theta = -r\sin(\Theta-\Theta_f) \tag{29}$$

$$p_{\gamma E} = -\frac{1}{\sigma}[\cos(\gamma_{Ef}-\Theta_f)-\cos(\gamma_E-\Theta_f)]\text{sign}[\sin(\gamma_E-\Theta_f)] \tag{30}$$

$$p_{\gamma P} = r\sin(\Theta-\Theta_f) - p_{\gamma E} \tag{31}$$

$$p_m = -[\mu_m + r\cos(\Theta-\Theta_f) - r_f]/m \quad (\text{for } p_{\gamma E} \equiv 0) \tag{32}$$

where $\mu_m > 0$ is a constant of integration, as defined by the transversality condition.

The case of singular trajectories for both evader and pursuer can now be considered. As previously stated a singular arc for the pursuer can exist only if

$$p_{\gamma P} \equiv 0 \tag{33}$$

from where it follows, $\gamma_p \equiv \Theta_f$ and

$$\alpha^* = 0 \tag{34}$$

Along a singular trajectory

$$a_1 = p_v/m - p_m/c \equiv 0 \tag{35}$$

Substituting now (28)-(35) into H as defined in (115)

$$H = -v + \cos\phi_{Ef} - p_v v^2(k-1)/m = 0 \tag{36}$$

where $\phi_{Ef} = \gamma_{Ef} - \Theta_f$

Differentiating (35) with respect to time, taking into account (36) and rearranging,

$$\dot{a}_1 = -[v^2+(c-\cos\phi_{Ef})v-2c\cos\phi_{Ef}]/cmv = 0 \tag{37}$$

Differentiating (37) with respect to time, taking into account (36), (37) and (12) and rearranging it is obtained,

$$\ddot{a}_1 = (v^2+2c\cos\phi_{Ef})[T'-(k-1)v^2]/cm^2v^2 = 0 \tag{38}$$

From (37) and (38) it follows that along a singular arc

$$T_s = (k-1)v_s^{\,2} \tag{39}$$

and

$$v_s = [\cos\phi_{Ef} - c + (\cos^2\phi_{Ef} + 6\cos\phi_{Ef}c + c^2)^{1/2}]/2 \tag{40}$$

<u>On a singular arc the pursuer flies with a constant velocity v_s along a straight line ($\alpha^*=0$).</u>

The possible existence of a singular control strategy for the evader is associated with $p_{\gamma E}=0$ for some finite period of time. For $\gamma_{Ef}=\Theta_f$ it follows that $p_{\gamma E}(t'_f)=p_{\gamma E}(t_f)=0$, which leads to $p_{\gamma E}(t')\equiv0$ for at least some finite period of time in the neighbourhood of t'_f.

The singular strategy can be obtained differentiating twice $p_{\gamma E}$ with respect to time,

$$\ddot{p}_{\gamma E} = \sigma\cos(\gamma_E-\Theta_f)u_E \tag{41}$$

and equating $\ddot{p}_{\gamma E}$ to zero along the singular trajectory. It leads to the singular evader control,

$$u_E^{\,s} = 0 \tag{42}$$

For all the other final conditions,

$$u_E^{\,*} = -\text{sign}[\sin(\gamma_E-\Theta_f)] \tag{43}$$

4. BARRIER TRAJECTORIES

The trajectories on the barrier can be obtained by a backward integration of the equations of motion.

At $t' = t'_f$, $a_1 = \mu_m / c > 0$, from where it follows, $T^* = 0$.

The pursuit ends with a coasting arc.

The integration begins at the Boundary of the Useable Part defined by,

$$v = v_f = \cos\phi_{Ef} / \cos\phi_{Pf},$$

$$r = r_f,$$

$$m = 1.$$

where $\phi_{pf} = \gamma_{pf} - \Theta_f$.

The evader singular control $u^{*s}_E = 0$ gives origin to a "Universal Surface". It reaches the target set and serves itself as the focus of attractive optimal trajectories.

REFERENCES

1. Prasad, U.R., Rajan, N., Rao, N.J., "Planar Pursuit Evasion with Variable Speeds, Part 1, Extremal Trajectory Maps", Journal of Optimization Theory and Applications, Vol. 33, No. 3, March 1981, pp. 401-418.

2. Prasad, U.R., Rajan, N., Rao, N.J., "Planar Pursuit Evasion with Variable Speeds, Part 2, Barrier Sections", Journal of Optimization Theory and Applications, Vol. 33, pp. 419-432, 1981.

3. Guelman, M., Shinar, J., Green, A., "Qualitative Study of a Planar Pursuit Evasion Game in the Atmosphere", AIAA paper No. 88-4158, Guidance and Control Conference, Minneapolis, August 1988.

4. Shinar, J., Guelman, M., Green, A., "An Optimal Guidance Law for a Planar Pursuit Evasion Game of Kind", Computers and Mathematics with Applications, Vol. 18, No. 1-3, 1989, pp. 35-44.

5. Blaquiere , A., Gerard, F., Leitmann, G., Qualitative and Quantitative Games, Academic Press, New York, 1969.

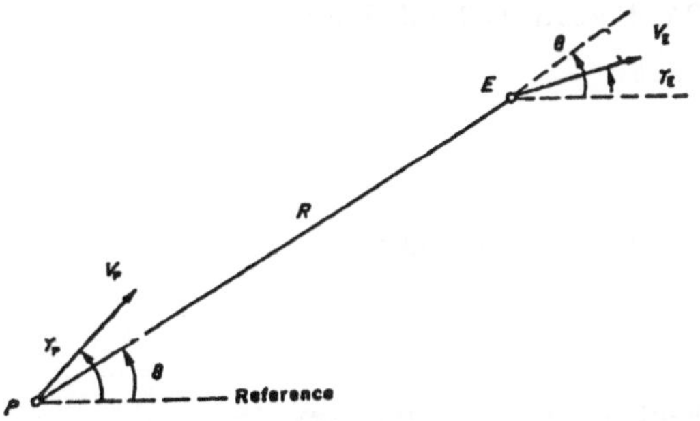

Fig. 1: Pursuit Geometry

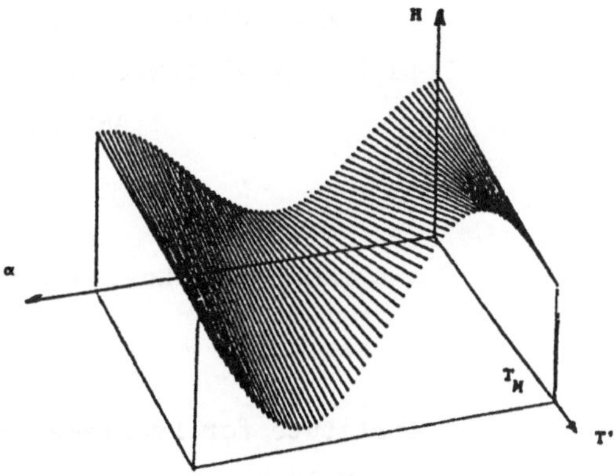

Fig. 2: The Hamiltonian H vs. α and T

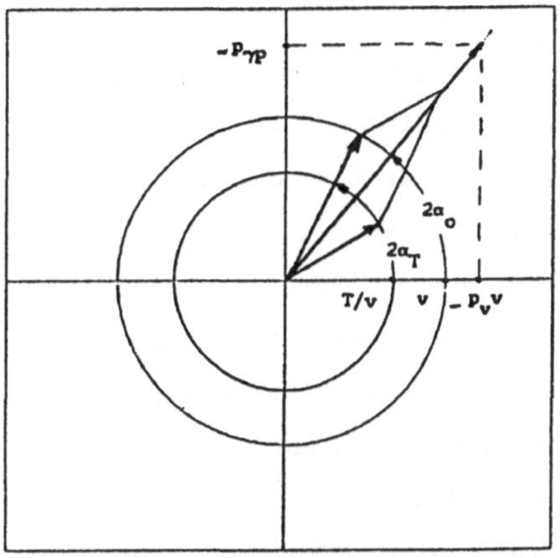

Fig. 3: The α_o and α_T values

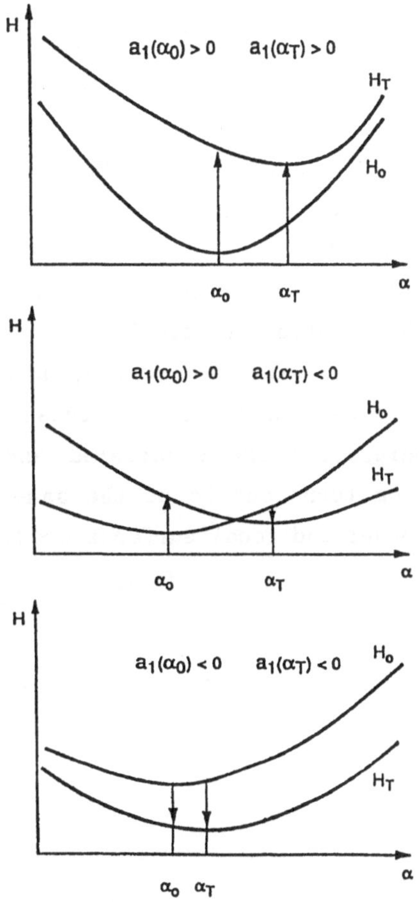

Fig. 4: H_o (T'=0) and H_T (T'=T_M) vs. α

A CONSTRUCTION OF THE VALUE FUNCTION IN SOME DIFFERENTIAL GAMES OF APPROACH WITH TWO PURSUERS AND ONE EVADER

A.Y.Levchenkov, A.G.Pashkov, S.D.Terekhov

Institute for Problems in Mechanics USSR Academy of Sciences, prospect Vernadskogo 101, Moscow, USSR 117526

Abstract

Problems of optimal approach with two pursuers and one evader are investigated under different assumptions about the player's dynamics. The problem's formulation is the same as in [1]. The duration of the game is fixed. The payoff functional is the distance between the evader and the pursuer, closest to it when the game terminates. Three following problems are solved here. The first is the problem of approach when the motion of all players is caused by the control and friction forces. The second one is the problem with simple motions and different pursuers. The third one is the problem with two inertial pursuers and non-inertial evader.
The value functions are constructed in the entire game space. Arising singularities are considered and analyzed. The game space of the last problem turns out to be the same as in [2], where the proximity of the dispersal and focal surfaces occurs [3].

1. The first problem

Let the motion of the pursuers $P_i(x_p^i)$, $x_p^i \in R^2$, (i=1,2) be described by the equations

$$\ddot{x}_p^i + \alpha \dot{x}_p^i = U^i \tag{1.1}$$

The control vectors $U^i = (U_1^i, U_2^i)$ are bounded by the inequality

$$((U_1^i)^2 + (U_2^i)^2)^{1/2} \le \mu \tag{1.2}$$

The pursued object $E(x_e)$ moves in accordance with the equation

$$\ddot{x}_e + \beta \dot{x}_e = V \tag{1.3}$$

with its control bounded by the inequality

$$(V_1^2 + V_2^2)^{1/2} \le \nu \tag{1.4}$$

We suppose, that μ and ν are related by the inequalities

$$\nu \ge \mu \ , \ \nu\beta^{-1} > \mu\alpha^{-1} \tag{1.5}$$

The termination time ϑ of the game is fixed. The game payoff is the distance between the object being pursued and the pursuer closest to it at the instant ϑ, i.e.

$$\begin{aligned} &\sigma(x(\vartheta)) = min[\sigma_1(x(\vartheta)), \sigma_2(x(\vartheta))] \\ &\sigma_i(x(\vartheta)) = \| x_p^i(\vartheta) - x_e(\vartheta) \| \quad (i=1,2) \end{aligned} \tag{1.6}$$

Assuming, that

$$y^i = x_p^i + a(t)\dot{x}_p^i \ , \ a(t) = (1 - exp(-\alpha(\vartheta - t)))\alpha^{-1} \qquad (i=1,2)$$

$$z = x_e + b(t)\dot{x}_e \ , \ b(t) = (1 - exp(-\beta(\vartheta - t)))\beta^{-1}$$

we can reduce the original differential game (1.1)-(1.6) to an equivalent R^6-game

$$\begin{aligned} &\dot{y}_1^i = a(t)U_1^i \ , \ \dot{y}_2^i = a(t)U_2^i \qquad (i=1,2) \\ &\dot{z}_1 = b(t)V_1 \ , \ \dot{z}_2 = b(t)V_2 \end{aligned} \tag{1.7}$$

with the payoff defined as

$$\sigma(y^i(\vartheta), z(\vartheta)) = \min_{i=1,2} \{ \| y^i(\vartheta) - z(\vartheta) \| \} \tag{1.8}$$

The region of attainability of objects P_i (i=1,2) by time ϑ will be a circle of radius $r(t) = \mu\alpha^{-1}[\vartheta-t-a(t)]$. Similarly, the region of attainability of object E will be a circle of radius $R(t) = \nu\beta^{-1}[\vartheta-t-b(t)]$.
We introduce open domains D_j (j=1,2,3,4):

$$D_1 = \{(t,x):\ t_0 \le t < \vartheta,\ \eta > 0\} \cap W,$$

$$D_2 = \{(t,x):\ t_0 \le t < \vartheta,\ \eta < 0\} \cap W,$$

$$D_3 = \{(t,x):\ t_0 \le t < \vartheta,\ \delta > 0\} \setminus W,$$

$$D_4 = \{(t,x):\ t_0 \le t < \vartheta,\ \delta < 0\} \setminus W$$

where $W = \{(t,x):\ t_0 \le t < \vartheta,\ |\delta| < R(t),\ Q|\zeta| > (Q + |\eta|)|\delta|\}$,

$$Q = [sup\ \{0,(R(t))^2 - \delta^2\}]^{1/2},$$

$$\zeta = (1/2)[(y_1^1 - y_1^2)^2 + (y_2^1 - y_2^2)^2]^{1/2},$$

$$\delta = [(z_1-(y_1^1+y_1^2)/2)(y_1^2-y_1^1)+(z_2-(y_2^1+y_2^2)/2)(y_2^2-y_2^1)]/(2\zeta),$$

$$\eta = [(y_2^1-y_2^2)z_1+(y_1^2-y_1^1)z_2+y_1^1y_2^2-y_2^1y_1^2]/(2\zeta)$$

It is not hard to establish that the programmed maximin function for the game (1.2),(1.4),(1.5),(1.7),(1.8) has the form

$$\varepsilon_0(t,x) = max\{\varepsilon_1(t,x),\varepsilon_2(t,x),\ min\{\varepsilon_3(t,x),\varepsilon_4(t,x)\}\} \qquad (1.9)$$

where

$$\varepsilon_{1,2}(t,x) = [(Q \pm \eta)^2 + \zeta^2]^{1/2} - r(t),$$

$$\varepsilon_{i+2}(t,x) = [(z_1 - y_1^i)^2 + (z_2 - y_2^i)^2]^{1/2} + R(t) - r(t) \quad (i=1,2)$$

We note that $\varepsilon_0(t,x)=\varepsilon_j(t,x)$ for $(t,x) \in D_j$ (j=1,2,3,4).

Proposition. The function $\varepsilon_0(t,x)$ is the value of differential game (1.2), (1.4), (1.5), (1.7), (1.8).

The proof of this statement is based on examination of the Bellman-Isaacs equation for $\varepsilon_0(t,x)$ at points of smoothness. On the singular surface $S = \{(t,x):\ t_0 \le t < \vartheta,\ \eta = 0,\ |\delta| \le |\zeta|\}$, which separates domains D_1 and D_2, the function $\varepsilon_0(t,x)$ is not differentiable, but satisfies the necessary and sufficient conditions [4,5].

2. The second problem

This problem was named in [6] as "two motor boats against the ship". Suppose the motions of the fast pursuer $S(X)$, the slow pursuer $Q(Y)$ and the evader $E(Z)$ are described by the equations

$$\dot{X}_i = U_i, \quad \dot{Y}_i = V_i, \quad \dot{Z}_i = W_i, \quad i=1,2 \tag{2.1}$$

The pursuer's and evader's control vectors satisfy the bounds

$$(U_1^2 + U_2^2)^{1/2} \le \mu, \quad (V_1^2 + V_2^2) \le \lambda, \quad (W_1^2 + W_2^2) \le \nu \tag{2.2}$$

the pursuers' superiority being guaranteed by the condition

$$\nu \le \mu < \lambda \tag{2.3}$$

The game is considered in the time interval $[t_0, \vartheta]$. The payoff functional is the Euclidean distance between the evader and the nearest pursuer at the time the game ends, $t = \vartheta$, i.e.

$$\sigma = \min \{\|Z(\vartheta) - X(\vartheta)\| \ , \ \|Z(\vartheta) - Y(\vartheta)\|\} \tag{2.4}$$

The phase space of the system is in fact three-dimensional. For full description of the game position at any given time, it suffices to know the triple of numbers (x,y,z), where $z = \|SQ\|$, and (x,y) are the Cartesian coordinates of E in a coordinate frame attached to the pursuers. The players' domains of attainability are represented in the game plane by circles G_s, G_q, G_e of radii $R_s = \mu(\vartheta - t_0)$, $R_q = \lambda(\vartheta - t_0)$, $R_e = \nu(\vartheta - t_0)$, centered at the points S, Q and E, respectively. Let m be the locus of all points equidistant from the circles G_s and G_q. Denote the points at which the circle G_e cuts m by A^* and A_* (Fig.1). In the cases $G_e \cap m = \{\emptyset\}$ or $A^* = A_*$ it can be shown that the original game degenerates into a two-person game with value ρ^{11} equal to the programmed maximin γ^{11}. Let $A^* \neq A_*$. If $E \notin int\ SA^*QA_*$, the game again reduces to a two-person game (the domain in which $\rho^{21} = \rho^{11}$ is denoted by D_{11}. The domain in which $E \in int\ SA^*QA_*$ is denoted by D_{21}. Anticipating, we remark that in the domain D_{21} the pursuers can increase their chances of success by interacting. The programmed maximin function γ^{21} is defined as follows:

a) $\gamma^{21} = 0$ if $G_s \cup G_q \subset G_e$;

b) $\gamma^{21} = max\{\gamma^*,\gamma_*\}$, if $G_e \setminus (G_s \cup G_q) \neq \{\emptyset\}$

where $\gamma^* = \rho(A^*,G_s) = \rho(A^*,G_q)$ and $\gamma_* = \rho(A_*,G_s) = \rho(A_*,G_q)$, ($\rho(\cdot,\cdot)$ is the Euclidean distance from a point to a set in the plane). It can be seen from the definition that the function γ^{21} is piecewise- smooth in D_{21}, it may fail to be smooth only on the surface $y = 0$ (when $\gamma^{21} = \gamma^* = \gamma_*$).

Let us define an extremal motion of players S and Q as a motion with maximum velocity at the point A^* (or A_*). Denote the corresponding control constants of the players by U^0 and V^0. It will be shown later that the function γ^{21} is u-stable in D_{21}. The examination of γ^{21} to be u-stable at points of it's smoothness leads to verification of Bellman-Isaacs equation and carries out ordinarily. We will now investigate the behavior of the function γ^{21} on the surface $y = 0$. Let us assume that either $\gamma > 0$, or $\gamma = 0$ but $A^*,A_* \in (G_s \cup G_q)$. The stability of the function $\gamma = 0$ when $A^*,A_* \in int(G_s \cup G_q)$ is obvious.

It is convenient to conduct our investigation in a special coordinate frame. Let $a = \|SE\|$, $b = \|EQ\|$, $c = a + b$. Relative to this frame, the value of γ^{21} is determined from the formula

$$\gamma = (R_e^2 + ab(1 - ((R_s - R_q)/c)^2))^{1/2} - (R_s b + R_q a)/c \tag{2.5}$$

As system (2.1) is linear in the controls, it will suffice to verify the u-stability provided maximal velocities of players. Suppose that player E's position is to the right of the perpendicular A^*O ($a > \|SO\|$). Introduce angles as new controls (Fig.2):

$$U_1 = \mu \cos\Phi,\quad V_1 = -\lambda \cos\psi,\quad W_1 = -\nu \cos\chi$$
$$U_2 = \mu \sin\Phi,\quad V_2 = \lambda \sin\psi,\quad W_2 = \nu \sin\chi$$

(if $a \le \|SO\|$ we replace $\chi \mapsto -\chi$). Note that the angles corresponding to controls U^0 and V^0 in the new notation are Φ^0 and ψ^0.

Let $0 \le \chi \le \pi$ (for $-\pi \le \chi \le 0$ the reasoning is analogous). Divide the remaining part of the vectogram of player E into two subsets:

1) controls $\chi(t)$ which, together with Φ^0 and ψ^0, generate a motion with $\dot{y} > 0$; these controls constitute the set K_1;

2) all other controls - the set K_2.

For controls in the first subset, u-stability of γ^{21} can be proved by

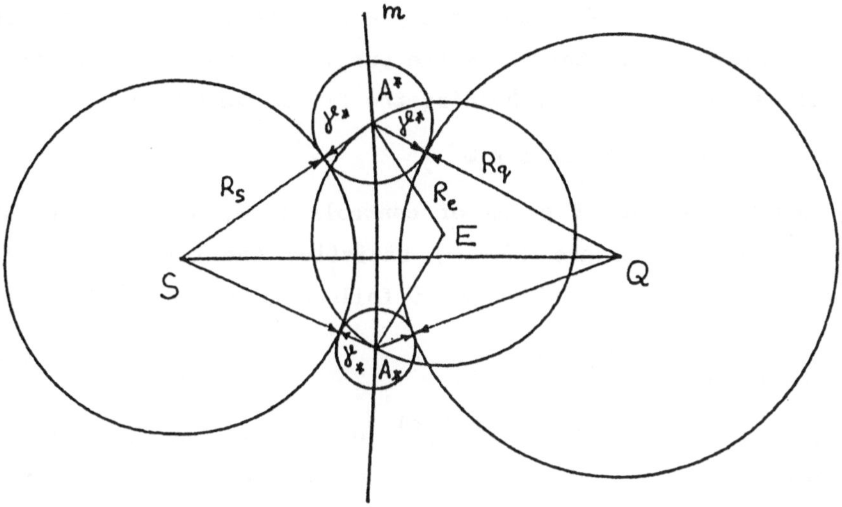

Fig.1

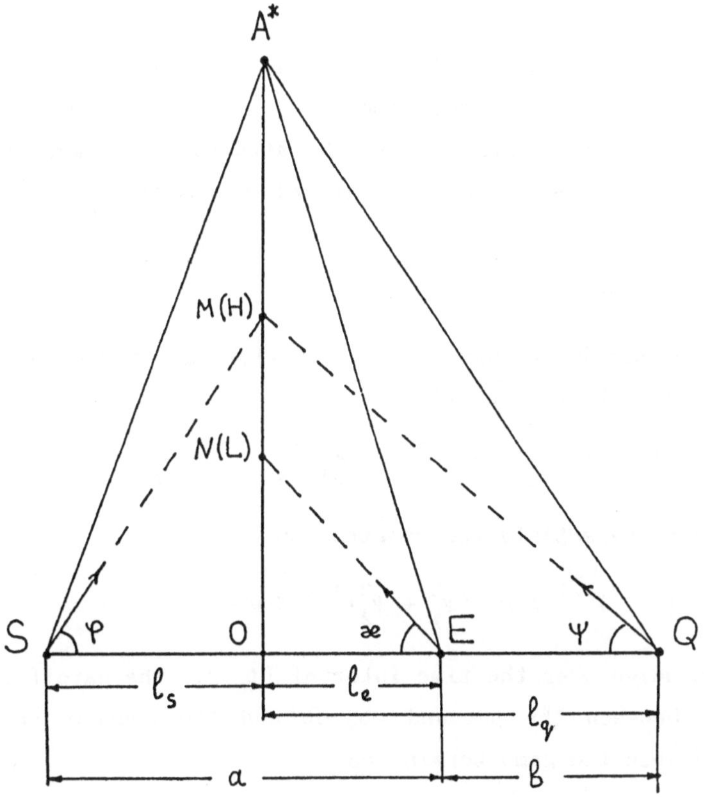

Fig.2

arguments similar to those in the case $y \neq 0$. Consider the second subset of controls for player E. In this case the pursuers can cause phase trajectory to slide along the plane $y = 0$ holding the finite condition

$$c\nu \sin \chi = b\mu \sin \phi + a\lambda \sin \psi \tag{2.6}$$

Let $\Phi(\chi)$ denote the set of pairs of controls (ϕ,ψ) such that $0 \leq \phi \leq \phi^0$, $0 \leq \psi \leq \psi^0$ and the triple (ϕ,ψ,χ) satisfies the condition (2.6). The u-stability of γ^{21} on the surface $y = 0$ follows from the in equality

$$\max_{\chi \in K_2} \min_{(\phi,\psi)\in\Phi(\chi)} (d\gamma^{21}/dt) \leq 0 \tag{2.7}$$

which expresses the fact, that γ^{21} does not increase on sliding trajectories. As it is u-stable, the programmed maximin γ^{21} is identical with the value ρ^{21} of the differential game (2.1)-(2.4); moreover, in the domain D_{21} interaction of players S and Q is essential in order to attain the optimum result.

We now consider problem (2.1)-(2.4) without fixing the final instant of the game, and determine the minimum time T at which the attainability domains of players S and Q completely cover the attainability domain of E. It is easy to show, that T is the minimum time of pointwise capture.

3. The third problem

Let the motions of the pursuers $P_i(X^i)$ (i=1,2) and the evader $E(Z)$ on the plane be described by the equations

$$\dot{X}_1^i = X_3^i,\ \dot{X}_3^i = U_1^i,\ \dot{X}_2^i = X_4^i,\ \dot{X}_4^i = U_2^i,\ \dot{Z}_1 = V_1,\ \dot{Z}_2 = V_2 \tag{3.1}$$

The control vectors satisfy the constraints

$$((U_1^i)^2 + (U_2^i)^2)^{1/2} \leq \mu,\ (V_1^2 + V_2^2)^{1/2} \leq \nu \tag{3.2}$$

The game is studied over the time interval $[t_0,\vartheta]$. The payoff functional is the distance between the pursued object and the nearest pursuer at the instant $t = \vartheta$ when the game terminates

$$\gamma = \min\ [((Z_1(\vartheta) - X_1^i(\vartheta))^2 + (Z_2(\vartheta) - X_2^i(\vartheta))^2]^{1/2} \tag{3.3}$$

Later we redenote the symbols P_i as the centers of attainability regions of the inertial objects. In the reduced space the equations of the motions take the form ($z = \|P_1P_2\|$, (x,y) - Cartesian coordinates of E in the frame related with P_i)

$$\begin{aligned}
\dot{x} &= V_1 - (\vartheta - t)[U_1^1 + U_1^2]/2 + y(\vartheta - t)[U_2^2 - U_2^1]/2z \\
\dot{y} &= V_2 - (\vartheta - t)[U_2^1 + U_2^2]/2 - x(\vartheta - t)[U_2^2 - U_2^1]/2z \\
\dot{z} &= (\vartheta - t)[U_1^2 - U_1^1]/2
\end{aligned} \qquad (3.4)$$

The constraints on the controls kept the form (3.3). The payoff functional is defined by the formula

$$\gamma = [((z(\vartheta) - |x(\vartheta)|)^2 + y^2(\vartheta)]^{1/2} \qquad (3.5)$$

The value function was obtained for all possible positions of the game. The typical optimal path consists of three parts (Fig.3):

1. Trajectory of extremal guidance to point N^* on the time interval $[t_0, T_f]$;
2. The time interval $[T_f, T]$ - trajectory of proportional pursuit (where the relations $z/x = const$, $y = 0$ hold);
3. The path of extremal aiming when $T \leq t \leq \vartheta$.

The particularity of the solution of this problem is the proximity of two singular surfaces - the dispersal S_d and the focal S_f. Dispersal surface is the part of the plane $y = 0$, which is close to the terminal set. From each point of S_d the two trajectories go out, one moving to the positive subspace $y > 0$ (related to the "extremal aiming strategy" at the "upper" point A_1), the other - to the negative subspace (symmetrical "lower" point A_2 - see Fig.3).

When stepping back from the terminal hyperplane $t = \vartheta$ the dispersal surface transforms to the focal one. Focal surface contains optimal trajectories itself and "attracts" trajectories from the symmetrical halfspaces. Unlike the universal surface the trajectories come on to the focal one tangentially, i.e. the condition $dy/dt = 0$ is satisfied.

On Fig.4 one can see isocost lines when $\mu = \nu = 1$, $\vartheta = 3$, $z = 4$. The curve S_1 and the symmetrical curve separate the positions of E, where pursuit is

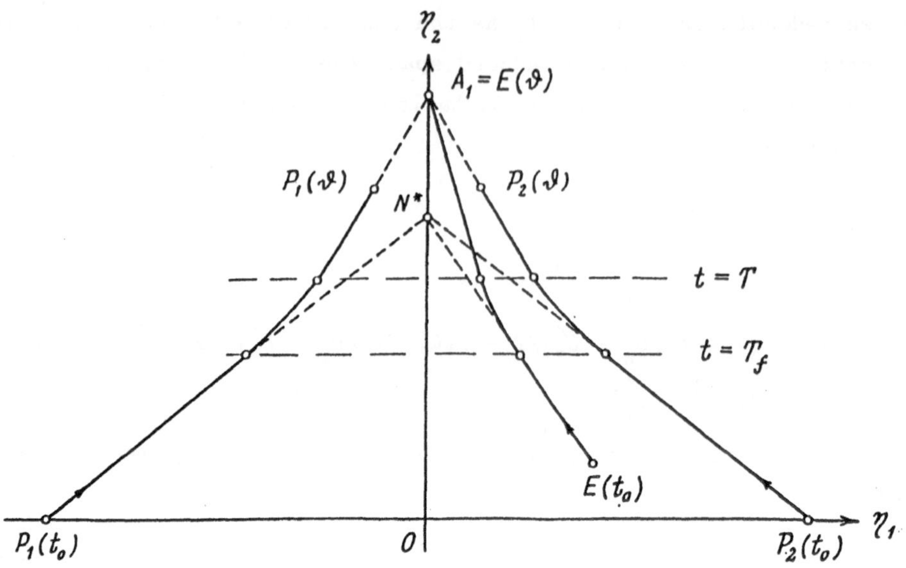

η_2
$A_1 = E(\vartheta)$
$P_1(\vartheta)$
$P_2(\vartheta)$
N^*
$t = T$
$t = T_f$
$E(t_0)$
$P_1(t_0)$
0
$P_2(t_0)$
η_1

Fig.3

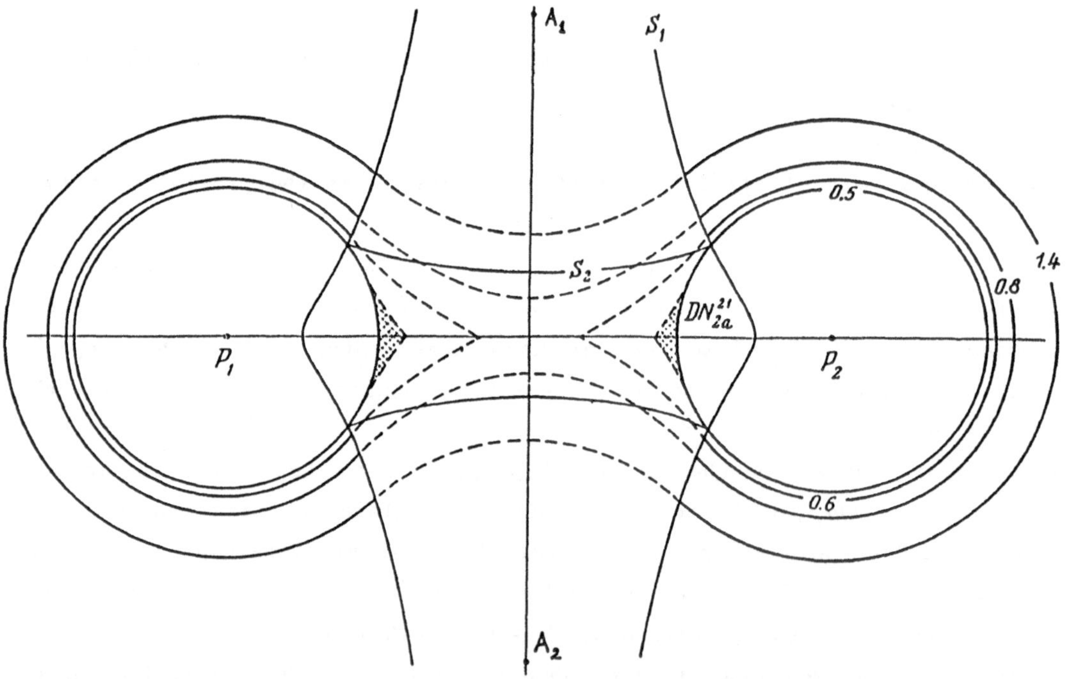

A_1
S_1
0.5
S_2
1.4
0.8
DN^{21}_{2a}
P_1
P_2
0.6
A_2

Fig.4

"one to one" (external region), from the positions, where pursuit is collective (internal region). The circles with the cost 0.5 and the centers in P_1, P_2 and touched domains are the regions of constant cost. The curve S_2 and the symmetrical one separate the positions of E, where the optimal trajectory consists of three parts (see above), from the region, in which the optimal paths don't reach singular surfaces. All indicated boundaries are calculated effectively by the numerical solution of transcendental equations.

References

[1] Pashkov A.G., Terekhov S.D. A Differential Game of Approach with Two Pursuers and One Evader // J.Optimiz.Theory&Appl., 1987, v.55, No 2, pp.303-311.

[2] Breakwell J.V., Hagedorn P. Point capture of Two Evaders in Succession // J.Optimiz.Theory & Appl., 1979, v.27, No 1, pp.89-97.

[3] Levchenkov A.Y., Pashkov A.G. Differential Game of Optimal Approach of Two Inertial Pursuers to a Non-inertial Evader // J. Optimiz. Theory & Appl., 1990, v.65, No 3, pp.501-518.

[4] Krasovskii N.N., Subbotin A.I. Game Theoretical Control Problems - Springer, New York etc., 1988, 518 p.

[5] Subbotin A.I. Generalization of the Main Equation of Differential Game Theory // J.Optimiz.Theory&Appl., 1984, v.43, No 1,pp.103-133.

[6] Isaacs R. Differential Games. - John Wiley, New York, 1965.

BARRIER TRAJECTORIES OF A REALISTIC MISSILE/TARGET PURSUIT–EVASION GAME

Breitner, M.[1), Grimm, W.[2)], Pesch, H.J.[3)]

1. Introduction

Presently a computer based pilot's decision aid is developed for future fighter aircraft. The objective is to increase the probability of success in the case of own attack and to improve the chance of survival in a situation of hostile attack. In the first case the firing range of a missile must be estimated by the onboard computer to pick the right launch time. In the second case the aircraft becomes the target of an adversary missile. Now, a favourable evasive maneuver must be initiated in time and controlled by an autonomous guidance algorithm.

For both purposes the pursuit–evasion game concept (Isaacs [8]) applied to the missile/target encounter provides the suitable mathematical framework. The game solution indicates both the firing range of the missile and the optimal evasive maneuver of the target.

First, game solutions were obtained for simplified models (e.g. linearized equations of motion (Shinar, Gutman [14]), simplified dynamics in the vertical plane (Guelman et al. [7])) or approximated with the help of singular perturbation technique (Shinar, Gazit [13]).

The game solution also shows the optimal missile guidance. If one is only interested in this aspect a one sided optimal control formulation is sufficient. This approach was applied to the complete point mass model in the vertical plane to maximize the missile's range subject to the condition that there remains enough energy for the final pursuit of the target (Kumar et al. [11]). The final pursuit phase itself was not explicitly considered in the study. The optimal trajectories of [11] served as reference flight paths in a closed–loop missile guidance law (Kumar et al. [10]).

The intention of the present paper is to resume the differential game approach and to combine it with a dynamic model containing realistic approximations for thrust and drag. The objective is to determine barrier trajectories in the vertical plane under the assumption that complete state and model information is available to each vehicle. Initial speed and altitude of the target aircraft are systematically varied within the flight envelope. For the missile two different launch positions are considered. The barrier trajectories are computed numerically by solving multipoint boundary value problems derived from the necessary conditions for the barrier. As a by–product, the dependence of the firing range on the altering initial values is obtained. Thus, the results draw a detailed picture of the firing envelope for the underlying vehicle models.

---

1) Graduate student, Mathematical Institute, Munich University of Technology.
2) Research scientist, Institute for Flight Systems Dynamics, German Aerospace Research Establishment DLR, D–8031 Oberpfaffenhofen.
3) Privatdozent, Mathematical Institute, Munich University of Technology, Arcisstr. 21, P.O. Box 20 24 20, D–8000 München 2.

2. Dynamic Model

The flight path of both vehicles (missile and target) is the solution of the following ODE–system:

$$\dot{x} = V \cos\gamma \tag{1}$$

$$\dot{h} = V \sin\gamma \tag{2}$$

$$\dot{V} = (T - D_o - \cos^2\gamma\ D_i) / m - g \sin\gamma \tag{3}$$

For abbreviation let

$$\dot{z} = f(z, \gamma, t) \tag{4}$$

with $z := (x, h, V)^T$ denote the ODE–system above. (1) – (3) describes the motion of a vehicle in the vertical plane with the flight path angle γ as control variable. This is already a model reduction in the sense of singular perturbation theory, since γ is a state variable in reality. The authors are aware of the falsifying effects of the simplification, as described by Katzir et al. [9].
The symbols in (1) – (3) have the usual meaning:

x:	range	h:	altitude
V:	velocity	γ:	flight path angle
T:	thrust	D_o:	zero–lift drag
D_i:	induced drag	m:	mass
g:	gravitational acceleration (constant)		

The structure of the drag components is

$$D_o(h, V) = q\, S\, C_{D0}(M), \qquad D_i(h, V) = C_{Di}(M)\, (m\, g)^2 / (q\, S)$$

with $q(h, V) = \rho(h)\, V^2 / 2$ and the following meaning of the additional symbols:

q:	dynamic pressure	M:	Mach number, = V / speed of sound
S:	reference wing area	C_{D0}:	zero–lift drag coefficient
ρ:	air density	C_{Di}:	induced drag coefficient

$\rho(h)$ is a realistic model function for air density.
In the generic <u>missile</u> model thrust is a decreasing step function of time. It is divided into the boost phase (the first 3 sec), the march phase (the following 12 sec) and the coasting phase (T = 0). In accordance with the thrust profile the mass is a piecewise linearly decreasing function of time which remains constant in the coasting phase. Therefore, the system (4) is explicitly time dependent.
In the <u>aircraft</u> model $T(h, M)$, $C_{D0}(M)$ and $C_{Di}(M)$ are analytic functions which are carefully adapted to realistic tabular data of a fighter aircraft. We assume maximum thrust to be optimal for the aircraft – an assumption to be verified a posteriori. Mass is assumed to be constant. The aircraft is constrained by its dynamic pressure limit q_{max}:

$$Q(z) := q - q_{max} \leq 0 \tag{5}$$

Differentiation of $Q(z) = 0$ w.r.t. time yields the control along the constraint (5):

$$0 = Q^{(1)}(z, \gamma) := \frac{d}{dt} Q(z) \qquad \Longrightarrow \qquad \gamma = \gamma^q(h) \tag{6}$$

Since h and V are coupled by $Q(z) = 0$ along the q_{max}–boundary, h is actually the only argument of γ^q.

3. Problem Formulation

In the sequel the subscripts P and E denote pursuer (=missile) and evader (=aircraft), respectively. We consider a game of kind with the terminal manifold

$$0 = R(z_P, z_E) = d^2 - (x_P - x_E)^2 - (h_P - h_E)^2 . \tag{7}$$

There is a barrier separating the capture zone of the pursuer from the escape zone of the evader. Since the pursuer (=missile) has a limited amount of fuel the existence of the barrier is clear from physical intuition. The task is to find the barrier trajectory determined by prescribed initial values for x_P, h_P, V_P, h_E and V_E. Without loss of generality one can set $x_P(0) = 0$. The resulting value of $x_E(0)$ (≥ 0) may be thought of as the maximum range from which a hit can be achieved. $h_E(0)$ and $V_E(0)$ are varied all over the flight envelope of the evader aircraft. $z_P(0)$ is restricted to two cases:

case 1:	$h_P(0) = 5000$ m	$V_P(0) = 250$ m/s
case 2:	$h_P(0) = 12000$ m	$V_P(0) = 400$ m/s

The main interest of the present paper is to evaluate the optimal control along a barrier trajectory for a representative set of initial conditions.

4. Necessary Conditions for the Barrier Trajectories

The necessary conditions for the barrier involve the adjoint vector $\lambda_z := (\lambda_{z,P}, \lambda_{z,E})$ corresponding to the state vector (z_P, z_E). Together with the state vector they form the expression

$$H := H_P + H_E + \lambda_t$$

with $\qquad H_P := \lambda_{z,P}{}^T f(z_P, \gamma_P, t)$

and $\qquad H_E := \lambda_{z,E}{}^T f(z_E, \gamma_E, t) + \mu_E Q^{(1)}(z_E, \gamma_E) . \qquad (8)$

λ_t is an adjoint variable accounting for the fact that the dynamic system is explicitly time dependent. The multiplier μ_E vanishes as long as $Q(z_E) < 0$ and is nonpositive on constrained arcs:

$$\begin{aligned} \mu_E &= 0 \qquad \text{for} \quad Q(z_E) < 0 \\ \mu_E &\leq 0 \qquad \text{for} \quad Q(z_E) = 0 \end{aligned} \tag{9}$$

The optimal control along the barrier is determined by the semipermeability condition:

$$0 = \min_{\gamma_P} \max_{\gamma_E} H . \tag{10}$$

This interpretation is based on the fact that λ_z is the normal vector on the barrier surface. Since H is separable in γ_P and γ_E the controls can be determined by one sided optimization of H and the min– and max–operations in (10) can be exchanged (minmax–assumption):

$$\min_{\gamma_P} H_P , \ \max_{\gamma_E} H_E \ \Rightarrow \ 0 = \partial H_i/\partial \gamma_i , \quad i = P, E. \tag{11}$$

Along the dynamic pressure boundary γ_E is determined by (6). In this case $0 = \partial H_E/\partial \gamma_E$ essentially serves to evaluate μ_E.
The terminal value of the adjoint vector is given by the transversality conditions

$$\lambda_{z_i}(t_f) = \partial R/\partial z_i^{\,T} , \quad i = P, E, \qquad \lambda_t(t_f) = 0 . \tag{12}$$

Along the barrier trajectories λ_z must satisfy the adjoint differential equations

$$\dot{\lambda}_{z_i} = - \partial H/\partial z_i^{\,T} , \quad i = P, E, \qquad \dot{\lambda}_t = - H_t . \tag{13}$$

Combining (7) – (12) yields the condition for the intersection of the terminal manifold and the barrier, the "Boundary of the Usable Part":

$$\begin{aligned} Q(z_E(t_f)) < 0 \quad &\Rightarrow \quad V_P(t_f) = V_E(t_f) \\ Q(z_E(t_f)) = 0 \quad &\Rightarrow \quad V_P(t_f) = V_E(t_f) \cos(\theta(t_f) - \gamma^q(h_E(t_f))) \end{aligned} \tag{14}$$

where θ is the aspect angle from the pursuer on to the evader:

$$\theta = \arctan \frac{h_E - h_P}{x_E - x_P} \tag{15}$$

At the entry point t_{q1} on to a constrained evader subarc the jump condition

$$\lambda_{z,E}(t_{q1}^+) = \lambda_{z,E}(t_{q1}^-) - \mu_E(t_{q1}^+) \, \partial Q/\partial z_E^{\,T} \tag{16}$$

must be satisfied. An eventual exit point $t_{q2} < t_f$ is determined by

$$\mu_E(t_{q2}^-) = 0 . \tag{17}$$

The treatment of the constraint is tacitly adopted from optimal control theory (e.g. [4], [5]). State constrained differential games have not yet received much attention in literature. Some publications on this topic confirming the conditions above are Blaquiere et al. [2], Bernhard [1] and Taylor [15]. A solid justification for the present example is given by Breitner [3].

5. Formulation and Numerical Solution of Multipoint Boundary Value Problems

The conditions of the physical problem (section 3) and the necessary conditions for barrier trajectories (section 4) constitute a multipoint boundary value problem (MPBVP) for 13 dependent variables: z_P, z_E, $\lambda_{z,P}$, $\lambda_{z,E}$, t_f. In the unconstrained case (no interior point conditions) the number of boundary conditions matches the number of dependent variables:

at $t = 0$: prescribed initial values for x_P, h_P, V_P, h_E, V_E (5 conditions)
at $t = t_f$: (7), (12), (14) (1 + 6 + 1 conditions)

λ_t does not affect the other variables and is therefore omitted. To cope with the free final time the independent variable is normalized in the usual manner. The jump condition (16) suggests backward integration of the trajectory: The limit value of μ_E for $t \to t_{q1} + 0$ is known on backward integration; it would represent another unknown parameter in the opposite direction. Thus, the roles of initial and terminal conditions are reversed from the view of the MPBVP.
Dependent on the "switching structure", i.e. the sequence of constrained and unconstrained arcs, the problem is augmented by switching points and switching conditions:

Switching structure A: $Q < 0$ / $Q = 0$. The unknown switching point t_{q1} is determined by $Q(t_{q1}) = 0$. Note that the jump (16) can be performed during numerical integration.
Switching structure B: $Q < 0$ / $Q = 0$ / $Q < 0$. In addition to A the unknown exit point t_{q2} and the interior point condition (17) is added.
Switching structure C: unconstrained case. MPBVP as described above.

The MPBVP's above are solved with the program package BNDSCO (see Bulirsch [6], Oberle, Grimm [12]), an implementation of the multiple shooting method taylored for the above kind of problem.

6. Discussion of the Numerical Results

The above problem is solved for case 1 and 2 for all pairs $(h_E(0), V_E(0))$ within the evader's flight envelope (see Fig. 1). Regions, where $h_E < 0$ or $\lambda_{V,E} < 0$ is encountered along the barrier trajectory, are omitted. Trajectories with $h_E < 0$ only occur for an evader starting in the right lower corner of his flight envelope (region I in Fig. 1). $\lambda_{V,E} < 0$ indicates that maximum thrust is not optimal for the evader; this happens for case 1 in a small domain along the upper right edge of the flight envelope (region II in Fig. 1). Switching structure A is found to be valid for most of the barrier trajectories (region A in Fig. 1). B and C only occur in case 1 (regions B and C in Fig. 1, respectively).
With the help of the multiple shooting algorithm also the boundaries of the different regions are computed. E.g. the boundary points of region I can be determined by solving a modified version of the MPBVP belonging to switching structure A: Let $h_E(0)$ be a free parameter of the problem and also the time t_h, where h_E takes its minimum. In return, two additional switching conditions are added: $h_E(t_h) = 0$ and $\gamma_E(t_h) = 0$. The resulting $h_E(0)$ together with the prescribed value $V_E(0)$ constitutes a point on the upper boundary of region I.

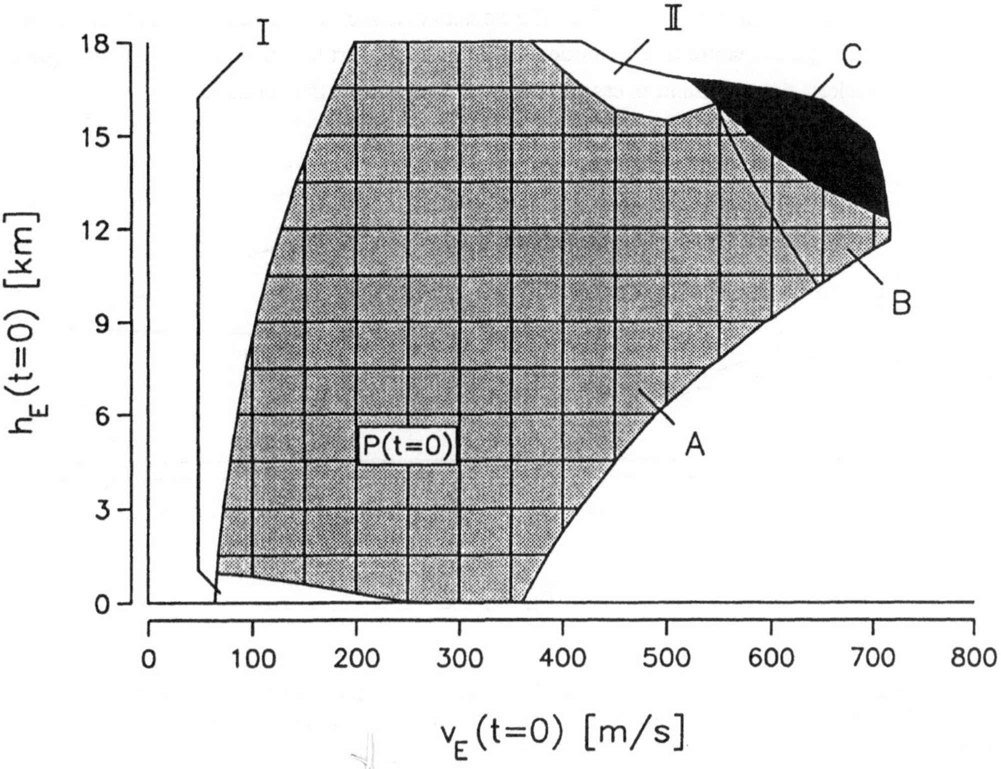

Figure 1: Switching structures of the evader's control in case 1. Regions A, B, C are the validity domains of switching structures A, B and C, respectively. In region I the evader would dive below sea level. In region II maximum thrust is not optimal for the evader. P(t = 0) indicates the initial state of the pursuer ($h_P(0) = 5$ km, $V_P(0) = 250$ m/s).

The whole set of MPBVP's is divided in suitable continuation series. Within each series the computational process is accelerated by generating start values extrapolated from already obtained solutions. Due to the high accuracy of the solutions this method causes an enormous speedup.

Figure 2 shows the the influence of the q_{max}–constraint on the game solution. The trajectories are an example for case 1. The dashed flight paths result on neglecting the constraint, leading to the typical dive of the evader ($q_E(t_f) \approx 150$ kN/m^2 , $q_{max,E} = 80$ kN/m^2). The constraint prevents the dive and extends the flight time considerably ($t_f \approx 33$ sec $\longrightarrow$ $t_f \approx 86$ sec). Of course, the firing range $x_E(0)$ increases (from about 7.2 km to about 11.0 km) when the evader is constrained by his q_{max}–boundary. The pursuer uses the longer duration for a climb up to about 18.5 km starting with $\gamma_P(0) > 60^0$. The small drag at this altitude brings him a distinct gain of range. This "lofting" is typical for all examples, where t_f is not too small. However, the flight path requires a missile with an all aspect seeker. Note that the "look angle" $\omega_P := \gamma_P - \theta$ (for θ see Equ. (15)) increases from -80^0 to 0^0 within a few seconds at the end of the trajectory. This is similar for all solutions obtained. In Fig. 3 the trajectories are depicted in the altitude/velocity diagram of the evader. The constrained subarc is a small portion of the whole flight path only – a typical feature of all solutions which do not start too close to the dynamic pressure boundary. Nevertheless, the constraint is essential for the character of the solution.

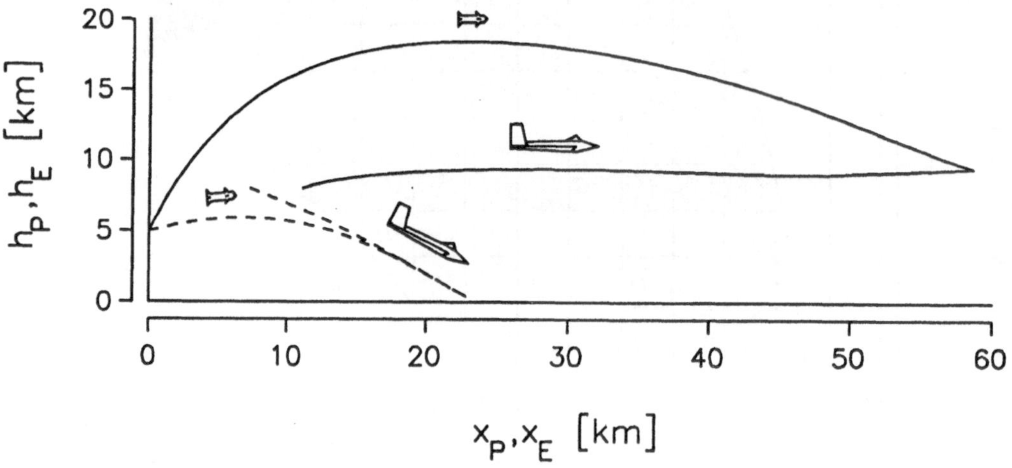

Figure 2: Barrier trajectory for case 1 in the (x, h)–plane. The picture shows the difference between the solutions with and without the dynamic pressure constraint for the evader (solid and dashed lines, respectively). $h_P(0) = 5$ km, $V_P(0) = 250$ m/s, $h_E(0) = 8$ km, $V_E(0) = 500$ m/s.

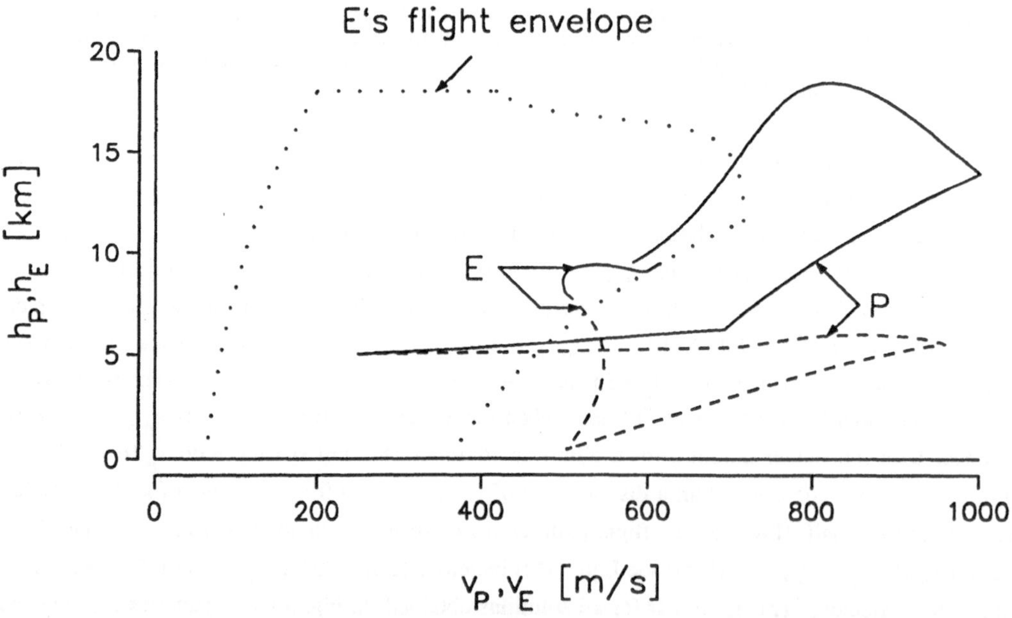

Figure 3: Barrier trajectory for case 1 in the (h, V)–plane. The picture shows the difference between the solutions with and without the dynamic pressure constraint for the evader (solid and dashed lines, respectively). $h_P(0) = 5$ km, $V_P(0) = 250$ m/s, $h_E(0) = 8$ km, $V_E(0) = 500$ m/s.

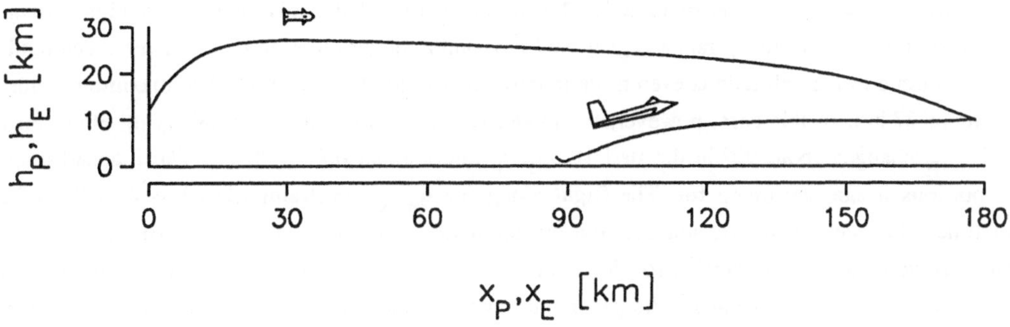

Figure 4: Barrier trajectory for case 2 in the (x, h)–plane.
$h_E(0) = 2$ km, $V_E(0) = 100$ m/s, $h_P(0) = 12$ km, $V_P(0) = 400$ m/s.

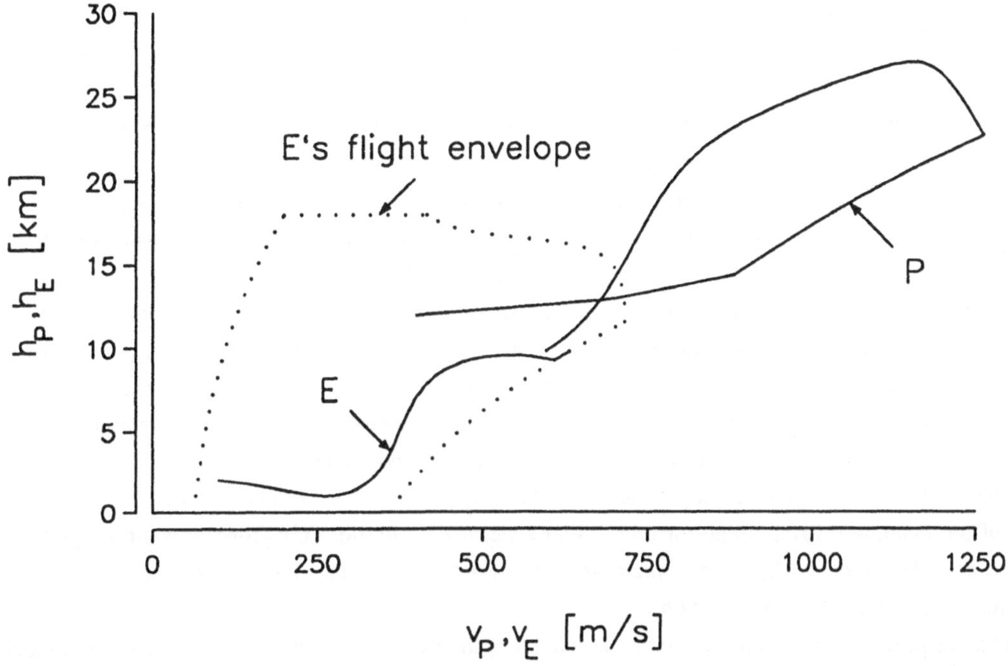

Figure 5: Barrier trajectory for case 2 in the (h, V)–plane.
$h_E(0) = 2$ km, $V_E(0) = 100$ m/s, $h_P(0) = 12$ km, $V_P(0) = 400$ m/s.

Figs. 4 and 5 show an example of case 2. The initial energy of the missile is much higher than in case 1 leading to a larger firing range ($x_E(0) \approx 87$ km) and a longer duration ($t_f \approx 201$ sec). The tendency of the missile to gain altitude is even more marked than in the first example. The maximum value of h_P is about 27 km. In this case, however, the model reduction probably has a falsifying effect on the solution. γ_E changes about 90^0 in the first 18 sec, γ_P alters about 60^0 in 30 sec. The induced drag in these portions is not accounted for. The flight along the q_{max}–constraint takes less than 9% of the flight time. One has the impression that the evader really wants to avoid the constraint as long as possible. Nevertheless, it is decisive for the solution. In spite of the low initial energy of the aircraft a dive below sea level is prevented. Altogether the evader's trajectory reminds one to maximum range extremals in former studies.

7. Conclusions

This paper presents barrier trajectories of a pursuit–evasion game formulation of a missile/target encounter. Two practically important observations are made:

- For sufficiently long flight time the missile climbs up to 20 km altitude or more to exploit the small drag for enlarging its range. Accordingly, the optimal flight path angle at launch time is extremely high (up to 70^0).
- The optimal target evasion maneuver generally ends on the dynamic pressure constraint, which prevents the aircraft from diving.

The multiple shooting algorithm turns out to be a useful tool for solving differential games:

- For the solution of (one sided) optimal control problems there are competing nonlinear programming algorithms. This is not true in the area of differential games.
- The solutions of the typically ill–conditioned multipoint boundary value problems are highly accurate and provide full information about the "switching structure" of the optimal control.
- The multiple shooting technique and the associated software package is an excellent tool to generate a series of solutions for continuously varying boundary conditions ("continuation" or "homotopy") and to demarcate the validity domains of switching structures.

The reduced vehicle model (flight path angle as control) permits a two dimensional tabular representation of the maximum firing range for each initial state of the missile. This property could be preserved in the complete model (flight path angle as state variable) if initial elevation was treated as a free parameter also subject to optimization.

Results are presented for a missile, which is guided optimally in the sense of differential game theory. To quantify the loss of firing range for a conventionally guided missile would be an interesting task.

References

[1] Bernhard, P., "New Results about Corners in Differential Games, Including State Constraints", Proc. of the 6th Triennial IFAC World Congress, Boston, Ma., 1975.

[2] Blaquiere, A., Gerard, F., Leitmann, G., "Quantitative and Qualitative Games". Academic Press: New York and London, 1969.

[3] Breitner, M., "Numerische Berechnung der Barriere eines realistischen Differentialspieles", Diploma Thesis, Mathematical Institute, Munich University of Technology, 1990.

[4] Bryson, A.E., Denham, W.F., Dreyfus, S.E., "Optimal Programming Problems with Inequality Constraints I, Necessary Conditions for Extremal Solutions", AIAA Journal, Vol. 1, 1963.

[5] Bryson, A.E.jr., Ho, Y.–C., "Applied Optimal Control", Waltham, Ma., Ginn & Co., 1969.

[6] Bulirsch, R., "Die Mehrzielmethode zur numerischen Lösung von nichtlinearen Randwertproblemen und Aufgaben der optimalen Steuerung", Report of the Carl–Cranz–Gesellschaft, Heidelberg, 1971.
Reprint: Munich University of Technology, Mathematical Institute, Munich, 1985.

[7] Guelman, M., Shinar, J., Green, A., "Qualitative Study of a Planar Pursuit Evasion Game in the Atmosphere", Proc. of the AIAA Guidance, Navigation and Control Conference, Minneapolis, Minnesota, August 15 – 17, 1988.

[8] Isaacs, R., "Differential Games", Wiley, New York, 1965.

[9] Katzir, S., Cliff, E.M., Lutze, F.H., "A Comparison of Dynamic Models for Optimal Midcourse Guidance", Proc. of the AIAA Guidance, Navigation & Control Conference, Boston, Ma., August 1989, pp. 1685–1691.

[10] Kumar, R.R., Seywald, H., Cliff, E.M., "Near Optimal 3–D Guidance Against a Maneuvering Target", Proc. of the AIAA Guidance, Navigation & Control Conference, Boston, Ma., August 1989, pp. 482–495.

[11] Kumar, R.R., Seywald, H., Cliff, E.M., Kelley, H.J., "3–D Air–to–Air Missile Trajectory Shaping Study", Proc. of the AIAA Guidance, Navigation & Control Conference, Boston, Ma., August 1989, pp. 470–481.

[12] Oberle, H.J., Grimm, W., "BNDSCO – A Program for the Numerical Solution of Optimal Control Problems", Internal Report No. 515–89/22, Institute for Flight Systems Dynamics, DLR, Oberpfaffenhofen, FRG, 1989.

[13] Shinar, J., Gazit, R., "Optimal No–Escape Envelopes of Guided Missiles", Proc. of the AIAA Guidance, Navigation and Control Conference, Paper No. 85–1960, Snowmass, Colorado, USA, 1985.

[14] Shinar, J., Gutman, S., "Three–Dimensional Optimal Pursuit and Evasion with Bounded Controls", IEEE Transactions of Automatic Control, Vol. AC–25, No. 3, June 1980, pp. 492 – 496.

[15] Taylor, J.G., "Necessary Conditions of Optimality for a Differential Game with Bounded State Variables", IEEE Transactions on Automatic Control, Vol. AC–20, December 1975, pp. 807–808.

NONLINEAR EFFECTS IN A VARIABLE SPEED PURSUIT-EVASION GAME

by
Hezi Atir
RAFAEL, MOD, POB 2250, Haifa, Israel

This work is based on the author's Ph.D thesis, completed at Stanford University, CA, under the supervision of Prof. J.V. Breakwell.

ABSTRACT

A three state pursuit-evasion differential game with variable speed, nonlinear control and open barrier is studied as a game of kind. A phenomenon of accumulating paths was found. The high accuracy requirements inherent in the game prevent retrogressive integration of the long optimal paths. An approximation is suggested, based on the characteristic behaviour of the controls and of the adjoints, for suboptimal completion of the game.

1. INTRODUCTION

Effects of realistically modeling the pursuer as a variable speed fighter plane in a simplified three state game are the concern of this paper. When solving differential games, one is restricted to a small number of state variables. Therefore games, such as the Two Cars game [7] and games with state reductions by energy formulation [6] [9], are widely studied. Games with one variable speed player are also treated [8]. When both players have variable speed, the game is often solved to satisfy the necessary conditions only [5]. There are no practical restrictions on the complexity of the model regarding the state equations, the constraints and the cost function. Simple games can be manipulated for more involved effects such as in the two target set air combat games [1] [3], and in the double threat effect [2]. Both the nonlinear effects and the open barrier characteristics are sought. An open barrier may be a property associated with the long range weapons in air combat as long as one assumes predetermined roles. In more involved games, when this is not assumed, only the portion where the evader cannot turn pursuer, should be considered the capture zone. When a swerve maneuver is called for, the evader may have the chance to switch roles. Therefore the barrier termination becomes of interest.

2. GAME DESCRIPTION

The game is a planar pursuit. The pursuit geometry is described in a relative coordinate system, attached to the pursuer, shown in Fig. 1. the game is described by three state variables:

- R - the range along the line of sight (L.O.S).
- Θ - the bearing, or angle between the pursuer's velocity directions and the line of sight.
- V_p - the pursuer's speed.

The states R and Θ are equivalent to the Relative X-Y coordinates. The pursuer controls his turn rate, ω_p, which is algebraically related to his load factor, n. The load factor is bounded by the maximum available lift and by the structural limit. The evader flies at a constant speed, V_e, smaller than the maximum pursuer's speed, and controls his heading, h. The target set is a circle of radius R_{cap} around the pursuer. The pursuer wishes to minimize the final range, and the evader - to maximize it. We seek the barrier, such that optimal paths starting at the barrier just graze the target set.

3. PROBLEM FORMULATION

3.1 The Game Dynamics

The time rate of change of R is the difference between the projections of the evader's velocity and the pursuer's velocity on the range vector. Using Fig. 1 we obtain:

$$\dot{R} = - V_p \cos \Theta + V_e \cos (h-\Theta) \qquad (1)$$

The time rate of change of the angle Θ is the rotation rate of the L.O.S. minus the rotation rate of the coordinate system, ω_p. The rotation of the L.O.S. is created by the components of V_p and Ve normal to it, thus:

$$\dot{\Theta} = \frac{V_p}{R} \sin \Theta + \frac{V_e}{R} \sin (h-\Theta) - \omega_p \qquad (2)$$

The time rate of change of the pursuer's speed is obtained from the excess thrust over the drag divided by the airplane's mass. Constant thrust and aerodynamic coefficients are used, and a coordinated horizontal turn is assumed. The detailed derivation is presented in [2]. The non-dimensional result is:

$$\dot{V}_p = A(1 - \gamma(V_p^2 + \frac{\gamma_1}{V_p^2}) - \beta\omega_p^2) \qquad (3)$$

The maximum turn rate is given by:

$$\omega_{max}(V_p) = \frac{K\omega}{V_p}\sqrt{n_{max}^2(V_p) - 1} \qquad (4)$$

where:

$$n_{max}(V_p) = \min\{(\frac{V_p}{V_{stall}})^2, n_{limit}\} \qquad (5)$$

The coefficient A represents the normalized thrust. The coefficients γ, γ_1, β and $K\omega$ are related to the following nondimensional quantities (see also [2]):

- V_{max} - the maximum speed sustained by the thrust.
- V_{min} - the minimum speed sustained by the thrust.
- V_{stall} - the minimum speed sustained, using the maximum lift coefficient.
- n_{limit} - the structural limiting load factor.

The nondimensional parameters were chosen to fit a fighter plane (except for the coefficient A). The values are presented in table 1 along with typical related dimensional quantities. The pursuer's performance is depicted in Figs. 2 and 3.

The cost function for the game is:

$$\phi = R_f \qquad (6)$$

where $\phi = R_{cap}$ represents the target set.

3.2 The Variational Formulation

Isaac's main equation [4] is:

$$\min_{\omega_p}\max_{h} H = \min_{\omega_p}\max_{h}(J_r\dot{R} + J_\Theta\dot{\Theta} + J_v\dot{V}_p) = 0 \qquad (7)$$

The controls h and ω_p are obtained from:

$$J_r \cos(h-\Theta) + \frac{J_\Theta}{R} \sin(h-\Theta) \to \max \qquad (8)$$

$$-(A\beta J_v \, \omega_p{}^2 + J_\Theta \omega_p) \to \min \qquad (9)$$

thus:

$$h = \arctan \frac{J_\Theta}{RJ_r} + \Theta \qquad (10)$$

$$\omega_p = \begin{cases} \omega_{max} \operatorname{sgn}(J_\Theta) & J_\Theta \neq 0 \quad J_v \geqq 0 \\ \min\{\,|J_\Theta/2A\beta J_v|,\ \omega_{max}\,\} \operatorname{sgn}(J_\Theta) & J_\Theta \neq 0 \quad J_v < 0 \\ +\omega_{max} \text{ or } -\omega_{max} & J_\Theta = 0 \quad J_v > 0 \\ 0 & J_\Theta = 0 \quad J_v < 0 \\ \min\{(V_p/2A\beta) \tan\Theta,\ \omega_{max}\,\} & J_\Theta = J_v = 0 \quad t = t_f \\ \text{any} & J_\Theta = J_v = 0 \quad t \neq t_f \end{cases} \qquad (11)$$

The adjoint equations are obtained by the following expression due to the dependence of the pursuer's control limit on the states:

$$J_x = \begin{cases} -\dfrac{\partial H}{\partial X} & |\omega| < \omega_{max} \\ -\dfrac{\partial H}{\partial X} - \dfrac{\partial H}{\partial \omega_p} \dfrac{\partial \omega_{max}}{\partial X} & |\omega| = \omega_{max} \end{cases} \qquad (12)$$

The terminal adjoints are:

$$J_{rf} = 1 \quad J_{\Theta f} = J_{vf} = 0 \qquad (13)$$

The termination conditions: $\phi = R_{cap}$, $d\phi/dt = 0$ yield:

$$R_f = R_{cap} \qquad \Theta_f = \arccos(V_e/V_{pf}) \qquad (14)$$

This leads to a one parameter family of optimal paths. This parameter is the final pursuer's velocity V_{pf}. The boundary of the usable part (BUP) of the target set, as obtained from Eqs. 14 is depicted in Fig. 4. Actually, the BUP is larger due to special paths described in section 4.1.

4. THE RESULTS

4.1 The Optimal Paths

Optimal paths in the state space, with different values of V_{pf}, are depicted in Fig. 5. There seems to be a central path that most of the paths merge into. Actually, we obtain a bundle of paths that terminate along a section of the BUP, rather than at a single point. Yet this section is extremely short compared to the whole range of V_{pf} (0.4 to 1.0). Between path 1 and path 12 there are still many more paths which were omitted for clarity. Details of such paths are presented in Fig. 6. The most "inner" paths cannot be computed by retrogressive integration. We distinguish between the "left paths", such as paths 1 through 6, that are "seen" on the left hand side upon advancing retrogressively along the central path; and the "right paths", such as paths 12 through 17.

Representative "left" and "right" paths are given in Fig. 6, showing state space and real space trajectories, and time histories of the pursuer's velocity and control. The "right path" starts at $V_p=1$. The pursuer turns at his maximum available rate towards the evader, letting his velocity drop rapidly, and gradually eases off his turn. Then, for a long period of time the pursuer turns at a rate just about the sustained one, with almost unnoticable change in speed. At this speed, his turn radius is near its minimal value. During this maneuver the pursuer performed a tight turn, shown in the real space. A reduced turn, with acceleration concludes the chase. This path may be divided into three phases: The initial phase, the "almost steady turn" phase, and the terminal phase. The evader flies in a straight line directly towards his best terminal point throughout the entire chase.

The "left path" includes quite the same "almost steady turn" and terminal phases. The initial phase (of which only the last part is shown in the figure) starts at $V_p \cong V_{min}$. The turn is postponed until enough speed is gained to enable maneuver capabilities.

The time histories of ω_p, shown for the initial phase of the right and left paths, are typical of all the right and left paths close to the central path. The duration of the almost steady turn phase becomes longer as the central path is approached. Note paths 9 through 11 of Fig. 5 that start and end on the target set, similar to the paths in the Isotropic Rocket Game [4, pages 244-250].

4.2 The Central Path and Its Tributaries

The central path phenomenon was found in all studies of this game with various parameters, including different shapes of the target set. Long paths were also found in other studies [8].

Pictorially, this phenomenon seems to be a nonlinear equivalence of the Universal Line. Another similarity is to a focal line, where the tangency assumes the form of merging bundle due to the numerical process. Affinity to a Universal Line, in terms of a steady turn, can be shown by linearizing Eq. 3 using: $\omega_p = \omega_o + \delta\omega$. The linear switch function resembles Eq. 11 for the almost steady turn phase where: $J_\Theta \neq 0$, $J_v < 0$ and $\omega_p < \omega_{max}$.

From a practical point of view, the central path phenomenon yields a rule of thumb for the pursuer's control, that applies to most starting points. The pursuer should first attempt to reach the central path using the initial phase strategy, and then follow this path. We extrapolate the central path assuming a sustained turn for its extended portion, and use the initial phase strategy to obtain its tributaries. The results are depicted in Fig. 7. A one sided semipermeable close surface is obtained. It represents an ensured capture zone that does not require swerve maneuvers. It turns out that optimality does not require the evader to fly directly towards the pursuer.

ACKNOWLEDGEMENTS

I would like to express my deepest gratitude to Prof. J.V. Breakwell for his invaluable direction during my Ph.D. studies and his enthusiasm. I am lucky to have been exposed to his influence which was a very special and enriching experience.

DEDICATION

I am dedicating this work with love in memory of my father, Eng. Oskar Reichenthal. He was fond of my academic involvement and encouraging. He died a month before this paper was submitted.

REFERENCES

[1] Ardema, M.D., Heymann, M., and Rajan N., Combat Games. Journal of Optimization theory and Applications. Vol 46, No. 4, Aug 1985.

[2] Atir, H., Double Threat in Pursuit-Evasion Games, Ph.D Thesis, Stanford University July 88. Supervised by Prof. J.V. Breakwell.

[3] Davidovitz, A., and Shinar J., Eccentric Two Target Model for Qualitative Air Combat Analysis. AIAA Journal of Guidance, Control and Dynamics. May-June 1985.

[4] Isaacs, R., Differential Games. Krieger Publishing Co. 1975 (First Published by John Wiley, 1965).

[5] Jarmark, B.S.A., A Missile Duel Between Two Aircraft. Journal of Guidance. Vol. 8, No. 4, July-Aug 1985.

[6] Kelley, H.J., and Lefton, L., Differential Turns. AIAA Journal. Vol. 11, No. 6, June 1973.

[7] Merz, A.W., The Game of Two Identical Cars. Journal of Optimization Theory and Applications. Vol. 9, No. 5, 1972.

[8] Prasad, U.R., Rajan, N., and Rao, N.J., "Planar Pursuit-Evasion with Variable Speed, Part 1, Extremal Trajectory Map" and "Part 2, Barrier Sections." Journal of Optimization Theory and Applications, Vol. 3, 1981.

[9] Rajan, N., and Ardema, M.D., Interception in Three Dimensions: An Energy Formulation. Journal of Guidance. Vol 8, No. 1, Jan-Feb 1985.

Table 1: The Parameters of the Problem

NONDIMENSIONAL PARAMETERS		
INDEPENDENT	DEPENDENT	PURSUER'S PERFORMANCE
$A = 1.0$	$\beta = 3.60577$	$V_{max} = 1.0$
$V_{stall} = 0.2$	$\gamma = 0.96159$	$\max \{\omega_{max}\} = 1.0$
$V_{min} = 0.2$	$\gamma_1 = 0.04000$	at $V_p = 0.4$
$n_{limit} = 4.0$	$K\omega = 0.10328$	$\max \{\omega_{sustain}\} = 0.4131$
		at $V_p = 0.4472$
$V_c = 0.4$		
$R_{cap} = 0.5$		

TYPICAL UNITS		
Velocity :	308.7	m/sec
Turn Rate :	.3077	rad/sec
Time :	3.250	sec
Distance :	1003	m

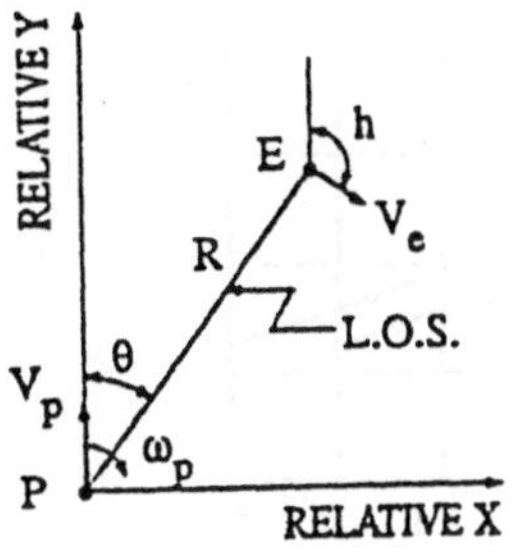

Fig. 1: Pursuit Geometry

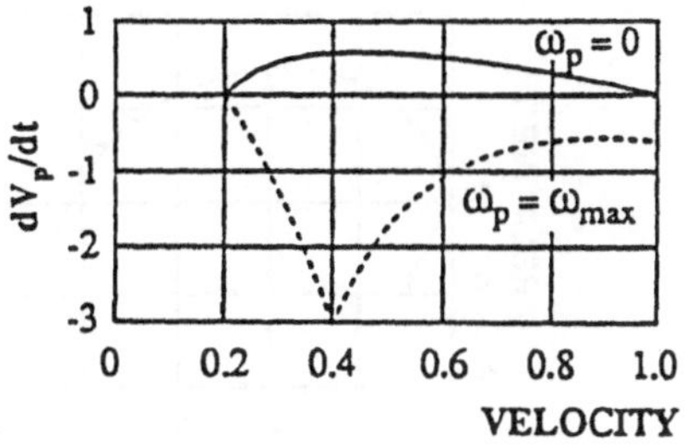

Fig. 2: Longitudinal Acceleration

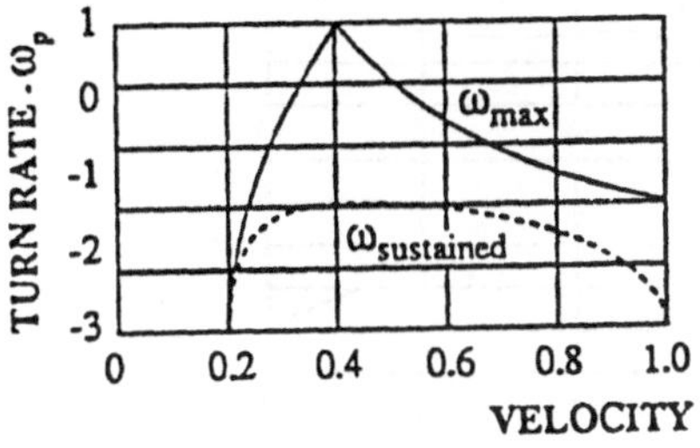

Fig. 3: Turn Rate

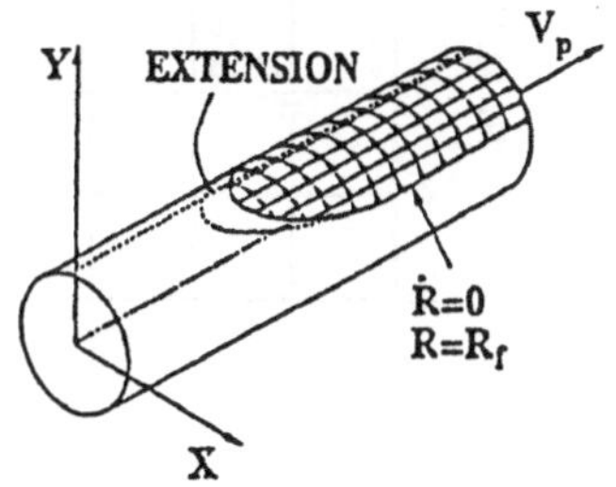

Fig. 4: The Useable Part

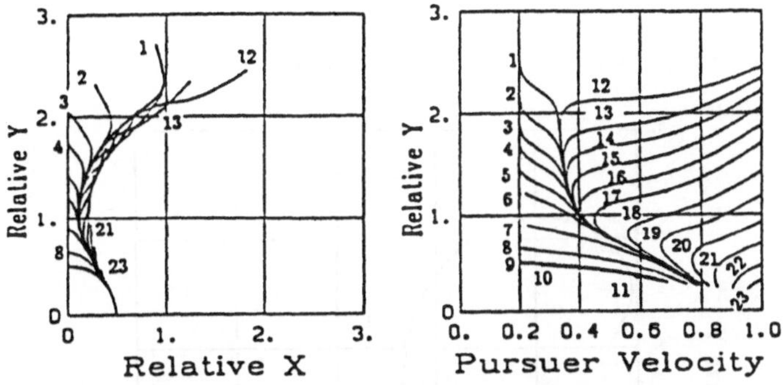

V_{pf} - TERMINAL VALUES OF THE PURSUER'S VELOCITY			
LEFT PATHS		RIGHT PATHS	
(1) 0.78538272	(7) 0.78	(12) 0.78538273	(18) 0.79
(2) 0.7853825	(8) 0.75	(13) 0.785383	(19) 0.80
(3) 0.78538	(9) 0.6889	(14) 0.785385	(20) 0.82
(4) 0.78537	(10) 0.67	(15) 0.7854	(21) 0.85
(5) 0.7853	(11) 0.60	(16) 0.7855	(22) 0.90
(6) 0.785		(17) 0.786	(23) 0.95

Fig. 5: Map of Optimal Path

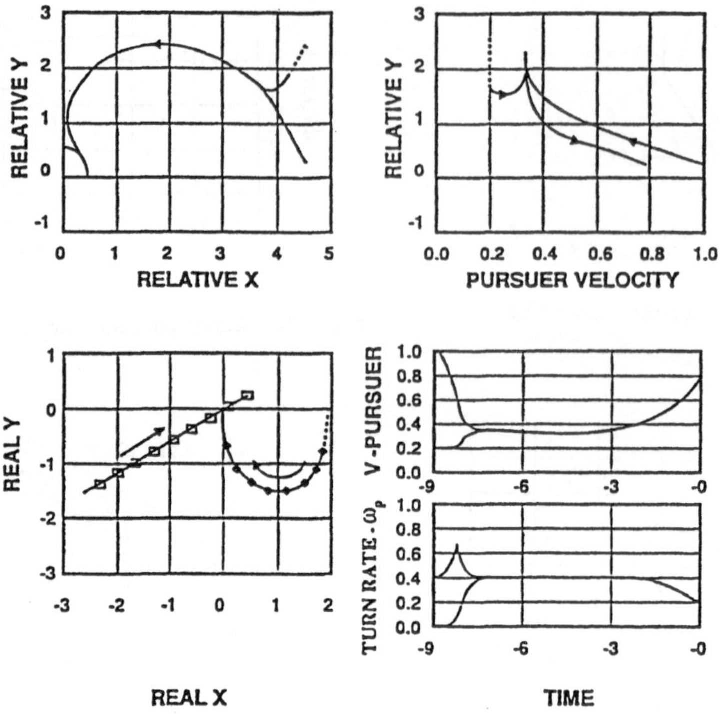

Fig. 6: "Long" Left and Right Paths

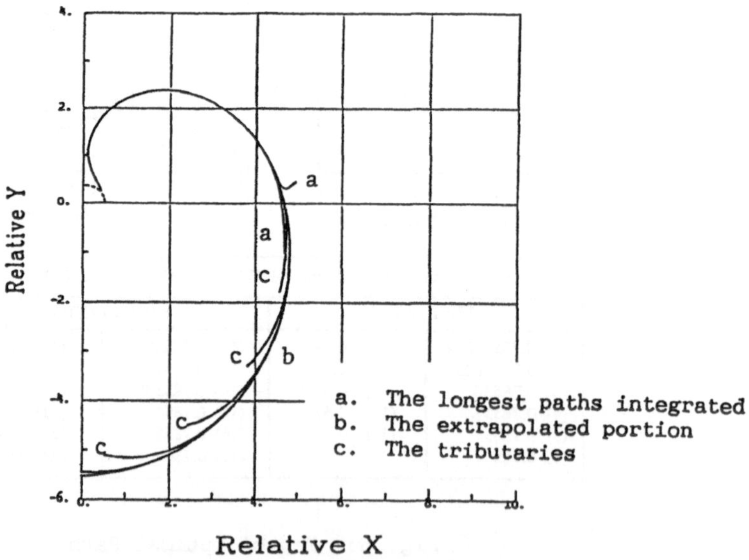

Fig. 7: The Extended Central Path and its Tributaries

THE DIRECT PONTRJAGIN METHODS IN THE THEORY OF DIFFERENTIAL GAMES

M.S.Nikolskii
Steklov Institute of Mathematics, USSR, Academy of Sciences
Vavilova 42, Moscow 117966, GSP-1, USSR

L.S.Pontrjagin constructed two direct methods for effective solving of linear differential game of pursuit (see [1-3]) which is described by the equation:

$$\dot{z} = Cz - u + v \, , \qquad z(0) = z_0 \, , \tag{1}$$

where $z \in R^n$, C is $n \times n$ -matrix, u is control vector of pursuer, $u \in P$ (P is compactum from R^n), v is control vector of evader (Q is compactum from R^n). The control functions $u = u(t) \in P$, $v = v(t) \in Q$, $t \geqslant 0$, are measured functions in Lebesgue sense. The terminal closed set M is fixed in R^n , $z_0 \notin M$.

The pursuer tends to lead the point $z(t)$ (see (1)) on the set M . The evader resists to the pursuer. It is supposed that the pursuer knows the equation (1), z_0 , P , Q , M and the measured function $v(s) \in Q$, $0 \leq s \leq t$, for every $t \geqslant 0$.

We shall consider the problem of quality: to find initial states $z_0 \notin M$ from which the pursuer can guarantee the leading of point $z(t)$ on the set M within some finite time $T(z_0)$ for every measured $v(t) \in Q$, $t \geqslant 0$.

The direct methods of Pontrjagin give the sufficient conditions for that the given initial point $z_0 \notin M$ belongs solution of the problem of quality.

In the first method M is linear subspace from R^n . In the second method M is arbitrary closed set.

L.S.Pontrjagin invented the important example of linear differential game of pursuit.

<u>The checking example of Pontrjagin.</u>

The equation (1) has such appearance:

$$\dot{z}_1 = z_2$$
$$\dot{z}_2 = -\alpha z_2 - \tilde{u}$$
$$\dot{z}_3 = z_4$$
$$\dot{z}_4 = -\beta z_4 + \tilde{v}$$

where $z_1, z_2, z_3, z_4, \tilde{u}, \tilde{v} \in R^k$, $k \geq 1$, $\alpha > 0$, $\beta > 0$,

$|\tilde{u}| \leq \rho$, $|\tilde{v}| \leq \sigma$, $\rho > 0$, $\sigma > 0$. Let us

$$M = \{z \in R^{4k}: \ z_1 = z_3\}.$$

With help of first method of Pontrjagin it can be proved that in the case $\rho \geq \sigma$, $\frac{\rho}{\alpha} > \frac{\sigma}{\beta}$ from an arbitrary $z_0 \notin M$ pursuit can be finished for finite time $\tau(z_0)$. For this time $\tau(z_0)$ the first method gives transcendental equation. If in this example $M = \{z \in R^{4k}: |z_1 - z_3| \leq \ell\}$, where $\ell > 0$, then the second method of Pontrjagin (method of alternating integral) can be applied.

Let us give a survey of some researches connected with the direct methods of Pontrjagin.

The little book [4] contains systematic modern account of the first direct Pontrjagin method on the base of theory of set-valued multifunctions.

In [5] it is developed the regularization of computing of Pontrjagin set

$$W(t) = \int_0^t \hat{w}(\tau)\, d\tau \qquad \text{where} \quad \hat{w}(\tau) = \pi e^{\tau C} P \stackrel{*}{-} \pi e^{\tau C} Q,$$

π is matrix of orthogonal projection on L which is orthogonal supplement to the linear space M, $\stackrel{*}{-}$ denotes the Minkowski difference (geometrical difference), the integration is understood in Aumann's sense.

The first method of Pontrjagin was generalized by author (see [6]) on linear differential games of pursuit:

$$\dot{z} = Cz - Au + Bv \,, \qquad z(0) = z_0$$

where $z \in R^n$, $u \in R^p$, $v \in R^q$, A is $n \times p$-matrix, B is $n \times q$-matrix,

$$\int_0^{+\infty} |u(s)|^2 ds \leq \mu^2 \,, \qquad \int_0^{+\infty} |v(s)|^2 ds \leq \nu^2 \,. \tag{2}$$

In [6] also more general than [2] integral restrictions were considered.

Important results in the theory of alternating integral of L.S.Pontrjagin were got in the paper [7].

In [8] the second method of Pontrjagin was generalized on nonlinear differential games of pursuit.

For applications it is very important to know how to construct positional winning strategy $u(t,z)$ for pursuer.

N.N.Krasowskii and A.I.Subbotin developed in [9] very important theory of positional differential games. With help of this theory it is possible to construct the winning positional strategy in conditions of both methods of Pontrjagin.

The methods of Pontrjagin were generalized on differential games of pursuit where pursuer and evader are teams (see [10]).

The systematic account of the theory of linear differential games with integral restrictions on base of Pontrjagin ideas contains in [11].

References

1. Pontrjagin, L.S. About linear differential games. I. Soviet Math. Doklady, Vol. 174, Nr. 6, 1976.
2. Pontrjagin, L.S. About linear differential games. II. Soviet Math. Doklady, Vol. 175, Nr. 4, 1976.
3. Pontrjagin, L.S. Linear differential games of pursuit. Math. Sbornik, Vol. 112. Nr. 3, 1980.

4. Nikolskii M.S. The First direct method of L.S.Pontrjagin in differential games. Publ. House of Moscow University, Moscow, 1984.
5. Nikolskii, M.S. Some computing aspects of the first direct method of Pontrjagin in differential games. Vestnik of Moscow Univ., ser. computing math. and cybernetics, Nr. 4, 1980.
6. Nikolskii, M.S. The direct method in linear differential games with general integral restrictions. Diff. Uravnenija, Vol. 8, Nr. 6, 1972.
7. Polovinkin, E.S. Nonautonomous differential games. Diff. Uravnenija, Vol. 15, Nr. 6, 1979.
8. Pshenichnii, B.N., Sagaidak, M.I. About differential games with fixed time. Kibernetika, Nr. 2, 1970.
9. Krasovskii, N.N., Subbotin, A.I. Positional differential games. Nauka, Moscow, 1974 (English translation in Springer-Verlag, 1988).
10. Grigorenko, N.L. Differential games of pursuit by some objects. Publ. House of Moscow University, Moscow, 1983.
11. Mezencev, A.V. Differential games with integral restrictions on control. Publ. House of Moscow University, Moscow, 1988.

THE PROBLEM OF PURSUIT BY SEVERAL OBJECTS

Nikolai L.Grigorenko
Optimal Control Department
Faculty of Cybernetics & Computational Mathematics
M.V.Lomonosov Moscow State University
MSU,Faculty VMK,Leninskye Gory,119899,Moscow,USSR

Abstract

We consider a linear differential game with a group of pursuers and a single evader.Sufficient condition for the solvability of the pursuit problem are obtained for the case when there are three kinds of players in the pursuers group:those with equal,those with poorer and those with greater dynamic capability than the evader, for the case when the pursuers are inferior to the evader in control resources, and for the case when there are players in the pursuers group with greater inertia then the evader. The methods can be applied to a large class of linear nonstationary and nonlinear game.

Keywords: Differential games with many players, pursuit problem, dynamic non-cooperative game of kind,combat games, multicriteria games.

1 Introduction and Problem Formulation

The fundamental works [1-5] are devoted to various classes of pursuit and evading problem for differential games of two players:one evader and one pursuer.They have served as a basis for investigation of problem of pursuit of one evader by several players [6-11]. It turned out that one can arrange the pursuit process, in dependence on the composition of the pursuers group, in such a way that the whole group of pursuers can terminate the pursuit game, whereas separere players or separete subgroups of players cannot catch the evader. In case that one of pursuer can terminate of pursuit, the group of pursuers can terminate of pursuit quickly, as a rool.

Suppose that the motion of the vectors $z_1,\dots,z_m$ in n_i-dimentional Euclidean space R^{n_i} is described by the equations

$$\dot{z}_i=C_i z_i+u_i-v, \qquad i=1,2,\dots,m, \tag{1}$$

where C_i is a constant matrix of dimension n_i*n_i, $u_i\in P_i\subset R^{p_i}$, $v\in Q\subset R^q$, and P_i and Q are convex sets. Terminal sets M_i, $i=1,\dots,m$ are given in R^{n_i}, where $M_i=M_i^1+M_i^2$, with M_i^1 a linear subspace of R^{n_i} and M_i^2 a compact convex set in the orthogonal complement L_i^1 of M_i^1 in R^{n_i}.

The information given above describes a differental game with several players (1) in which the particlpiants are a group of pursuers having at their disposal a control vector $u=(u_1,\dots,u_m)$, and a pursued player having the vector v at his disposal. We consider the pursuit problem for the differential game (1).

A *counterstrategy* for the *i*th pursuer $u_i(t)=U_t^i(z^0,v(t))$ is defined to be a mapping on the set of arbitary measurable controls $v(t)\in Q$ and the set of arbitary vectors $z^0=(z_1^0,\dots,z_m^0)\in R^{\bar{n}}$, $\bar{n}=\sum_1^m n_i$, dependent on the set M_i, and having the following property: $u_i(t)=U_t^i(z^0,v(t))$ is measurable as a function of t for any measurable $v(t)\in Q$ and $z^0\in R^{\bar{n}}$, $u_i(t)\in P_i$, and $U_t^i(z^0,v_1(t))=U_t^i(z^0,v_2(t))$ almost everywhere if $v_1(t)=v_2(t)$ almost everywhere.

A *program strategy* for the *i*th pursuer (the evader) is defined to be an arbitary measurable function $u_i(t)$ $(v(t))$ such that $u_i(t)\in P_i$ $(v(t)\in Q)$. A *pursuit strategy* is defined to be a vector $u_i(t)=(u_i(t),\dots,u_m(t))$, where $u_i(t)$, $i=1,\dots,l$, is a counterstrategy for the *i*th pursuer, $u_i(t)$, $i=l+1,\dots,m$, is a program strategy for the *i*th pursuer, and $0\leqslant l\leqslant m$ (if $l=0$, then $u_i(t)$,

$i=1,\dots,m$, are program strategies for the respective pursuers).

The *pursuit problem* is formulated as follows: find a condition on the parameters of the game (1) such that for givem initial states $z_i^0, i=1,\dots,m$, there exists a pursuit strategy $u^*(t)=(u_1^*(t),\dots,u_m^*(t))$, for which at least one vector $z_i(t), i=1,\dots,m$, solving the equation

$$\dot{z}_i=C_i z_i+u_i^*(t)-v(t), \quad z_i(0)=z_i^0,$$

arrives at the corresponding terminal set M_i by some finite time.

2 Sufficient Conditions For Terminating The Game Of Pursuit

Let π_i be the orthogonal projection of R^{n_i} onto L_i^1, $\overset{*}{-}$ the geometric difference of sets, $N=\{1,\dots,m\}=N_1\cup N_2\cup N_3$, $N_1=\{1,\dots,k\}$, $N_2=\{k+1,\dots,l\}$, $N_3=\{l+1,\dots,m\}$, $\nu_i=\dim L_i^1$, and T a positive constant.

Assumption 1. There exist $\nu_i\times\nu_i$ matrices $A_i(\tau)$, $i\in N_1\cup N_2$, depending in a measuarble way on τ such that for $0\leqslant\tau\leqslant t$ the following sets are nonempty:

$$\hat{W}_i(\tau)=\pi_i e^{\tau C_i}P_i \overset{*}{-} A_i(\tau)\pi_i e^{\tau C_i}Q, \qquad i\in N_1\cup N_2,$$

$$M_i^3(t)=M_i^2 \overset{*}{-} \int_0^t (A_i(t-\tau)-E)\pi_i e^{(t-\tau)C_i}Q\,d\tau, \qquad i\in N_1.$$

Assumption 1 and the conditions on the game (1) imply the existence of a measurable function $\gamma_i(\tau)\in\hat{W}_i(\tau)$, $i\in N_1\cup N_2$, for all $\tau\geqslant 0$. Fix such a function. For $t_0\leqslant\tau\leqslant t$, $m_i\in L_i^1$, $z_i\in R^{n_i}$, and $i\in N_1\cup N_2$ let

$$\lambda(i,t,\tau,t_0,v(\tau),m_i,z_i)$$

$$=\max\left\{\lambda:\lambda\geqslant 0,-\lambda\left[\pi_i e^{(t-t_0)C_i}z_i-m_i+\int_0^t\gamma_i(\tau)d\tau\right]\right.$$

$$\in \pi_i e^{(t-\tau)C_i} P_i - A_i(t-\tau)\pi_i e^{(t-\tau)C_i} v(\tau) - \gamma_i(t-\tau)\Big\},$$

$$\lambda(i,t,\tau,0,v(\tau),M_i^3(t),z_i^0) = \max_{m_i^3 \in M_i^3(t)} \lambda(i,t,\tau,0,v(\tau),m_i^3,z_i^0), \quad i \in N_1, \qquad (2)$$

$$E(t,T,z^0)$$

$$=\Big\{v(\tau): v(\tau)\in Q, 0\leqslant\tau\leqslant t\leqslant T, \int_0^t \lambda(i,T,\tau,0,v(\tau),M_i^3(T),z_i^0)d\tau<1, i\in N_1\Big\}.$$

Assumption 2. $\nu_i=\nu$ for $i\in N_2$, and for the position $z^0=(z_1^0,\dots,z_m^0)$ there exist a positive integer p, nonnegative constants δ_i^j, T_j, $j=1,\dots,p$, $i\in N_1$, $0=T_0\leqslant T_1\leqslant\dots\leqslant T_p=T$, vectors $\alpha_i^j\in L_i^1$, $i\in N_2$, $j=1,\dots,p$, and one-to-one linear mappings F_i: $L_i^1\to R^\nu$ such that the following conditions hold:

a) $$\min_{v(\cdot)} \min_{z_i(T_{j-1})\in D_i(T_{j-1})} \min_{i\in N_2} \int_{T_{j1}}^{T_i} \lambda(i,T_j,\tau,T_{j-1},v(\tau),\alpha_i^j,z_i(T_{j-1}))d\tau\geqslant 1$$

there $D_i(0)=z_i^0$, $D_i(T_{j-1})=\{z_i(T_{j-1}),\ \dot z_i(t)\in C_i z_i(t) + P_i - Q,$ $t\in[T_{j-2},T_{j-1}],\ z_i(T_{j-2})\in D_i(T_{j-2})\}$, $j=2,3,\dots,p$, and the minimum is taken over all measurable functions on $[T_{j-1},T_j]$ taking values in Q.

b) The vectors

$$\int_{T_{j-1}}^{T_j} F_i(A_i(T_j-\tau)-E)\pi_i e^{(T_i-\tau)C_i} v(\tau)d\tau$$

do not depend on i, $i\in N_2$, for all $v(\tau)\in E(t,T,z^0)$.

c) For any $v(\tau)\in\Pi(T_{j-1},T_j)$, where

$$\Pi(T_{j-1},T_j)= \Big\{v(\tau):\ v(\tau)\in Q,\ v(\tau)\in E(T,T,z^0),\ \tau\in[T_{j-1},T_j],$$

$$\int_{T_{j-1}}^{T_j} F_i(A_i(T_j-\tau)-E)\pi_i e^{(T_j-\tau)C_i}v(\tau)d\tau \cap \left(\bigcup_{i\in N_2}(F_i\alpha_i^j+F_iM_i^2)\right)=\emptyset, i\in N_2\Big\},$$

we have

$$\int_{T_{j-1}}^{T_j} \lambda(i,T,\tau,0,v(\tau),M_i^3(T),z_i^0)d\tau>\delta_i^j$$

for at least one $i\in N_1$.

Assumption 3. $\nu_i=\nu$ for $i\in N_3$, and there exist one-to-one linear mappimgs $F_i\colon L_i^1 \to R^\nu$ such that the vectors $\int_0^t F_i\pi_i e^{(t-\tau)C_i}v(\tau)d\tau$, $i\in N_3$, do not depend on i for all

$$v(\tau) \in E(t,T,z^0) \cap \cup_{j=1}^p \Pi(T_{j-1},T_j),\ 0\leqslant\tau\leqslant t\leqslant T.$$

Let
$$j^*(t)=\max\{j: j\in\{1,2,\dots,p\}, T_j\leqslant t\},$$

$$\hat{T}(t)=\begin{cases}\max\{T_j : T_j \leqslant t, j\in\{1,2,\dots,p\}\} & \text{if } N_2\neq\emptyset,\\ 0 \quad \text{if } N_2=\emptyset.\end{cases}$$

For $i\in N_3$ and $t\in[0,T]$ we consider the set

$$Q(t,\hat{T}(t),T,z^0)=\Big\{-\int_0^t F_i\pi_i e^{(t-\tau)C_i}v(\tau)d\tau : v(\tau)\in Q, v(\tau)\in E(t,T,z^0),$$

$$\int_{T(t)}^{t} \lambda(\eta,T,\tau,0,v(\tau),M_\eta^3(T),Z_i^0)d\tau$$

$$< \Big(1-\sum_{j=1}^{j=j^*(t)}\int_{T_{j-1}}^{T_j} \lambda(\eta,T,\tau,0,v(\tau),M_\eta^3(T),z_i^0)d\tau\Big), \eta\in N_1\Big\}.$$

Let $u_i(\tau)$, $i\in N_3$, $0\leqslant\tau\leqslant T$, be arbitrary program controls. We use the notation

$$M_i(t)=-F_i\pi_i e^{tC_i}z_i^0-\int_0^t F_i\pi_i e^{(t-\tau)C_i}u_i(\tau)d\tau+F_iM_i^2,\ i\in N_3,$$

$$M(t)=\bigcup_{i\in N_3} M_i(t),\qquad t\leqslant T.$$

Assumption 4. For a position z^0 there exist a positive constant T and admissible controls $u_i(t)$, $0\leqslant t\leqslant T$, $i\in N_3$, such that $M(t_*)\supset Q(t_*,\hat{T}(t_*),T,z^0)$ for some time $t_*\leqslant T$.

Condition B. For the sets $M(t)$ and $Q(t,\hat{T}(t),T,z^0)$ we say that Condition B holds on the interval $[T^1,T^2]$, $T^1\leqslant t\leqslant T^2\leqslant T$, (the condition of combing of $Q(t,\hat{T}(t),T,z^0))$ by $M(t)$ on interval $[T^1,T^2]$, if there exist a continuos functions $\xi(x,t)$: $R^{\nu}\times[T^1,T^2]\to R^1$ such that:

a) $Q(T^1,\hat{T}(T^1),T,z^0)\in\{x:\xi(x,T^1)\leqslant 0\}$,
$Q(T^1,\hat{T}(T^1),T,z^0_i)\cap\{x:\xi(x,T^1)=0\}\neq\emptyset$;
b) $M(t)\ni(Q(t,\hat{T}(t),T,z^0)\cap\{x:\xi(x,t)=0\})$ for all $t\in[T^1,T^2]$;
c) $M(T^2)\ni(Q(T^2,\hat{T}(T^2),T,z^0)\cap\{x:\xi(x,T^2)\leqslant 0\})$.

Assumption 5. For a position z^0 there exist controls $u_i(t)$, $i\in N_3$, $t\geqslant 0$, and positive constants T^1,T^2,T, $0\leqslant T^1\leqslant T^2\leqslant T$, such that Condition B holds for the sets $M(t)$ and $Q(t,\hat{T}(t),T,z^0)$ on $[T^1,T^2]$.

Theorem. *Suppose that for the game* (1) *in the position* z^0 *there exist sets* N_1,N_2, *and* N_3 *such that one of the following conditions holds:*

1) $N_1\neq\emptyset$, *Assumption* 1 *holds, and there exists a positive constant* T *such that* $E(t^*,T,z^0)=\emptyset$ *for* $t^*\leqslant T$.

2) $N_2\neq\emptyset$, *Assumption* 1 *and Conditions* a) *and* b) *of Assumption* 2 *hold, and* $\Pi(T_{j-1},T_j)=\emptyset$ *for at least one* j.

3) $N_1\neq\emptyset$, $N_2\neq\emptyset$, *Assumptions* 1 *and* 2 *hold, and* $\sum_{j=1}^{p}\delta_i^j\geqslant 1$ *for some* $i\in N_1$.

4) $N_1 \neq \emptyset$, $N_2 \neq \emptyset$, $N_3 \neq \emptyset$, *and Assumptions 1-4 hold.*

5) $N_1 \neq \emptyset$, $N_2 \neq \emptyset$, $N_3 \neq \emptyset$, *and Assumption 1-3 and 5 hold.*

6) $N_2 \neq \emptyset$, $N_3 \neq \emptyset$, *and Assumption 1,3, and 5 hold.*

Then for the position z^0 *the pursuit problem is solvable by time* T.

Remark. The magnitude T of the guaranteed pursuit time is influenced by the form of the partition of N into the sets N_1, N_2, and N_3, and this form can change in the process of the game.

3 Game Examples

The effectiveness of the results we would demonstrated by several examples.

Example 1. Suppose that the differential game is described by the equations

$$\dot{z}_{i1}=z_{i2}-v, \quad \dot{z}_{i2}=u_i, \quad i=4,5, \quad \dot{z}_i=u_i-v, \quad i=1,2,3,6,7,$$

where $z_{ik},u_i,v \in R^2$, $\|u_i\| \leqslant \rho_i$, $\|v\| \leqslant \sigma$, $k=1,2$, $\rho_1=\rho_2=\ldots=\rho_5=\sigma=1$, $\rho_6=\rho_7=0.5$, the terminal sets are $M_i=\{\|z_{i1}\| \leqslant l_i\}$, $i=4,5$, $l_4=l_5=\sqrt{3}/4$, $M_i=\{\|z_i\| \leqslant l_i\}$, $i=1,2,3,6,7$, $l_1=l_2=l_3=0$, $l_6=1.125$, $l_7=0.75$. We have $N_1=\{1,2,3\}$, $N_2=\{4,5\}$ and $N_3=\{6,7\}$. Assumption 1 holds if $A_i(\tau)=\mu_i(\tau)E$, $i=4,5$, where

$$\mu_i(\tau)=\begin{cases}\tau, & 0\leqslant\tau\leqslant 1,\\ 1, & \tau>1,\end{cases} \qquad i=4,5, \qquad A_i(\tau)=E, \qquad i=1,2,3.$$

According to (2),

$$\lambda(i,t,\tau,0,v,0,z_i^0)=\left[(z_i^0,v)+\sqrt{(z_i^0,v)^2+\|z_i^0\|^2(\sigma^2-(v,v))}\right]/\|z_i^0\|^2,$$

$$i=1,2,3,$$

$$Q(t,0,T,z_0)\in\left\{-\int_t^0 v(s)ds,\ \left(z_i^0,-\int_0^t v(s)ds\right)>\frac{\|z_i^0\|^2}{2}\right\}.$$

We give a typical case of interaction of the pursuing players. Let

$z_1^0=(0;-4)$, $z_2^0=(4;0)$, $z_3^0=(0;4)$, $z_{41}^0=(1;0)$, $z_{42}^0=(0;0)$,
$z_{51}^0=(0;1.125)$, $z_{52}^0=(0;0)$, $z_6^0=(6;-2.125)$, $z_7^0=(6;-2.75)$.

Assumptions 2 and 5 will hold if, for example, $T_1=3$, $T=19$, $\alpha_4^1=(\sqrt{3}/8;1/8)$, $\alpha_5^1=(-\sqrt{3}/8;1/8)$, $\delta^1=0.25$, $1=4.5$, $\alpha_4^j=(-1/8;\sqrt{3}/8)$,

$\alpha_5^j=(-1/8;-\sqrt{3}/8)$, $j\geq 2$, $u_6(t)=(0;-0.5)$, $u_7(t)=(0;0.5)$, $0\leq t\leq 3$, $u_6(t)=(-0.5;0)$, $u_7(t)=(-0.5;0)$, $t\geq 3$, $\xi(x,t)=x_1-7.5+0.5t$, $t\geq 3$, $\nu=2$.

Moreover,

$$Q(t,0,19,z^0) \in \{-2<z_2<2,\ z_1>-2\},$$

$$Q(t,3,19,z^0) \in \{-2<z_2<1.75,\ z_1>-2\},$$

the players with indices $l\in N_2$ force the evader to approach the player with index 1 (otherwise the evader is caught) by the time $T^1=3$, and for $t>3$ they force him to approach the player with index 2. For $t>3$ the players with indices 6 and 7 comb the region $Q(t,3,19,z^0)$.

Example 2. Suppose that the differential game is described by the equations

$$\dot z_{i1}=z_{i2}-z_3,\quad \dot z_{i2}=-\alpha_i z_{i2}+u_i,\quad \dot z_3=-\beta z_3+v,\quad i=1,2,\dots,m,$$

where $z_{ik},u_i,v\in R^{\nu}$, $\|u_i\|\leq\rho_i$, $\|v\|\leq\sigma$, $k=1,2$; the terminal sets are $M_i=\{z_{i1}=0\}$, $\nu\geq 0$, $m\geq 1$. For this game $N_1=\{1,2,\dots,m\}$ and if $\rho_i\geq\sigma$, $\rho_i\beta>\sigma\alpha_i$ or $\rho_i>\sigma$, $\rho_i\beta\geq\sigma\alpha_i$ then the pursuit problem is solvable for all initial states z_{i1}^0,z_{i2}^0,z_3^0. If $\rho_i=\sigma,\beta=\alpha_i$ then for initial states z_{i1}^0, z_{i2}^0,z_3^0 the pursuit problem is solvable if exist pozitive constant θ that next two inequality make up simultaneous:

$$\delta=\min_{\|\psi\|=1}\ \max_{i=1,\dots,m}\ (\eta_i(\theta)/\|\eta_i(\theta)\|,\psi)>0,$$

$$(2\delta\sigma/\beta^2)\,(e^{-\beta\theta}+\beta\theta-1)\geq m\max_{i=1,\dots,m}\|\eta_i(\theta)\|$$

where $\eta_i(t)=z_{i1}^0+((1-e^{-\alpha_i t})/\alpha_i)z_{i2}^0-((1-e^{-\beta t})/\beta)z_3^0$.

Example 3. Suppose that the differential game is described by the equations

$$\dot z_{i1}=z_{i2}-v,\ \dot z_{i2}=u_i,\ \dot z_3=u_3-v,\ i=1,2,$$

where $u_i,v,z_{i1},z_{i2},z_3 \in R^2, \|u_j\|\leqslant\rho_j, \|v\|\leqslant\sigma, j=1,2,3, \rho_1=\rho_2=\rho, \rho_3=\sigma$, the terminal sets are $M_i=\{z:\|z_{i1}\|\leqslant l\}, i=1,2; M_3=\{z:z_3=0\}$. For this game $N_1=\{1,2\}, N_2=\{3\}$ and the pursuit problem is solvable for all initial states z_{i1}^0,z_{i2}^0,z_3^0 $(l>0,\rho>0,\sigma>0)$.

Example 4. Suppose that the differential game is described by the equations

$$\dot z_{i1}=z_{i2}-v,\ \dot z_{i2}=-\alpha z_{i2}+u_i,\ i=1,2,\dots,m,$$

where $u_i,v,z_{i1},z_{i2} \in R^2, \|u_i\|\leqslant\rho, \|v\|\leqslant\sigma$, the terminal sets are $M_i=\{z:\|z_{i1}\|\leqslant l\}$. For this game $N_1=\emptyset, N_2=\{1,\dots,m\}$ and the pursuit problem is solvable for all initial states z_{i1}^0,z_{i2}^0 for next m and l: $m=2, l=\beta; m=3, l=(\sqrt{3}/2)\beta; m=4, l=\beta/\sqrt{2}; m=5, l=2\beta/(1+\sqrt{5}); m=6, l=\beta/\sqrt{3}; m=7, l=\beta/2$; $\beta = (\sigma/\alpha)+((\rho/\alpha)-\sigma)((1/\alpha)\ln(1-(\alpha\sigma/\rho))), \alpha>0, \rho>0, \sigma>0$.

4 Conclusion

For the proof of the theorem,algorithms for finding parameters and initial states for solvability pursuit games,and calculation time T and ith pursuer control see [10,11].The results presented in this paper can be extended to nonstationar and some nonlinear case [10].Other related references on the pursuit problem for differential games with several pursuers and one evader are [6-11]

References

[1] L.S.Pontryagin. *Linear differential games of pursuit*.Mat.Sb. 112(154) 1980,307-332; English transl. in Math. USSR-Sb. 41 1981.

[2] N.N.Krasovskii and A.I.Subbotin, *Positional differential games*, "Nauka", Moscow, 1974; French transl., *Jeux differentiels*, "Mir", Moscow, 1977.

[3] A.I.Subbotin and A.G.Chentsov, *Optimization of the guarantee in control problems*, "Nauka", Moscow, 1981.(Russian)

[4] E.F.Mishenko, M.S.Nikol'skii and N.Yu.Satimov, Trudy Mat. Inst. Steklov **143** (1977); English transl. in Proc. Steclov Inst. Math. **1980**, no 1(143).

[5] L.A.Petrosyan, *Differential games of pursuit*, Izdat. Leningrad. Gos. Univ., Leningrad, 1977. (Russian)

[6].B.N.Pshenichnyi, A.A.Chikrii and I.S.Rappoport, Dokl. Akad. Nauk SSSR **259** (1981), 785; Englich transl. in Soviet Math. Docl. **24** (1981).

[7] N.Satimov.Pursuit and evasion problems for the class of linear differential n-person games.Tashkent.Gos.Univ.Sb.Nauchn.Trudov. No.670 Prikl.Mat.Mekh.:54-64,1981.

[8] N.L.Grigorenko, Dokl. Akad. Nauk SSSR **258** (1981), 275; English transl. in Soviet Math. Dokl. **23** (1981).

[9] N.L.Grigorenko, Vestnik Moskov. Univ. Ser. XV Vichisl. Mat. Kibernet. **1982**, no.1,49; English transl. in. Moscow Univ. Comput. Math. and Cybernet. **6** (1982).

[10] N.L.Grigorenko.*Mathematikal methods of control several dynamic processis*.Izdat.Moskov.Gos.Univ.,Moskov,1990.(Russian)

[11] N.L.Grigorenko.*The problem of pursuit by several objects*. Proceedings of the Steklov Institute of Mathematics,63-77,1986, Issue 1.

THE METHOD OF CHARACTERISTICS FOR CONSTRUCTING SINGULAR PATHS AND MANIFOLDS IN OPTIMAL CONTROL AND DIFFERENTIAL GAMES

Arik A.Melikyan
The Institute for Problems in Mechanics, USSR Academy of Sciences, prospect Vernadskogo, 101, 117526, Moscow, USSR

Singular hypersurfaces in phase space of a problem of optimal control or differential game are under study, which consist of singular paths. The analysis of such surfaces is important for development of methods for constructing solutions in inverse time. The neighbourhood Γ_0 of some point x^* on a singular surface Γ_1 in the state space R^n can be represented as $\Gamma_0 = D_0 + \Gamma_1 + D_1$, where D_i are half-neighbourhoods. The function of optimal result (Bellman function) will be denoted $S(x)$, $x \in \in \Gamma_0$, $S_i(x)$, $x \in D_i$ in the domains Γ_0, D_i. Two main types of singular hypersurfaces, containing paths, are shown on Fig.1. The first type (eqivocal surface) can be met in game problems, the second (universal surface) is also usual for optimal cotrol. As it's clear from the figure the backward constructing of Bellman's function is done on the sets D_i, Γ_1 in the order $D_1 \to \Gamma_1 \to D_0$ for the first case, and $\Gamma_1 \to D_i$ for the second, while the surfaces themselves are also sought for. Under the appropriate descriptions of surfaces Γ_1 the backward constructions including the surface itself can be reduced to the following standard mathematical problem, the solution of which is based on the generalization of the classical method of characteristics.

1. Cauchy problem on the manifolds. Let in the space R^{2n+1} of vectors $z = (x, p, S)$; $x, p \in R^n$, the manifold W_{2m+1} of odd codimension $2m + 1$, $0 \le m < n$ is given:

$$W_{2m+1} = \{z \in R^{2n+1}: G_i(z) = 0, \quad i = -m,\dots,-1,0,1,\dots,m\} \tag{1}$$

with scalar functions $G_i \in C^2$. Let I_m -be the set of all smooth integral manifolds (surfaces) Σ of 1-form $dS-pdx$, contained in the domain (1), $\Sigma \subset W$. Let π be the projection operator from R^{2n+1} into R^n.

Problem 1 . A manifold $\Sigma_{m+1} \in I_m$ - of dimension $n-(m+1)$ is given. Find manifold of dimension $n-m$ and it's projection $\Gamma_m = \pi\Sigma_m \subset \subset R^n_x$, such that $\Sigma_{m+1} \subset \Sigma_m \subset W_{2m+1}$.

For $m = 0$ the Problem 1 is the classic Cauchy problem for non-linear differential equation $F(x, S_x(x), S(x)) = 0$ with respect to $S(x)$. It's solution is constructed using the integral curves of characteristic field $\xi_F = (F_p, -F_x - pF_S, (p, F_p)) \in R^{2n+1}$ tangent to W_1. Here $(.,.)$ is the scalar product. When $m = 1$ the projection $\Gamma_1 = \pi\Sigma_1$ of the problem's solution is hypersurface.

Similarly in terms of Hamiltonian H construct the field ξ_H on W_{2m+1} and the corresponding characteristic system:

$$\dot{x} = H_p, \quad \dot{p} = -H_x - pH_S, \quad \dot{S} = (p, H_p); \quad H = \sum_i \lambda_i(z)G_i(z), \ |i| \le m \quad (2)$$

For the tangent field ξ_H the functions G_i must be the first integrals of the system (2). This yields to linear homogenious equation $C\lambda = 0$ with respect to Lagrange multipliers vector $\lambda \in R^{2m+1}$. The elements of matrix C are Jacobi brackets $c_{ij} = \{G_i, G_j\}$ (Poisson brackets, if there is no dependence on S). The solution $\lambda(z)$ of this system can be represented as Pfaffians of 2m-submatrixes of C .

For $m = 1$, in particular, we have:

$$H = \{G_0G_1\}G_{-1} + \{G_1G_{-1}\}G_0 + \{G_{-1}G_0\}G_1 \quad (3)$$

Theorem 1. Let the vector $\xi_H(z^*)$, $z^* \in \Sigma_{m+1}$, be transversal to the manifold Σ_{m+1}, i.e.particularly, $\xi_H \neq 0$. Then in the neighbourhood of the point z^* there exists unique solution Σ_m of the problem 1 of the class C^1. If vector $H_p(z^*) \in R^n$ is transversal to the surface Γ_{m+1}=

$= \pi\Sigma_{m+1}$, then exists also the smooth projection $\Gamma_m = p\Sigma_m$. The manifold Σ_m consists of integral curves of the system (2) passing through the points of Σ_{m+1}.

Thus for the construction using theorem 1 in the domains D_i it is necessary to know only one function G_0, determining the equation of Bellman-Isaacs. For the construction of the hypersurfaces Γ_1, as it can be seen from (2),(3), it's necessary to give three functions, three equalities $G_i(z) = 0$, $|i| \le 1$, and the initial manifold $\Sigma_2 \in I_1$. These three equations, one of which is Bellman's equation, are necessary conditions of optimality. In general for the construction of the manifold Γ_m of codimension m, 2m + 1 conditions of the form $G_i = 0$ are needed. Thus for constructing the surface Γ_1 one has to find two more necessary conditions. The functions G_i, $|i| \le 1$, must produce vector $\dot{x} = H_p$, which equals to the optimal phase speed in R^n_x. Note that Hamiltonian H in (2), (3) is defined up to homogenuity multiplier $\mu(z) > 0$. The manifold Σ_2 is constructed using the terminal set of the problem or the results of intermediate constructions.

This approach allowed to obtain the equations of singular characteristics for the case of equivocal surface, and for universal surface to obtain also necessary conditions of optimality, generalizing particularly Kelley's and Kopp-Moyer's conditions.

2. Equivocal surfaces. Let the differential game of two players P and E with controls u and v be given by:

$$\dot{x} = f(x, u, v), \quad t \in [0, T], \quad u \in U, \quad v \in V$$
$$x(0) = x^0, \quad x(T) \in M \subset R^n; \quad T + \Phi(x(T)) \to \min_u, \max_v \tag{4}$$

Define Hamiltonian (assuming commutativeness of minimax)

$$F(x, p) = \min_u \max_v (p, f(x, u,v)) + 1, \quad u \in U, \quad v \in V \tag{5}$$

Assume that in the neighbourhood of the point $x^* \in R^n$ in the problem (4) the singularity of the type 1, equivocal hypersurface, exists [1]. Equivocal surface is the switching surface for both players.

If the player, definite for the surface (refered as the player, which governs the surface), doesn't switch his control, then singular optimal motion on the surface is realized. On the eqivocal surface the game value is continuous and it's gradient is discontinuous, i.e. $p_0 \neq p_1$, $x \in \Gamma_1$, where $p_i = S_{ix}(x)$, $x \in D_i$. Both extrema in (5) are reached on unique vectors, while the uniqueness may fail only for one player (who doesn't govern on the surface) in the points of Γ_1 from the D_0 - side, i.e. for $p = p_0$, $x \in \Gamma_1$ in (5). In the domain Γ_0 the following necessary conditins of optimality are true

$$F_0(x, p_0) = 0, \quad x \in D_0; \qquad F_1(x, p_1) = 0, \quad x \in D_1,$$
$$\min_u \max_v \dot{S} \geq -1 \geq \max_v \min_u \dot{S}, \quad u \in U, \quad v \in V, \quad x \in \Gamma_1 \tag{6}$$

Here F_i are functions (5), corresponding to half-neighbourhoods D_i, $\dot{S} = \dot{S}(x, u, v)$ denotes the derivative of the game value $S(x)$ in the direction $f(x, u, v)$, i.e. the total time derivative. If the vector $p_0 - p_1$, $x \in \Gamma_1$, is directed to $D_1(D_0)$, then we have $\dot{S} = \min [\dot{S}_0, \dot{S}_1]$ ($\dot{S} = \max [\dot{S}_0, \dot{S}_1]$); left (right) equality takes place in the second line of (6), the surface is governed by player P(E), minimum over u (maximum over v) in (5) is always unique. It's proved [1], that if the extrema in (5) are always unique then the trajectories from region D_0 approach to Γ_1 tangently. Thus, there exist four types of equivocal surfaces, depending on the player, governing on the surface, and whether the surface is envelope of the paths coming from D_0.

Assume that $F_1(x, p)$ can be defined in the neighbourhood of the point (x^*, p^*), $p^* = S_{1x}(x^*)$ and trajectories in D_1 leave the Γ_1 with non-zero angle. Let S_1, $F_1 \in C^2$. Then the function $S_1(x)$ can be defined in the whole domain Γ_0 as the solution of the equation (6). In the backward procedure the function $S_1(x)$ is the first to be determined. Considering it known in Γ_0, write the condition of continuity of the game value as one of the necessary conditins of optimality determining the manifold W_3: $G_1(x, S) \equiv S - S_1(x) = 0$. As G_0 take the function F_0 in (6). If Γ_1 is an envelope-surface, then function G_{-1} is defined by the tangency condition

$$G_{-1}(x, p) \equiv \{G_1 G_0\} \equiv (F_{0p}, p - p_1(x)) = 0 \tag{7}$$

If the trajectories from D_0 approach to Γ_1 with non-zero angle, and the surface is governed by player P(E), then the uniqueness of maximum (mimimum) is failed in (5). The uniqueness is lost usually when a relation of type $Q(x, p_0) = 0$, $x \in \Gamma_1$, holds. The function Q may be taken as G_{-1}. Thus three functions are defined to write Hamiltonian (3) and equations (2) for equivocal motion. In the case of envelope-surface the equalities can be expressed in terms of two functions F_0 and S_1 ($F_1 \equiv S - S_1$):

$$\dot{x} = F_{0p}, \quad \dot{p} = -F_{0x} - (\{F_0\{F_1F_0\}\}/\{F_1\{F_0F_1\}\})(p - S_{1x}(x)) \tag{8}$$

The last equation $\dot{S} = (p, F_{0p})$ is separated. The system (8) from the regular one differs only by second term in the second equation.

3. Universal surface. Let the dynamic equations in (4) do not contain parameter v , and in the corresponding optimal control problem the cost function and terminal condition are the same. Singularity of type 2 is considered, Fig. 1. For the sake of simplicity we assume Γ_1 to be the plane $x_1 = 0$, where x_1 is the first component of the vector $x \in R^n$. The main assumptions are the following ones. Smooth Bellman function $S(x) \in C^1(\Gamma_0)$ exists, while $S_i(x) \in C^2(D_i)$. Hamiltonian $F(x, p) = \min_u(p, f(x, u)) + 1$, $u \in U$, can be represented in the neighbourhood of the point (x^*, p^*), $p^* = S_x(x^*)$, in the form $F = \min [F_0, F_1]$, where functions $F_i \in C^2$ are defined in the above mentioned neighbourhood. Minimum over u is always unique vector, except of $p = p(x)$, $x \in \Gamma_1$. Hamiltonian F_i corresponds to the domain D_i, that is the function S_i satisfies the equation $F_i(x, S_{ix}(x)) = 0$, $x \in D_i$, which allows to extend it on the whole domain Γ_0. Here we assume that paths approach to Γ_1 with non-zero angles ,i.e. $F_{0p_1} > 0$, $F_{1p_1} < 0$. (The substitution $p = -\psi$ gives the Hamiltonian of the maximum principle). Let the functions S_i be as smooth as it's required. The following necessary conditions of optimality are true [2].

Theorem 2. The difference $\Delta(x) = S_1(x) - S_0(x)$ as a function of x_1 is non-decreasing function in the neighbourhood of 0, i.e. if $\Delta^{(k)} \equiv \partial^k\Delta/\partial x_1^k = 0$, $k = 0,1,\ldots,m$, $\Delta^{(m+1)} \neq 0$ for $x_1 = 0$, then m is an even integer and $\Delta^{(m+1)} > 0$, Bellman's funcnion is represented in the form $S = \max [S_0, S_1]$.

It follows from here, particularly, that $S(x) \in C^2(\Gamma_0)$. The following mathematical statement reduces these conditions to an invariant form, irrespective of the particular form of Γ_1.

Theorem 3. Let the pair of functions $V_i(x) \in C^m(\Gamma_0)$, $m \geq 2$, have equal partial derivatives of order $k = 0,1,\ldots,m-1$ on the surface $\Gamma_1 = \{x \in \Gamma_0 : x_1 = 0\}$ and satisfy the equations $G_i(x, V_{ix}(x), V_i(x)) = 0$, $x \in \Gamma_0$, $G_i \in C^{m-1}$, $i = 0, 1$. Then in the points of surface Γ_1 the following equality holds

$$\{\ldots\{G_0G_1\}G_{j_3}\}\ldots G_{j_m}\} + (G_{0p_1})^r(G_{1p_1})^{m-r}\partial^m(V_0 - V_1)/\partial x_1^m = 0$$

Here the indices j_k can be either 0 or 1, r is the number of zeroes in the set $(0, 1, j_3, \ldots, j_m)$.

From the theorems 2, 3 it follows.

Theorem 4. Let $S_i \in C^{2d+1}$, $F_i \in C^{2d}$. Then on the surface Γ_1 for some $d \geq 1$ the following necessary conditions of optimality hold

$$(d = 1) \quad \{F_0F_1\} = 0, \quad \{\{F_0F_1\}F_0\} \leq 0, \quad \{\{F_1F_0\}F_1\} \leq 0$$

$$(d \geq 2) \quad \{F_0F_1\} = 0, \quad \{\ldots\{\{F_0F_1\}F_{j_3}\}\ldots F_{j_k}\} = 0, \quad k = 3,\ldots, 2d \qquad (9)$$

$$(-1)^r\{\ldots\{\{F_0F_1\}F_{j_3}\}\ldots F_{j_m}\} \leq 0, \quad m = 2d + 1$$

Note, that the number of equalities (inequalities) in (9) is equal to the number of integer vectors of dimension not more than 2d (equal to 2d + 1) with components 0, 1, while the the first two components are fixed and equal to 0, 1. For $d = 1$ in (9) we have two inequalities; Kelley's condition (one inequality) is one of the consequences of (9). For constructing the surface with the help of Hamiltonian (3) one has to take functions F_0, F_1 and as the third one of the Poisson brackets in (9) with maximal k .

4. Game problem of simple pursue on the manifold. Let the points (players) P and E move in n-dimensional manifold N (in the physical space of game) according to the equations:

$$P: \dot{z} = u, \quad u \in E_1(z); \quad E: \dot{y} = v, \quad v \in E_\nu(y), \quad \nu < 1 \tag{10}$$

Here $z, y \in R^n$ are the local coordinates of the points P and E, $E_\alpha(z)$ denotes ellipsoid $\{u \in R^n: (G(z)u, u) \le \alpha^2\}$, where $G(z)$ is the positive defined matrix, metric tensor of the manifold. Let $L(z, y)$ be the global minimum of the distance between P and E in the metric G. The game starts at the moment $t = 0$ and ends at $t = T > 0$, when at first the following condition holds

$$(z(T), y(T)) \in M = (z, y) \in R^{2n}: L(z, y) \le h , \quad (h \ge 0) \tag{11}$$

The pursuer P, having unity as maximal speed, minimizes time, and evador E , having speed $\nu < 1$, maximizes it. The main assumption about the geometry of the manifold N is, that if the global minimum L is reached not on a unique curve, then there exist smooth local minimums L^+, L^-, such that $L(z, y) = \min [L^+(z, y), L^-(z, y)]$. This condition always holds, for example, when N is two-dimensional euclidean plane with convex closed bounded obstacle D , the bound of which the players can not intersect, Fig.2 [3]. Euclidean space with obstacle and 2-dimensional surfaces in 3-dimensional space are the main examples of N. Let X denote the phase space of the game N x N, $x = (z, y) \in X$. It is obvious, that in some subset $X_1 \subset X$ the optimal behaviour of the players is to pursue and evade along the geodesic line, connecting the points P and E. In that domain the value of game has the form

$$S(x) = [L(x) - h]/(1 - \nu) \tag{12}$$

As far as the geodesic length satisfies the eikonal equations $(G^{-1}(z)L_z, L_z) = 1$, $(G^{-1}(y)L_y, L_y) = 1$ the function (12) in the points of smoothness satisfies Bellman-Isaac's equation ($p = (S_z, S_y)$):

$$F(x, p) = -\sqrt{(G^{-1}(z)S_z, S_z)} + \nu\sqrt{(G^{-1}(y)S_y, S_y)} + 1 = 0, \tag{13}$$

obtained by calculating mimimax (5) on the base of equations (10). Note that both extrema are reached on unique vectors for any $p \neq 0$. .On the surface $\Gamma = \{x \in R^{2n}: L^+ = L^-\}$ function (12) has the form $S = \min [S^+, S^-]$, so the condition in the second line of (6) ($S^{\pm}$ mean (12) with substitution $L = L^{\pm}$) holds. The calculations show, that the right (maximin) condition in (6) holds in all points of Γ, the left (minimax) condition holds in the form of equality only on the part of the surface Γ defined by [4]

$$L^+(x) = L^-(x), \quad F(x, R_x(x)) \geq 0, \quad R(x) = \frac{1}{2}(S^+(x) + S^-(x)) \tag{14}$$

Let $B \subset \Gamma$ be the set of points, for which in (14) equality holds. Then the manifold B can be represented as

$$|a^+ + a^-|_x - \nu|b^+ + b^-|_y = 2(1 - \nu), \quad L^+ = L^-, \quad |a|_x = \sqrt{(G(x)a, a)}$$

where $a^{\pm}$, $b^{\pm}$ are unit vectors,tangent to geodesics in it's extremities, Fig. 2. Analysis shows, that set B can be the origin of two branches of equivocal envelope-surfaces Γ^+, Γ^-, as shown on Fig. 3. On the surface Γ^+, (Γ^-) the necessary condititons are Bellman's equation (13), the continuity condition $S - S^+(x) = 0$ ($S - S^-(x) = 0$) and tangency condition (7) (in correspondence with the above mentioned uniqueness of extremal vectors). The corresponding functions are used to write the equations of singular characteristics (8). The initial manifold Σ_2 for both branches Γ^+ and Γ^- is the same suspension on $B = \Gamma_2$ in R^{4n+1}:

$$\Sigma_2 = \{(x, p, S): p = R_x(x), S = S^+(x), x \in B\} \tag{16}$$

If N is Euclidean space with obstacle, then the Hamiltonian (13) depends only on p and function $F(p) - 1$ is homogeneous. Then equations (8) are symplified and reduced to the form

$$\dot{x} = F_p, \quad \dot{p} = [(S_{xx}F_p, F_p)/(F_{pp}q, q)](p - q) \tag{17}$$

Figure 1:

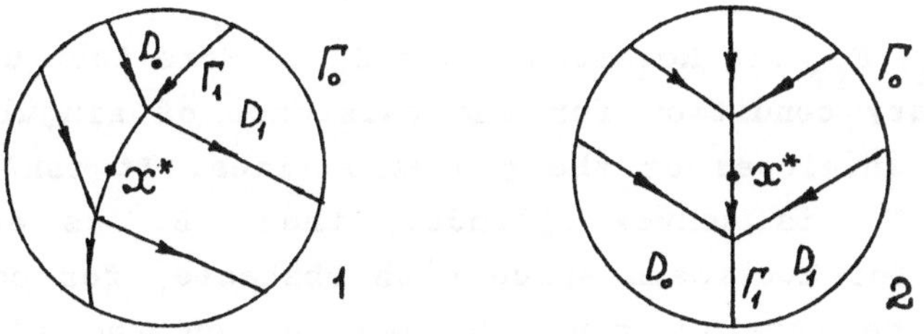

Figure 2:

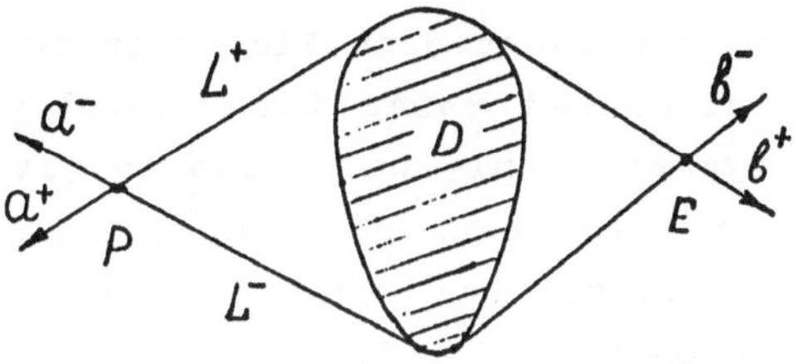

Figure 3:

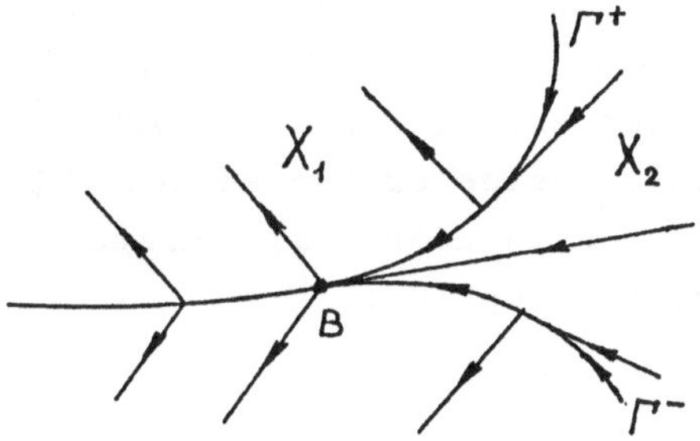

Figure 4:

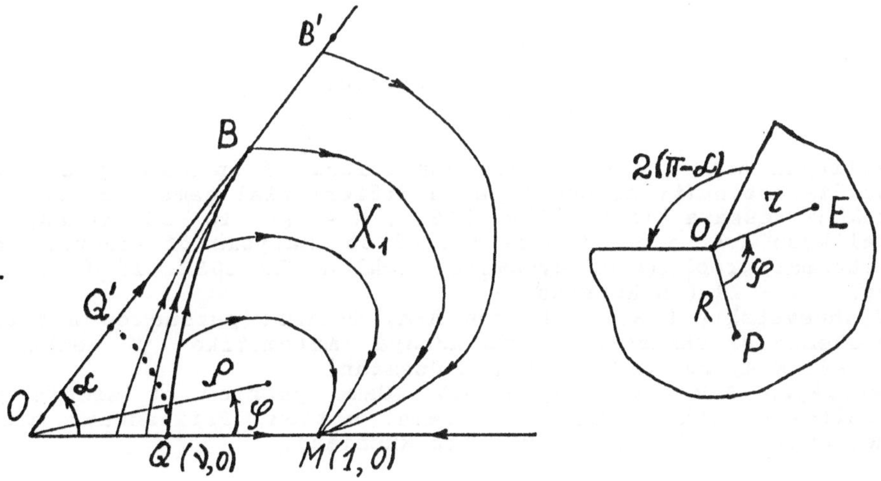

where S_{xx}, F_{pp} are Hessians, $q = S_x^{\pm}$. Note that unemptiness of set B is necessary condition for the existence of singular equivocal paths, which are envelopes of the geodesic lines. It isn't difficult to show that if N is convex cylinder, then B is empty; B is always non-empty for Euclidean space with obstacle, for convex cone.

For the case of cone the mapping on the plane two-sided angle allows to use locally euclidean coordinates, i.e. equations (17). The relations (10) - (17) have self-similarity, due to which the problem is reduced to two-dimensional one. The complete solution of the pursue problem on the cone is presented on Fig. 4 in terms of polar coordinates ρ, φ ($\rho = r/R$) introduced using the development of the cone on a plane. Because of symmetry only half-cone is given on Fig. 4. The point B is calculated from (16), equivocal curve BQ, the main element of the solution, which determins the whole pattern of optimal paths, is obtained by integration of (17). Primary domain X_1 is on the right side of BQ. The half-line BB′ is dispersal line, as well as the segment OQ for $\arcsin(1-\nu) < \alpha < \pi$, $0 < \nu < 1$. If $0 < \alpha \le \arcsin(1-\nu)$ a new singular curve, the circular arc QQ′ appears on which the value function is differentiable. Inside of the sector OQQ′ the value doesn't depend on the position of the evador E; pursuer P must move with maximal velocity towards the top O .

References

1. Melikyan A.A. The necessary conditions of optimality on the surface of discontinuity of one type in differential game. Izvestiya AN SSSR. Tekhnicheskaya kibernetika. 1981., N 4, pp. 10 -18 (in Russian).
2. Melikyan A.A. A positional method of analysis of singular motions in extremal problems of dynamics. Doklady AN SSSR. 1990, v. 311, N 1, pp. 22 - 27 (in Russian).
3. Vyshnevetskiy L.S., Melikyan A.A. Optimal pursue on a plane in the presence of obstacle. Prikladnaya matematika y mechanika. 1982, v. 46, N 4, pp. 613 - 620 (in Russian).
4. Melikyan A.A., Ovakimyan N.V. Game problem of simple pursue on manifolds with nonunique extremal motion. Prikladnaya matematika y mechanika. 1991, v. 55, N 1 (in Russian).

Differential games and totally risk-averse optimal control of systems with small disturbances

Martino BARDI and Caterina SARTORI
Dipartimento di Matematica Pura e Applicata
Dipartimento di Metodi e Modelli Matematici per le ScienzeApplicate
Università di Padova
via Belzoni,7 , I-35131 Padova - Italy

Abstract. We propose a differential game as a model for the optimal control of systems affected by an unpredictable disturbance, when the controller follows a policy of total risk aversion. We study the problem of small disturbances of a given deterministic system, and estimate the difference between the value functions of the unperturbed and of the perturbed problem.

0. Introduction.

In this paper we study the optimal control of a system which contains some parameters whose future behaviour is completely unpredictable. Assuming a totally risk-averse behaviour of the controller who seeks to minimize a given payoff, we model the problem by a differential game where the second player is a "hostile nature" who drives the unknown parameters towards the goal of maximizing the same payoff. For instance time-optimal problems where the system is affected by an unpredictable disturbance are modeled by games of pursuit-evasion type where the nature controlling the disturbance is the evader. The idea of using differential games to model similar situations goes back to a classical paper of Fleming [F]. A recent paper of Barron & Jensen [BJ] gives a rigorous foundation to this modelization within the theory of risk, at least for a finite horizon problem of maximization.

Our main result concerns the case when the unknown parameters cause a small perturbation of a given deterministic systems. We prove estimates on the difference between the value functions of the perturbed and the unperturbed system. In particular we show, roughly speaking, that the rate of convergence to zero when the perturbation vanishes is the Hölder exponent of the value function of the unperturbed problem.

The proofs are based on some general results on approximation and perturbation of differential games we obtained in [BS], where we considered the Hamilton-Jacobi-Isaacs equations of Dynamic Programming and worked in the framework of the theory of viscosity solutions for such equations. We limit ourselves here to time-optimal problems, but similar results can be given for infinite horizon problems following again [BS] (and they are indeed easier). Finite horizon problems can be treated by the same methods.

We refer to Crandall & P.L. Lions [CL], Crandall, Evans & Lions [CEL], Lions [L] and Elliott [E] for an introduction to the theory of viscosity solutions and its connection to optimal control and games. The applications of viscosity solutions to differential games were first made by Barron, Evans & Jensen [BEJ], Evans & Souganidis [ES] and Souganidis [S] for finite horizon problems, and by Bardi & Soravia [BSo1, BSo2] for pursuit-evasion problems, see also the references therein.

1. Total risk-aversion and differential games.

We consider a controlled dynamical system containing an unknown parameter

$$(1.1)\qquad \begin{cases} \dot{y} = f(y,a,b) & t > 0\,, \\ y(0) = x & x \in \mathbb{R}^N\,, \end{cases}$$

where a is the control function taking values in a given compact set $A \subset \mathbb{R}^M$, and b is the "random" disturbance taking values in a known compact set $B \subset \mathbb{R}^M$. The controller wants to minimize a given payoff functional $P(x,a,b)$, where a and b belong respectively to

$$\mathcal{A} := \{\, a : \mathbb{R}_+ \to A \text{ measurable} \,\}\ ,\quad \mathcal{B} := \{\, b : \mathbb{R}_+ \to B \text{ measurable} \,\}\,,$$

and the partial information about the behaviour of b makes it impossible to choose a reasonable probability distribution for b, so that the problem does not fit into the theory of stochastic control.

Instead, we assume that the controller is totally risk-averse, that is, he gives up possible better outcomes in case of a favorable behaviour of the disturbance b, which is unpredictable, and prefers to optimize his behaviour in the worst possible case. A model for this situation is a zero-sum differential game where b is chosen by a second player seeking to maximize the same payoff $P(x,a,b)$. To complete the model we assume that the hostile nature driving b is not so powerful to foresee the controller's behaviour in the future. More precisely we assume that the second player chooses from the set of nonanticipating strategies

$$\Gamma := \{\, \beta : \mathcal{A} \to \mathcal{B} : a(t) = a'(t) \text{ for all } t \le t' \text{ implies } \beta[a](t) = \beta[a'](t) \text{ for all } t \le t' \,\},$$

and therefore the value function of our totally risk-averse optimal control probem under uncertainty is the upper value in the sense of Varaya-Roxin-Elliott-Kalton [EK, E] of the differential game with dynamics (1.1) and payoff P, that is

$$(1.2)\qquad U(x) := \sup_{\beta \in \Gamma}\ \inf_{a \in \mathcal{A}} P(x,a,\beta[a])\,.$$

This model may be considered a little too pessimistic, in that the information pattern gives an advantage to the hostile nature. A more optimistic model is the lower value of the differential game, that is

(1.3) $$V(x) := \inf_{\alpha \in \Delta} \sup_{b \in \mathcal{B}} P(x,\alpha[b],b) ,$$

where Δ is the set of nonanticipating strategies for the first player, defined in an obvious symmetric way. This means that at each time t the controller of a can choose knowing the values of b up to time t included.

In a recent paper Barron & Jensen [BJ] studied a finite horizon problem of maximizing the utility of a given payoff, where the parameters b in the system were a Brownian motion reflected back into B upon reaching the boundary. They showed that, as the index of risk aversion of the utility function tends to infinity, the value function of the stochastic control problem tends to the upper value of the differential game where a second player controls b and seeks to minimize the payoff. They also show that if, roughly speaking, the controller cannot respond fast enough to the changes of the disturbance, then the stochastic value converges to the lower value of the same game as the index of risk aversion goes to infinity. Their results support our modelization and give more precise information for the choice between U and V.

A situation of uncertainty on the behaviour of the system (1.1) similar to ours was studied recently by Alberto Bressan [Br1, Br2], who proposed a notion of "likelihood" for the trajectories of the system which relies only on the metric structure and is entirely independent of probability theory. A theory of optimal control based on Bressan's likelihood functional might be an interesting future direction of research.

We are now going to describe the mathematical framework of our results. We will suppose the function $f : \mathbb{R}^N \times A \times B \to \mathbb{R}^N$ to be continuous and to satisfy

(1.4) $$| f(x,a,b) - f(y,a,b) | \leq L | x-y | \qquad \text{for all } a \in A,\ b \in B \text{ and } x,y \in \mathbb{R}^N.$$

We consider only time-optimal problems because they are simple to formulate and they contain all the main difficulties. Indeed our results for these problems are deduced from the study of rather general infinite horizon problems with exit time, via a nonlinear transformation (see [BS]). We are given a closed target $\mathcal{T} \subseteq \mathbb{R}^N$ and our payoff will be the first time the system hits the target, that is

(1.5) $$P(x,a,b) = t_x = t_x(a,b) := \inf \{ t : y(t) \in \mathcal{T} \} \leq +\infty ,$$

where y(.) is the trajectory of (1.1) corresponding to $a \in \mathcal{A}$ and $b \in \mathcal{B}$. The unperturbed problem occurs when B is a singleton, that is $B_0 := \{b_0\}$. The value function in this case is the minimal time function

(1.6) $$T(x) := \inf_{a \in \mathcal{A}} t_x(a,b_0) \leq +\infty , \quad x \in \mathbb{R}^N.$$

Let $\mathcal{R}$ be the set of starting points x from which the system can reach the target in finite time, i.e.

$$\mathcal{R} := \{x \in \mathbb{R}^N : T(x) < +\infty\}.$$

It is well known that the Dynamic Programming Principle implies that at each point of $\mathcal{R}$ where T is differentiable the Bellman partial differential equation is satisfied

$$(1.7) \qquad \max_{a\in A} \{-f(x,a,b_0) \cdot DT\} = 1 .$$

However T is not differentiable in general, its regularity being tipically Hölder continuity, provided some controllability condition is assumed. The discovery of a good concept of weak solution for Hamilton-Jacobi equations such as (1.7) [CL, CEL, L] allowed to revive the Dynamic Programming approach to the optimal control of continuous-time systems. P.L. Lions [L] indeed showed that T is a viscosity solution of (1.7) as soon as it is continuous. One of the authors [B] proposed recently the following singular boundary value problem for Bellman's equation

$$(1.8) \qquad \begin{cases} H(x,DS) = 0 & \text{in } \Omega\backslash\mathcal{T} ; \\ S = 0 & \text{on } \partial\mathcal{T} ; \\ S(x) \to +\infty & \text{as } x \to \partial\Omega ; \end{cases}$$

where the Hamiltonian is

$$H = H_0(x,p) := \max_{a\in A} \{ -f(x,a,b_0) \cdot p \} - 1 ,$$

and he proved that, if Ω is chosen to be $\mathcal{R}$, then $S = T$ is the unique continuous viscosity solution of (1.8) bounded from below. Later Bardi & Soravia [BS2] discovered that (1.8) is indeed a free boundary problem, in the sense that it singles out the pair (Ω, S). This is made precise in Theorem 1.1 below.

When the system is perturbed, i.e. B is not a singleton, the upper value function U given by (1.2) with the payoff (1.5) satisfies, where it is continuous, the "upper" Isaacs equation of Dynamic Programming in the viscosity sense. The Hamiltonian of this equation is

$$H^u(x,p) := \max_{a\in A} \min_{b\in B} \{ -f(x,a,b) \cdot p \} - 1 .$$

Similarly the lower value function V given by (1.3) with the same payoff (1.5) is a viscosity solution of the "lower" Isaacs equation whose Hamiltonian is

$$H^l(x,p) := \min_{b\in B} \max_{a\in A} \{ -f(x,a,b) \cdot p \} - 1 .$$

We need also to introduce the sets of starting points from which the system can be driven to the target in finite time, no matter what the evader (the disturbance) does:

$$\mathcal{P} := \{x \in \mathbb{R}^N : U(x) < +\infty\}, \quad \mathcal{Q} := \{x \in \mathbb{R}^N : V(x) < +\infty\}.$$

We are now ready to state precisely the main results on the well-posedness of the Dynamic Programming partial differential equations for time-optimal and pursuit-evasion problems.

Theorem 1.1. ([BSo1, BSo2]) Assume (1.4) and T, U, V continuous at each point of $\partial\mathcal{T}$. Then

(i) $\mathcal{R}, \mathcal{P}, \mathcal{Q}$ are open and contain $\mathcal{T}$;

(ii) T, U, V are continuous in $\mathcal{R}, \mathcal{P}, \mathcal{Q}$ respectively, and they satisfy in the viscosity sense (1.8) where H and Ω are H_0, H^u, H^l and $\mathcal{R}, \mathcal{P}, \mathcal{Q}$ respectively;

(iii) $\mathcal{P} \subseteq \mathcal{Q} \subseteq \mathcal{R}$ and $U(x) \geq V(x) \geq T(x)$ for all $x \in \mathcal{P}$;

(iv) if Ω is an open set containing $\mathcal{T}$, $S \in C(\Omega)$ is bounded below, and the pair (S, Ω) satisfies (1.8), then $\Omega = \mathcal{R}$ and $S = T$ if $H = H_0$, $\Omega = \mathcal{Q}$ and $S = V$ if $H = H^l$;

(v) if $H^u = H^l$ for all x, p (Isaacs condition), then $\mathcal{P} = \mathcal{Q}$ and $U = V$, i.e. the game has a value.

Note that (v) is a trivial consequence of (iv). This P.D.E. formulation of the problem is the basis for our results in next Section.

2. Estimates of the error for small disturbances.

In this section we study the problem of small disturbances to the minimum time problem. In our model this means that the set of values of the disturbance, B_h, is close to a singleton in the following sense

(2.1) $$d_H(B_h, B_0) \leq h ,$$

where $B_0 = \{ b_0 \}$ and d_H is the usual Hausdorff distance, that is

$$d_H(B_h, B_0) = d_H(B_h, \{b_0\}) := \max_{b \in B_h} |b - b_0| .$$

We call U_h and V_h respectively the upper and the lower value of the game when b takes value in B_h. Our main result is the following.

Theorem 2.1. Let $B_h \subseteq B$ and A be compact, assume (2.1) and $f : \mathbb{R}^N \times A \times B \to \mathbb{R}^N$ continuous and satisfying

(2.2) $\quad |f(x,a,b) - f(y,a',b')| \le L(|x - y| + |a - a'| + |b - b'|)$ for all x, y, a, a', b, b'.

Let T_h be either U_h or V_h, and suppose that, for $\alpha, E > 0$, and for all x in a neighbourhood of $\mathcal{T}$

(2.3) $\quad T_h(x) \le E\, d^\alpha(x) \quad$ for all $0 \le h \le 1$,

where $d(x) := \text{dist}\,(x, \mathcal{T})$. Then for every compact set $\mathcal{K} \subseteq \mathcal{R}$ there exist positive constants C and $\overline{h}$ such that

$$\sup_{x \in \mathcal{K}} |T(x) - T_h(x)| \le C h^\alpha \text{ for all } 0 < h \le \overline{h} . \qquad //$$

Before sketching the proof of this result, we give an example of its application, in the case $\alpha = 1$, where the assumptions are made only on the unperturbed problem, as is the rule in perturbation theory, and the perturbation is only supposed to be small in the sense that (2.1) holds with h small. For $\mathcal{T}$ bounded and smooth, namely the closure of an open set with C^2 boundary, we will assume

(2.4) $$\min_{a \in A} f(x,a,b) \cdot n(x) < 0 \ , \quad \text{for every } x \in \partial\mathcal{T},$$

where $n(x)$ is the exterior normal to $\mathcal{T}$ at x. This condition has been used by many authors beginning with Fleming [F1], and it is necessary and sufficient for the local Lipschitz continuity of T, see the recent work of Bardi & Falcone [BF] and the references therein.

Corollary 2.2. Assume $\mathcal{T}$ is bounded and smooth and (2.1), (2.2), (2.4) hold. Then for every compact set $\mathcal{K} \subseteq \mathcal{R}$ there exist positive constants C and $\overline{h}$ such that

$$\sup_{x \in \mathcal{K}} |T(x) - T_h(x)| \le C h \text{ for all } 0 < h \le \overline{h} ,$$

for $T_h = U_h$, V_h.

Proof of Corollary 2.2. It is easy to see that (2.4), (2.2) and (2.1) imply, for h small enough,

$$\min_{a \in A} \max_{b \in B_h} f(x,a,b) \cdot n(x) < 0 \ , \quad \text{for every } x \in \partial\mathcal{T},$$

which implies (2.3) with $\alpha = 1$, for $0 \le h \le \overline{h}$ and for $T_h = U_h$, V_h by Proposition 2 and the Remark in [BSo1]. Then Theorem 2.1 applies. //

Theorem 2.1 follows from a more general result proven in [BS] on approximation and perturbation of differential games of pursuit-evasion type, which we now recall. Let us introduce for $0 \le h \le 1$ new dynamical systems in $\mathbb{R}^N$ controlled by two players

$$(2.5) \qquad \begin{cases} \dot{y} = f_h(y,a,b) & t > 0, \\ y(0) = x & x \in \mathbb{R}^N, \end{cases}$$

where $a \in \mathcal{A}_h$, $b \in \mathcal{B}_h$. Recall the definiton of *Hausdorff distance* of two compact sets J and K :

$$d_H(J,K) := \max \{ \max_{j \in J} \operatorname{dist}(j,K), \max_{k \in K} \operatorname{dist}(k,J) \}$$

and define

$$\| g \|_\infty := \sup \{ | g(x,a,b) | : (x,a,b) \in \mathbb{R}^N \times A \times B \}.$$

Theorem 2.3. [BS] Given the system (2.5), suppose that $B_h \subseteq B$ and $A_h \subseteq A$, and assume $f_h : \mathbb{R}^N \times A \times B \to \mathbb{R}^N$ continuous and satisfying, for all $0 \le h \le 1$,

$$|f_h(x,a,b) - f_h(y,a',b')| \le L(|x - y| + |a - a'| + |b - b'|) \text{ for all } x, y, a, a', b, b',$$

$$d_H(B_h, B_0) \le Ch, \quad d_H(A_h, A_0) \le Ch, \quad \| f_h - f_0 \|_\infty \le Ch.$$

Moreover suppose that, for $\alpha, E > 0$ and for all x in a neighbourhood of $\mathcal{T}$

$$(2.6) \qquad V_h(x) \le E\, d^\alpha(x) \qquad \text{for all } 0 \le h \le 1,$$

where $d(x) := \operatorname{dist}(x, \mathcal{T})$. Then for every compact set $\mathcal{K} \subseteq \mathcal{R}$ there exist positive constants C and $\bar{h}$ such that

$$\sup_{x \in \mathcal{T}} | V_0(x) - V_h(x) | \le C h^\alpha \text{ for all } 0 < h \le \bar{h}. \qquad //$$

Remark 2.4. This Theorem is a particular case of a more general result on the rate of convergence of viscosity solutions of Hamilton-Jacobi when the Hamiltonians converge uniformly, at a given rate, to a limit Hamiltonian. It is obtained using comparison techniques for such weak solutions, and we can fit our control problem into this general theory thanks to Theorem 1.1 .

Remark 2.5. A result similar to Theorem 2.3 holds when V_h is replaced by U_h for $0 \leq h \leq 1$, by Remark 2.4.

Proof of Theorem 2.1. In view of Theorem 2.3 and Remark 2.5 it is enogh to show that when B_0 is a singleton the upper and the lower value functions both coincide with the minimal time function. We have that

$$V_0(x) := \inf_{\alpha \in \Delta_0} \sup_{b \in \mathcal{B}_0} t_x(\alpha[b], b) = \inf_{\alpha \in \Delta_0} t_x(\alpha[b_0], b_0)$$

where the set Δ_0 of nonanticipating strategies for the first player is in this case $\Delta_0 = \{\alpha : \{b_0\} \to \mathcal{A}\}$. It is obvious that Δ_0 is in one to one correspondence with $\mathcal{A}$, giving therefore $V_0 = T$, see (1.6). On the other hand the upper value of the game is

$$U_0(x) := \sup_{\beta \in \Gamma_0} \inf_{\alpha \in \mathcal{A}} t_x(a, \beta[a]) = \inf_{\alpha \in \mathcal{A}} t_x(a, b_0) = T(x)$$

because $\Gamma_0 = \{\beta : \mathcal{A} \to \{b_0\}\}$. //

References.

[B] M. Bardi: A boundary value problem for the minimum time function, SIAM J. Control Optim. 27 (1989), 776-785.

[BF1] M. Bardi, M. Falcone: An approximation scheme for the minimum time function, SIAM J. Control Optim. to appear.

[BS] M. Bardi, C. Sartori: Approximation and regular perturbation of optimal control problems via Hamilton-Jacobi theory, preprint Università di Padova 1990, submitted.

[BSo1] M. Bardi, P. Soravia: A PDE framework for differential games of pursuit-evasion type, in "Differential games and applications," T. Basar and P. Bernhard eds., pp. 62-71, Lecture Notes in Control and Information Sciences 119, Springer-Verlag 1989.

[BSo2] M. Bardi, P. Soravia: Hamilton-Jacobi equations with singular boundary conditions on a free boundary and applications to differential games, Trans. Amer. Math. Soc. to appear.

[BEJ] E.N. Barron, L.C. Evans, R. Jensen: Viscosity solutions of Isaacs' equations and differential games with Lipschitz controls, J. Differential Equations 53 (1984), 213-233.

[BJ] E.N. Barron, R. Jensen: Total risk aversion, stochastic optimal control, and differential games, Appl. Math. Optim. 19 (1989),313-327.

[Br1] A. Bressan: The most likely path of a differential inclusion, J. Differential Equations to appear.

[Br2] A. Bressan: Nonprobabilistic filtering, Proceedings of the Conference "New Trends in System Theory", Genova 1990, to appear.

[CEL] M.C. Crandall, L.C. Evans, P.L. Lions: Some properties of viscosity solutions of Hamilton-Jacobi equations, Trans. Amer. Math. Soc. 282 (1984), 487-502.

[CL] M.C. Crandall, P.L. Lions: Viscosity solutions of Hamilton-Jacobi equations, Trans. Amer. Math. Soc. 277 (1983), 1-42.

[E] R.J. Elliott: Viscosity solutions and optimal control, Longman, Harlow 1987.

[EK] R.J. Elliott, N.J. Kalton: The existence of value in differential games, Mem. Amer. Math. Soc. 126 (1972).

[ES] L.C. Evans, P.E. Souganidis: Differential games and representation formulas for solutions of Hamilton-Jacobi equations, Indiana Univ. Math. J. 33 (1984), 773-797.

[F1] W.H. Fleming: The convergence problem for differential games, J. Math. Anal. Appl. 3 (1961), 102-116.

[L] P.L. Lions: Generalized solutions of Hamilton-Jacobi equations, Pitman, Boston 1982.

[S] P.E. Souganidis: Max-min representations and product formulas for the viscosity solutions of Hamilton-Jacobi equations with applications to differential games, Nonlinear Anal. T.M.A. 9 (1985), 217-257.

DIFFERENTIAL GAMES AND ARTIFICIAL INTELLIGENCE - A NEW APPROACH FOR SOLVING COMPLEX DYNAMIC CONFLICTS*

Joseph Shinar
Professor, Faculty of Aerospace Engineering
Technion - Israel Institute of Technology
Haifa 32000, Israel.

ABSTRACT. Dynamic conflicts exhibit differential game characteristics and their analysis by any method which disregards this feature may turn out to be, by definition, futile. Unfortunately, most realistic conflicts have an intricate information structure and a complex hierarchy which don't fit in the classical differential game formulation. Moreover, in many cases even well formulated differential games are not solvable. In the recent years great progress has been make in Artificial Intelligence techniques, put in evidence by successful applications in Scientific Modeling, Automated Engineering Design Processes as well as for Fuzzy and Intelligent Control Systems. This progress has raised hopes that Artificial Intelligence methods can be also of help in solving complex dynamic conflicts. This paper outlines a feasible option which combines Artificial Intelligence techniques with concepts of Differential Game Theory for attaining such an objective. This approach is illustrated by an air combat game example.

1. INTRODUCTION

Activities of the human society have been always characterized by situations of conflict expressed in either political, economical or military terms. In all of these conflicts the decisions have been made by following some rational reasoning, called some times a "strategy". The search for optimal strategies in this sense has been an important objective of all decision makers. The need for some mathematical modeling of a conflict has been realized since the fifth century B.C., when the Chinese General Sun Tzu observed the correlation between victory in a battle and well calculated planning. In the last century a formal mathematical frame for conflict analysis was created [1]. It is the Theory of Games, a recognized branch of Applied Mathematics, that provides the language and the concepts for describing and

---

*A part of the work described in the paper was carried out during the author's Sabbatical leave at INRIA, Sophia Antipolis Research Center, Valbonne, France, as Visiting Scientist sponsored by the French Ministry of Research and Technology, Delegation of International Affairs.

analyzing strategies in a conflict characterized by several agents (players) with independent goals, each having only a partial control of the environment.

In most cases the conflict environment is not static. Its evolution in time can be described by some set of ordinary (or partial) differential equations or by their discretized equivalents. Such a mathematical model brings us to the domain of "dynamic" conflicts and to the theory of dynamic or Differential Games [2].

The study of differential games as a model for dynamic military conflicts started during the second world war. It emerged from the pioneering work of Rufus Isaacs on optimal pursuit and evasion attempting to model tactical air combat problems. Scientific investigation in this direction has continued since the early fifties without yielding (unfortunately) useful applications. In the last decade there has been a renewed research interest in differential games. It is evidenced by four International Symposia on Differential Game Applications (1984 in Haifa, Israel, 1986 in Williamsburg, Virginia; 1988 in Sophia Antipolis, France and the present one in Helsinki, Finland) as well as by the great number of special sessions devoted to this topic in the last years at several international and national meetings of the automatic control community (IFAC, CDC, AIAA GN&C). There is a clear tendency of efforts to overcome the difficulties which create the gap between theory and applications.

Anyone who wishes to carry out a meaningful analysis of a real-life conflict is facing two major challenges:

1. to formulate a relevant and insightful mathematical model of the situation,
2. to solve a well posed mathematical problem based on the model.

However, these steps are only prerequisites for the *real solution*, which consists of interpreting the mathematical results in terms of the original conflict.

The main difficulty is imbedded in the attributes "relevant and insightful" of the modeling challenge. If this task is not carried out properly it may be possible to solve the mathematical problem, but the formal results will be either irrelevant or even worse - misleading. The technical literature is full of intellectually intriguing mathematical game models, most of them being solved (or are at least solvable) by classical mathematical methods. Unfortunately, there is virtually no evidence of a practical application of these mathematical results. As the models become more realistic the solution of the mathematical problem involved becomes more and more difficult, time consuming and expensive. This is the main reason that in most cases decision makers prefer to rely on the alternative of "brute force" simulations using detailed physical models with arbitrary or at the best, heuristically determined control strategies.

In the last decade great progress has been made in the domain of Artificial Intelligence (AI). Both innovative concepts, efficient computation and search techniques were developed and put in evidence by successful applications for Scientific Modeling, Automated Engineering Design Processes as well as for Fuzzy and Intelligent Control Systems. This progress provides an indication that AI methods have the potential to be of help also in solving complex dynamic conflicts.

The objective of this paper is to outline a feasible option for the analysis of complex dynamic conflicts based on combining AI techniques with concepts of Differential Game Theory.

These seems to be a great potential of success for such an approach because the classical representation of a dynamic conflict (essentially a multi-stage decision making process) by a "decision tree" is equivalent to a game in *extensive* form [3]. The *pruning* of such a tree is one of the basic tasks of heuristic search techniques used in most AI programs. This observation indicates the compatibility of the two different scientific disciplines in the present context. A second aspect is related to the complementary and probably synergetic features of these disciplines. Differential games require an unambiguous and strict description of the problem to be solved and therefore complex *real-world* situations cannot be always introduced in such a frame. On the contrary, AI is suitable for a qualitative or a fuzzy modeling but cannot guarantee an optimal solution in the pure mathematical sense.

2. DIFFICULTIES IN A DIFFERENTIAL GAME SOLUTION

The difficulties in solving a well formulated differential game have the following origin. The classical solution of a differential game [2] is based on simultaneous backwards integration of the state and adjoint equations, starting at the target set (terminal manifold) of the game, in order to fill the entire game space by the ensemble of optimal trajectories. The backwards integration is a rather direct operation as long as no singular surface of the game, implying a discontinuity of the adjoint vector, is encountered. Experience has shown that several types of different singular surfaces are frequently encountered in differential game solutions. The game solution requires, once a singular surface is reached, to determine the type of singularity and to continue the backwards integration accordingly. As a first step, the existence of the singular surface has to be identified by observing the intersection of *retrograde* trajectories belonging to different end conditions. In a simple game, with no more than two state variables, such intersection can be easily visualized. For a differential game of three independent state variables the same process becomes very cumbersome and for dynamic models of higher dimension it is virtually impossible.

At the other end, the numerical solution of a "two-point boundary-value problem" associated with the differential game satisfies only the necessary conditions of optimality. Though this method can be applied to dynamic models of any order at the cost of increasing computational effort, it cannot identify the singular surfaces of the game and provides no tool to verify the sufficiency conditions of the game solution. For this reason the usefulness of this method is rather limited.

One can also encounter great conceptual difficulties in an attempt to formulate a dynamic conflict as a differential game. This can be well illustrated by an example of an air-to-air combat duel between two fighter aircraft carrying guided missiles. A simple pursuit-evasion game formulation is certainly not suitable for such a scenario. Rather, the conflict has to be viewed as an interaction of a "two-target differential game" (between the aircraft) with two independent missile-aircraft "pursuit-evasion games". The target sets in the two-target game are the respective missile firing zones, each being the "capture zone" of a missile-aircraft pursuit-evasion game of kind. The encounter between the two aircraft (say Blue and Red) exhibits a *threat reciprocity* and must terminate with one of the following outcomes:

a. Red alone is shot down = Blue wins

b. Blue alone is shot down = Red wins

c. Both are shot down = Mutual kill

d. Both survive = Draw

The solution of the two-target differential game with the given target sets is a qualitative one, namely the decomposition of the set of admissible initial conditions into the respective zones of fixed outcome. Inside the Blue and Red winning zones many zero-sum pursuit-evasion games of degree can be played with the winning player (the pursuer) minimizing and its opponent (the evader) maximizing the same cost function. Different cost functions may yield different optimal strategies, but all games have the same guaranteed outcome, namely the termination of the game (in some finite time) on the target set of the winning player. If the engagement starts in the "mutual kill" zone both players have to be aggressive, otherwise the state of the game may slip to the opponent's winning zone. However, cooperative strategies can drive the state of the game to the "draw" zone. In the "draw" zone each player can guarantee his own survival against any action of the opponent, but cooperative aggressive strategies can lead to a "mutual kill".

A version of this air combat duel was recently analyzed as a game of kind by a simplified dynamic model of the "game of two cars". In this game model the target sets represented the "capture zones" of advanced all-aspect, *fire and forget* air-to-air missiles [4]. This modeling assumption implies that game termination on one of the target sets is equivalent to a missile firing which guarantees the destruction of the opponent, even if it employs an optimal missile avoidance strategy. Curiously, the interpretation of the apparently satisfactory results obtained in this investigation raised the problem of a conceptual incompleteness of the two-target game formulation, as explained briefly in the sequel.

Inside the winning zone of Blue, contrary to the classical assumption, Red (the loser) may have no interest to behave defensively by playing the role of an evader. Though he cannot force the game to his own target set (the "capture set" of his missile), he can still fire a missile within the classical firing envelope assuming a straight flying opponent. Such a firing disrupts the original two-target game by starting an unexpected pursuit-evasion game between the Red missile and Blue aircraft. In order to survive, Blue must take evasive action. This action will lead to a successful escape from the missile, but may prevent him - at least temporarily - from reaching an effective firing opportunity, i.e., the victory guaranteed by the two-target game solution. Moreover, during this evasive maneuver Red may be able to escape from the Blue winning zone.

The insight gained by this interpretation presents a warning, which indicates that the frame of classical Game Theory is rather limited to accommodate even relatively simple models of a *real-world* dynamic conflict.

Another difficulty is related to the very nature of the solution concept. In future air combat most engagements will start at rather long (beyond visual) ranges, thus the initial conditions of the above described two-target game are generally in the "draw" zone. Therefore the only guaranteed outcome of such non-cooperative game is a "draw".

This result, which denies the very essence of an air-to-air combat, - and consequently the justification of the high cost of advanced aircraft and missile development, - is clearly unacceptable from an operational point of view. At the other end, cooperative strategies are also inadmissible in a hostile environment. The inherent non-cooperative nature of the scenario requires from each player to determine his *preference ordering* between a "mutual kill" and a "draw" and to act accordingly. Such *preference ordering* is one of the elements of the players' strategy and therefore not known by the opponent. This uncertainty implies a major difficulty in the proper mathematical formulation and consequently in a meaningful analysis of the dynamic conflict exhibited in an air-to-air combat duel.

If one of the players prefers a "draw" and plays the corresponding optimal strategy, the opponent cannot enforce his preferred outcome. For initial conditions in the "draw" zone an eventual victory of anyone of the players is not possible, unless the other player fails to act according to his optimal defensive strategy. Such a situation, which arrives frequently in real-life conflicts, can be exploited by the opponent by using an appropriate *reprisal* strategy.

Moreover, one has to remember that future air combat will be fought by groups of adversary fight aircraft and not by individual pilots. An air-to-air combat duel is, therefore, only the simplest example and eventually a building block of the more complex multiple scenarios. Successful solution of the air-to-air combat duel is certainly a prerequisite, but it is only an element of the complete conflict.

The points raised in this section jointly indicate that Differential Game Theory alone cannot provide a satisfactory solution to complex dynamic conflicts, such as a realistic air-to-air combat.

3. AN ALTERNATIVE SOLUTION BY ARTIFICIAL INTELLIGENCE

As a consequence of the difficulties outlined in the previous section, one is strongly tempted to search for an alternative approach for the analysis of dynamic conflicts. A priori, Artificial Intelligence seems to represent such an alternative. The example of a *chess-playing* computer program demonstrates the feasibility of the idea.

Every dynamic conflict is essentially a multi-stage decision making process and can be represented as such by a "tree". Such representation is indeed identical to a game in its *extensive* form [3]. In fact, a differential game is also a game in *extensive* form and equivalent to a "decision tree" of infinite nodes. By discretizing time a finite "game tree" is obtained. The nodes of the tree represent the state of the game, where the players can select their controls for a given period of time. The branches of the tree are the moves in the game space. The *pruning* of such a tree is one of the basic tasks of heuristic search techniques used in most AI programs. In the case where the players don't change their decision at every node (selected by arbitrary discretization of the time scale), some moves can be aggregated to a single branch and the resulting "game tree" becomes simplified. It is also possible that the players don't make their decision simultaneously and in this case Blue and Red moves can be distinguished. Such an approach was pursued recently with some success [5].

Using AI techniques for the analysis of dynamic conflicts allows a qualitative description of the problem, which is extremely important in complex situations. Another favorable feature of AI is related to the already mentioned heuristic search techniques, which provide a way of solving problems for which no more direct approach is available. In such a search process any other available computational technique can be easily embedded. AI techniques can facilitate the solution of complex problems, such as conflicts, by exploiting the knowledge on the structure of the objects involved. Moreover, by using *abstraction* AI provides a way of separating important features and variations from the many unimportant ones, that otherwise may overwhelm the entire solution process.

There is no formal guarantee that an AI program will find the *true* optimal solution of a complex dynamic conflict. There are, however, two positive AI features that compensate for this deficiency: interaction with a human operator and *learning*. Most AI programs are structured to allow a human operator to interact with them by asking questions, providing missing inputs or even by modifying goals and constraints. There are AI programs that can improve their performance by changing their structure as the consequence of previous operations.

In order to possess all the above mentioned favorable features AI programs are in general complex software. They require, in addition to the problem oriented *knowledge base*, a great amount of knowledge for own management. Moreover, the very selection of an efficient method for solving the problem may require a considerable amount of specific knowledge on the problem domain. This is the reason that, though all Expert Systems provide some general structure for knowledge representation and an inference engine to manipulate it, a system designed for a given problem domain cannot be used or even modified for other purposes. In other words, *expertise* includes the knowledge of the particular concepts and methods for obtaining the solution in a particular problem domain.

In order to obtain an efficient solution of complex dynamic conflicts, it seems to be necessary to include concepts of differential games in the *knowledge base* of an AI program designed for this purpose. Analysis of such conflicts by using only *classical* heuristic search methods may turn out to be not only cumbersome, but also inaccurate and unsatisfactory.

4. AN EXPERT SYSTEM FOR SOLVING DYNAMIC CONFLICTS

In this section the elements of an Expert System, designed to solve complex dynamic conflicts, and its functioning are outlined. In this system AI techniques and Differential Game Theory concepts are combined. The basic idea is to incorporate in the *knowledge base* of the system the elements of *expertise* necessary to solve differential games. The characterization of the respective elements will be as general as possible. However, as an illustrative example of a dynamic conflict the already mentioned air-to-air combat with missiles will be used.

Realistic dynamic conflicts, such as an air combat, are characterized by the following features:

a. the moves in the *real* conflict are irreversible,
b. the conflict is composed of several interacting elements,
c. the conflict environment is only partially predictable,
d. the optimal solution is not always obvious,
e. there is a need, in general, to interact with a human operator.

These features suggest that the Expert System of interest should have a set of particular properties. First, it must solve the problem by an *off-line planning* mechanism before the user is committed to any *real* action (a.). The solution can be based on eventual decomposition of the original conflict to well defined subproblems (b.). When the *planning* process is completed and the *plan* is accepted by the decision maker, the first actions can be taken. However, since the environment and the behavior of the opponents are not fully predictable (c.) it has to be verified *on line* that the original plane is still valid. Such a *real-time monitor* serves as a mechanism which either authorizes the execution of the plan, or activates a new *planning* process.

A very important functional element of the Expert System is an *intelligent interface* with the decision maker (e.). It is a human being who in most conflicts must make the critical decisions. It is him who has to accept the proposed plan and its predicted outcome as satisfactory, because the optimal solution is not always obvious (d.). Moreover, the conflict environment may change in a way that unpredictable readjustments of the system are required. Such an action can be performed only by a human being directly involved in the conflict.

Both the *real-time monitor* and the *intelligent interface* are, in spite of their great importance, only supporting elements of the *off-line planning* mechanism which has to resolve actually the dynamic conflict. An approach for obtaining the solution of a complex dynamic conflict by the combination of AI techniques and Differential Game Theory concepts is outlined in the sequel.

By using this approach the very solution of a complex dynamic conflict starts, as already mentioned, by a decomposition of the conflict to a set of well defined and numerically solvable *subgames*. This most critical phase requires deep knowledge in the problem domain, as well as formal knowledge of differential game concepts. A subgame can be considered numerically solvable if its solution is either known, or it can be simulated by approximating the respective optimal strategies.

There are in fact several types of solvable subgames. The first distinction is between *complete subgames*, which represent a possible way of playing the entire conflict, and secondary subgames describing only a phase of the conflict. A *complete subgame* is generally composed by a sequence of *secondary subgames*. These definitions provide a direct analogy to the classical "tree" structure. The branches of the "tree" are the respective *secondary subgames* and the nodes are the transitions between such *subgames*.

Another distinction is between *optimal subgames*, which guarantee a predictable outcome in the "saddle point" sense, and *reprisal subgames* based on exploiting an assumed nonoptimal adversary behavior. This distinction is very important, because strategies based on an optimal game solution are *security* strategies, while using a *reprisal* strategy always involves a *risk*. A *complete subgame* is an *independent subgame* if it is not interacting with any other *complete subgame*. Two *independent subgames* are, by definition, mutually exclusive options of playing the conflict.

The above given definitions can be illustrated by some air combat examples. An air-to-air combat duel is composed of a set of consecutive missile firing sequences. Each missile firing sequence can be decomposed into three phases: a *pre-launch* phase, a *post-launch* phase and an eventual *avoidance* of the adversary missile. Each phase can be considered as a well defined and numerically solvable *optimal secondary subgame*. If each aircraft has only one missile to launch than a firing sequence of each aircraft, composed of the above defined three phases, is a *complete subgame*. The transition from one phase to the following one depends on the pilot's decision when to launch his missile and when to start the *avoidance* maneuver from the missile launched by the opponent. Consequently, there are many possible "firing sequence" scenarios. The "firing sequence" *subgame* of an aircraft is not an *independent subgame*, because it interacts with the "firing sequence" of the adversary. The union of two adversary "firing sequences" becomes a more complex, but *independent complete subgame*. These examples show the hierarchical structure of the decomposed conflict.

The solution of the air-to-air duel conflict for each pilot consists of finding the optimal timing for own missile launch and for initiating the *avoidance* from the adversary missile. This problem doesn't have a known solution. It is the role of the *planning* mechanism of the Expert System to provide the respective pilot with an optimal, or at least satisfactory, proposition for the two critical decisions. This task may require a rather extensive search in the parameterized space of the pilot strategies. The search can be simplified by using some (or several) rational behavior models of the opponent. In this case the respective subgames are *reprisal subgames*. The heuristic search for the satisfactory solution (as defined by the pilot) can be made a very efficient one by exploiting knowledge on air combat and on differential games.

This *planning* approach with a probabilistic adversary behavior model was implemented recently [6,7] using a simplified dynamic model of an air-to-air combat duel. In this study the optimal strategies of the *secondary subgames* are known and a *complete* "firing sequence" *subgame* can be directly simulated for any given pilot decision on the timing of missile launch and the *avoidance* phase. The *knowledge base* of the Expert System includes the definition of differential game concepts, as well as a simulation module for *reprisal subgame* solution. The inference engine manipulates the subgame simulations and drives a rule based automated reasoning process to guide the search. In a prototype design, suitable for an exploratory study with a simplified model, only a relatively small set of rules were implemented. Nevertheless, the tools were prepared for a larger and consequently more complex system that will be required both for a realistic multiplied aircraft scenario, where the number of rules and the complexity of rule interaction may increase by orders of magnitude. Moreover, the prototype design allowed to demonstrate the feasibility and the usefulness of incorporating differential game concepts in the *knowledge base* of an Expert System for solving a dynamic conflict.

5. CONCLUSIONS

In this paper a feasible option is outlined for the analysis of complex dynamic conflicts. This option is based on incorporating in the *knowledge base* of an Expert System the basic definitions and solution concepts of Differential Game Theory. Such an approach benefits from the compatibility and complementary features of both disciplines: Artificial Intelligence and Differential Game Theory. Some recent university investigations [5-7] demonstrated the feasibility of such an approach, as well as its effectiveness in the analysis of yet unsolved complex dynamic conflicts such as an air-to-air combat. Though air combat is a particularly interesting example, the proposed approach is equally suitable to analyze other complex dynamic conflicts.

REFERENCES

1. Shubik, M., *The Mathematics of Conflict*, North Holland, Amsterdam, 1983.

2. Isaacs, R., *Differential Games*, Wiley, New York, 1965.

3. Kuhn, H.W., *Extensive games and the problem of information*, Annals of Mathematical Studies, Vol. 28, Princeton, pp. 193-216, 1953.

4. Shinar, J. and Davidovitz, A., *A qualitative analysis of future air combat with "fire and forget" missiles*, AIAA Paper No. 87-2315, AIAA Guidance, Navigation and Control Conference, Monterey, California, 1987.

5. Lirov, Y.V., *Artificial intelligence methods in decision and control systems*, D.Sc. Thesis, Washington University, Saint Louis, Missouri, 1987.

6. Shinar, J., Siegel, A.W. and Gold, Y.I., *On the analysis of a complex differential game using artificial intelligence techniques*, 27th IEEE Conference on Decision and Control, Austin, Texas, 1988.

7. Shinar, J., Siegel, A. and Gold, Y., "*A medium-range air combat game solution by a pilot advisory system*", AIAA Paper No. 89-3630, Guidance Navigation and Control Conference, Boston, MA. August 1989.

SIMULTANEOUS TWO-PURSUER DIFFERENTIAL GAME

Scott K. Meyer and William W. Trigeiro, Ph.D.
STR Corporation
10700 Parkridge Boulevard
Reston, Virginia 22091 USA

We present and discuss a proprietary methodology which approximates the solution to differential games. The methodology computes "capture regions" (i.e., the space in which one or the other player can capture or kill the other with some probability), computes the probability of capture or kill within the capture region, and computes and displays optimal trajectories for both players. Unlike traditional approaches to differential game theory, the methodology does not require an *a priori* specification of pursuer and evader. Both sides have the option to pursue or evade depending upon the payoff function, the relative capabilities of the players, and the current position and velocity vectors.

Like traditional differential game theory, it works backwards through time computing boundary surfaces and optimal trajectories over the entire state space. It numerically calculates points along the value boundary surfaces whose values are scalar quantities which are functions of the state space (i.e., position, velocity vector, etc.). These values contain information that can be used to construct optimal tactics for both sides from that point forward.

By using numerical techniques imbedded within the traditional approach to differential game theory, we believe the methodology:

1. is theoretically sound (although no proof has been developed),
2. is capable of handling arbitrarily complex equations of motion, and
3. is capable of letting either side pursue or evade as the situation dictates.

The last point, essential to air-to-air combat analysis, is unique to our methodology.

Background: Differential Game Theory

From its earliest days, differential game theory has held the promise of being able to properly treat the tactics problem (Isaacs 1975 and Hájek 1975). Given any set of aircraft characteristics and the proper payoff function, differential game theory promises to find optimal tactics for both sides. These tactics would be the very best that either side can do given the characteristics of the aircraft; they would be mutually enforceable tactics. Use of any other tactic by either side would result in a more favorable result for the opponent. Thus, whenever one analyzes the value of an alternative aircraft characteristic, game theory promises to find tactics which specifically respond to and react to the changed characteristics.

The problem with differential game theory, however, is that in actual practice it has failed to live up to its promises. While game theory can theoretically be used to find optimal tactics, the mathematical derivations become so complex that traditional game theory approaches can only solve extremely simplified games. Whenever game theory has attempted to address games of the complexity of tactical air combat, it has failed because the derivations become intractable.

Several years ago, the authors participated in the development of Optimal Marginal Evaluator-III (OME-III), a methodology to approximate the solution to the optimum allocation of tactical air forces over a multi-day theater campaign (Dresher 1961). In this methodology, we got around the difficulties of solving complex mathematical equations representing the value surface on each day by approximating that surface by numerically fitting a surface (Goodson 1988). This technique, originally proposed by George Dantzig of Stanford University for dynamic programming problems (Control Analysis Corporation 1972 and Lansdowne, et al. 1973), permits one to use arbitrarily complex descriptions of the outcomes of various engagements (transition equations) and is guaranteed to get arbitrarily close to the optimum solution.

Our current methodology to approximate the solution to differential game theory is based on the OME-III methodology. Like traditional differential game theory, it works backwards in time computing boundary surfaces and optimal trajectories over the entire space. Also like OME-III, the methodology gets around the difficulties of mathematically deriving complex equations for the boundary value surfaces on each move and the trajectories by numerically calculating points along the value boundary surfaces, thus approximating a grid of the surface. The values are scalar quantities which are functions of the state space (i.e., position, velocity vector, etc.) and which contain information that can be used to construct optimal tactics for both sides from that point forward.

Current Prototype Implementation

We have successfully completed two steps of the development of the methodology. First, a theoretical phase resulted in a conceptual methodology. Next, a computer program was written to test and evaluate the methodology at a very rudimentary level.

The copyrighted computer program, called a "Proof of Concept Implementation in Two Dimensions," is currently implemented on a Macintosh computer, written in Lightspeed Pascal. It uses a two-dimensional model of two aircraft engaged in one-versus-one (1-v-1) combat. It permits the user to define the characteristics of a Red and a Blue "aircraft" (velocity and turn rate) as well as the characteristics of a single "missile" (firing envelope and probability of kill within the envelope) carried by each aircraft. The program then finds the capture regions in two dimensions and displays the "optimum" trajectories for each aircraft. It has the following characteristics:

1. Movement is in two dimensions. The two-dimensional plane is tiled with hexagons (discussed in more detail below) and the "aircraft" are always located at the center of hexagons.
2. Players are limited to one Blue and one Red "aircraft", defined by their movement capabilities and their armament:
    a. Movement options -- which hexes can be reached in one move. Both the "velocity" and "turning rate" can be varied;

b. Armament options -- each aircraft carries one "missile" which is defined by a Pk (probability of kill) envelope. The envelope can be varied by changing the missile's range, angle off, lethality within the envelope, and aspect angle.

3. Time and space are discrete, with a fixed maximum time horizon.
4. Both aircraft can simultaneously be the pursuer and evader.
5. Capture regions are generated during backward play. A composite "score" is calculated and displayed as a result of the generation of the backward surfaces.
6. Problems can be played forward -- optimal paths can be shown or a human can play against the computer.

The methodology has been tested and has passed all "necessary" tests. Unfortunately, with the complexity of differential game theory, it has not been possible to create a "sufficiency" test of the methodology. However, the current methodology does allow a human to "play" against the computer. When exercised in such a manner, the tests have shown:

1. it is very difficult for a human to find "good" solutions,
2. when a human does find a good solution, he/she has little idea how that solution should change given a change in the aircraft characteristics, and
3. thus far, the human can not find a solution that is better than the solution developed by the software.

The user can first define the aircraft and missiles for both Red and Blue along with the number of moves in the game. The program then works backwards in time computing boundary surfaces, capture regions, and optimal trajectories from one time period to the next for both aircraft. When the program has computed the last surface, a score is calculated which is the integral of the game value over the entire surface. This score is used as the metric to evaluate one aircraft characteristic against another.

The program can also be run forward in one of two ways. The computer can display the optimal trajectories for both aircraft or it can permit a human to play against the computer. In this mode, the human selects, time period by time period, the path of either the Red or the Blue aircraft. The computer, without knowing the human's choice, selects the path of the other aircraft.

Because the current implementation breaks up time and space into discrete elements, solving the game at each move is a simple process. We have defined movement as a missile firing (if feasible and if chosen) followed by a physical movement of the aircraft. We assume that a game consists of at most T stages, each a simultaneous-move single-stage game. With n possible movement options for Blue and m for Red, a simple game matrix of size no greater than 2n x 2m describes the one-move game. This is easily solvable by applying a min-max criterion (for saddle point solutions) or with linear programming. The "score" for a move (for each cell in the game matrix) is based on the expected outcome of the one current move plus the expected value of the state space at the end of the move.

Problems are solved with a backward dynamic programming approach: one need only define the residual value of the final state space in order to initialize the problem. We have assumed a zero residual surface value. For each possible combination of points in the state space, a single-stage game must be solved at each of the T moves of the game. Although that sounds like a computationally insurmountable task, in practice there are only a limited number of combinations that one must solve.

For any set of state variable values for the game, only a limited number of stages remain until the end of the game. Also, only a limited set of movements is available to each player. Thus, only those moves which could lead to a non-zero state in the next move need be considered. Other computational savings have been exploited, such as assuming symmetry of movements, which cuts the number of solutions to explore in half. For example, if both players have identical maximum speeds, then one can never "run down" the other, so after a small number of moves the non-zero surface never grows larger. (Recall that neither side is assumed to make mistakes because a game is solved at each stage.) If a player is in a disadvantageous initial position, and he is sufficiently distant from his opponent, he will always choose to turn and run, avoiding any engagement whatsoever, since it can only worsen his score.

The geometry of the current implementation is not the Cartesian coordinate system, but instead a grid of hexagons. A hexagon is the most spherical regular polygon which fully tiles a two-dimensional surface. Each group of seven hexagons (the center plus its six surrounding hexes) form a higher level hex. This allows for varying the degree of detail of a surface, effectively allowing one to use successive approximations of the surface, providing greater detail only where necessary. The origin of this geometry and its corresponding algebra has been credited to workers at the Jet Propulsion Laboratory (The BDM Corporation 1976).

The Future of Our Methodology

In principal, it may be possible to use the methodology to find optimal air-to-air combat tactics, not only for 1-v-1, but also for when multiple players exist on each side. Such an extension of our implementation is at least two steps into the future.

The next objective is to develop an air-to-air, simultaneous-pursuer, differential game played in three dimensions with realistic aerodynamic modeling of the aircraft. This would permit analysts to study tradeoffs of aircraft and weapon system designs in engagements where both sides select tactics to optimize their outcome. It will be necessary to use a variable velocity vector, a delta energy difference, and movement in continuous time and space (though values will be maintained only on a discrete grid).

In this next implementation, it will likely be necessary to significantly improve the detail of the state space. For example, instead of limiting turns to multiples of 60 degrees, a finer grid will likely be necessary. If a hexagonal-based geometry is maintained, this will require interpolating off of surfaces in order to obtain an accurate measure of the next-stage residual value for each game.

Because the state space is likely to grow significantly larger than in the present implementation, we will need to further exploit various computer science techniques to minimize run time. There are primarily three areas which require significant computational effort: solving the game, computing the movement vector, and the actual number of grid points needed to be solved. Because one could easily construct a model so complex as to bring a supercomputer to its knees, one must be careful in the design and solution of problems. In addition to using a simplified aerodynamic model, we will pursue other methods to reduce computational burden without seriously hampering the quality of the solutions.

We will study alternatives to the solution technique currently used to solve the game in place of our current linear program, such as using a faster LP, a more extensive search for dominance, and a combination of techniques. We will also explore using branch and bound techniques for paring the tree of possible solutions to be investigated. Another area of potential gain is in the use of variable step sizes, because the necessary accuracy varies with distance between aircraft. Thus, a coordinate system for which variable resolution can be used would be very helpful, so that aggregation and deaggregation of grid points could be readily carried out. In addition, we will seek out ways to exploit symmetry in the problem in order to reduce the region which must be explicitly solved.

Bibliography

Control Analysis Corporation, Approximate Solutions of Multi-Stage Network Games, August 1972.

Dresher, Melvin, Games of Strategy: Theory and Applications (Project Rand), Prentice-Hall, 1961.

Goodson, W. Leon, Implementation of the Optimal Marginal Evaluator III for the U.S. Air Force---Analyst Manual, STR Corporation, January 13, 1988.

Hájek, Otomar, Pursuit Games, Academic Press, 1975.

Isaacs, Rufus, Differential Games, Krieger, 1975.

Lansdowne, Zachary F., George B. Dantzig, Roy P. Harvey, and Robert D. McKnight, Development of an Algorithm To Solve Multi-Stage Games, Control Analysis Corporation, May 1973.

The BDM Corporation, Preliminary Design and Resource Estimate for Combined Arms Simulation Model (CASM), March 1976.

DYNAMIC GAME APPLIED TO COORDINATION CONTROL OF TWO ARM ROBOTIC SYSTEM

M.D. Ardema
University of Santa Clara
Santa Clara, CA 95053

and

J.M. Skowronski
University of Southern California
Los Angeles, CA 90089-1453

Abstract

The previous kinematic study of a pick-and-place robot with two entirely independent arms, is extended to the dynamic model of such a robot. The arms are seen as two players avoiding collision in any event. The corresponding game of kind gives the all-option program to the designer. A barrier between dexterous regions for each of the arms is determined. The method can be applied to a large class of multi-arm robots.

Key Words: Coordination control, dynamic non-cooperative game of kind, Liapunov function.

Introduction

Present robotic manipulators rarely work alone on the manufacturing floor. Hence the coordination control between the arms operating in the same workspace has become an essential feature of recent design in automated manufacturing, and thus also in recent research investigations. Such coordination control must cover the objectives of particular arms and guarantee collision-free operation.

There are a number of studies on master-slave or follow-the-leader (hierarchical) techniques, in which the manipulator arms have a specified relationship to each other, see [1,2]. Other studies have used closed chain propositions, see [1,3-8] and symmetry assumptions, see [1,2]; for a review see [9]. In all of these approaches, the second arm motion is specified by constraints. It seems, however, that when both arms are controlled independently to operate on the same work piece that a differential game formulation is more appropriate for developing control algorithms. Two formulations are of interest. First, in normal operation it is desired that the manipulator arms operate cooperatively to optimize the manufacturing operation – a cooperative differential game. Second, it is desired that collision of the arms be avoided in any event, and this leads to a formulation of a noncooperative differential game. It is this latter case we consider here.

For example, in an unattended manufacturing space it would be desirable to design a controller for one manipulator arm such that it would continue to function safely, at least at some reduced level of effectiveness, even in the event of a catastrophic control failure of a nearby arm. This situation can be formulated as the problem of reaching one target set in the combined manipulator state space while avoiding another set (anti target). It has been called a game of combat in [10–12].

Such a scenario describes what we usually call a qualitative semi-game (i, j), for each arm $i, j = 1, 2,\ i \neq j$, and the interface of the semi-games (1,2) and (2,1) for the two arms concerned gives the designer of the coordinating controller a state space map of options for which the two-arm system must be prepared.

It is our purpose to illustrate this concept on a simple manufacturing scenario which leads to the known turret game, see [10,12,13]. The kinematic version of our study appeared in [14], we extend the case now to the dynamic model.

The Dynamical Model

Consider two single link robotic arms, shown in Figure 1, in the horizontal plane with inertial reference frame $(\xi,\ \eta)$. Arm 1 has a rigid link of length r and an end-effector e_1, rotating about the base B_1 fixed at (0, 0). Arm 2 has a rigid link of length r and end-effector e_2 rotating about the base B_2, which itself is fixed to a conveyor turn-table rotating about B_1 with angular speed $\dot{\beta}(t)$. The table provides the link C. The radius of the table is r. The rotation angles of the arms are $\theta_\sigma(t)$, $t \geq 0$. $\sigma = 1, 2$. The center of mass of the turn table is at B_1 and the center of mass of arm 2 is at B_2.

The gripper e_2 is supposed to pick-up an object at some point O_2 in inertial space outside the conveyor and deliver it to location B_1, by controlling the rotation of the turn-table and the rotation of gripper e_2 relative to it. Simultaneously and independently, gripper e_1 is supposed to pick-up an object at some point O_1 in inertial space and deliver

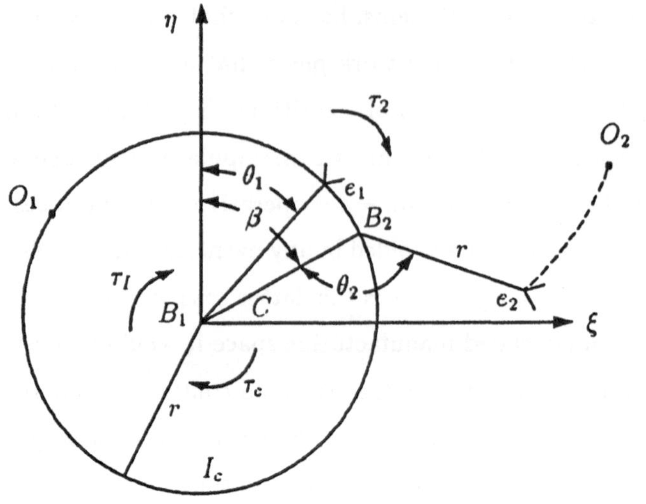

Figure 1. Two-arm robot.

it to the conveyor at location B_2. To prevent collision of the arms, both grippers must deposit their objects with zero relative velocity.

Our goal is to seek strategies or control programs that guarantee successful task completion of one arm, despite the actions of the other. Specifically, we seek two pairs of mutually dependent control programs. The first pair secures reaching the target point B_1 by arm 2 while avoiding collision with arm 1; if such a pair of programs exists at a point of the work space, we say that the system is controllable for arm 2 at that point. Conversely, the second pair of programs secures reaching the target point B_2 by arm 1 while avoiding collision with arm 2; if such strategies exist, we have controllability of the system by arm 1. It is clear that at a given point of the work space there are four. and only four possibilities [13]:

(1) System is controllable for only arm 1,

(2) System is controllable for only arm 2,

(3) System is controllable for both arms 1 and 2 and

(4) System is controllable for neither arm 1 nor 2.

Thus the work space is partitioned into four mutually exclusive regions by the controllability properties of the arms. Determining such regions and such pairs of programs is the purpose of the theory of differential games. Specifying the constraints and thus the regions is the goal of qualitative game analysis and specifying the (optimal) program is the goal of quantitative game analysis.

If all system elements are rigid, the system depicted in Figure 1 has three degrees of freedom. The kinetic equations of motion of such a system will be three dynamically coupled, nonlinear second order differential equations; if motors are used to drive the various angular motions, the motor torques would appear as the control variables. Denoting the moments of inertia of the arms by I_i, $i = 1,2$, that of the link C by I_c, the torques on the arms by τ_i, $i = 1,2$ and on the table by τ_c, and the mass of the arm 2 by M_2, we obtain the motion equations

$$\left.\begin{aligned} I_1\ddot{\theta}_1 &= \tau_1 \\ (I_c + M_2 r^2)\ddot{\beta} &= \tau_c - \tau_2 \\ I_2(\ddot{\beta} - \ddot{\theta}_2) &= \tau_2 \end{aligned}\right\} \tag{1}$$

or

$$\left.\begin{aligned} \ddot{\theta}_1 &= \frac{\tau_1}{I_1} \\ \ddot{\beta} &= \frac{(\tau_c - \tau_2)}{I'_c} \\ \ddot{\theta}_2 - \ddot{\beta} &= \left(\frac{\tau_2}{I_2}\right) \end{aligned}\right\} \tag{2}$$

where $I'_c = I_c + M_2 r^2$. Choosing the state variables $x_1 = \beta - \theta_1$, $x_2 = \theta_2$, $x_3 = \dot{x}_1$, $x_4 = \dot{x}_2$, and the control variables

$$u^1 = \left(\frac{\tau_1}{I_1}\right), \qquad u^{21} = \frac{(\tau_c - \tau_2)}{I'_c}$$

$$u^{22} = \left(\frac{\tau_2}{I_2}\right) - \frac{(\tau_c - \tau_2)}{I'_c}$$

we obtain the state equations

$$\left.\begin{aligned} \dot{x}_1 &= x_3 \\ \dot{x}_2 &= x_4 \\ \dot{x}_3 &= u^{21} - u^1 \\ \dot{x}_4 &= -u^{22} \end{aligned}\right\} \tag{3}$$

Since the problem is symmetric the angular state variables reduce to $x_i \epsilon [0, \pi]$, $i = 1, 2$. The described scenario requires x_3, $x_4 \geq 0$. Assuming the given upper values for the velocities x_3, x_4 as $\hat{x}_3$, $\hat{x}_4$, we obtain the state workspace i.e. the playing region for the game defined as the set

$$\Delta \geq \{(x_1, \ldots, x_4) | x_i \epsilon [0, \pi],\ x_{2+i} \epsilon [0, \hat{x}_{2+i}],\ i = 1, 2\}.$$

The scalar control function $u^1(t)$ is obtained from the actuator of arm 1. The responsibility of coordination lies with the contra vector $\bar{u}^2(t) = (u^{21}, u^{22})$ of arm 2. The component u^{21} influences coordination, affecting the relative dynamics between the conveyor and the arm 2, while u^{22} is the resultant of separate control actions on those two bodies.

Note from the equations or motion that negative torques may exist and in fact may be optimal. Considering the problem of control constraints, we must realize that player 2 (arm 2) has to share his control power between turning the table (coordination) to avoid arm 1, and his arm-turning to get his task accomplished. Thus there must be a constraint between τ_c and τ_2, the two motor torques at his disposal, see Fig. 2.

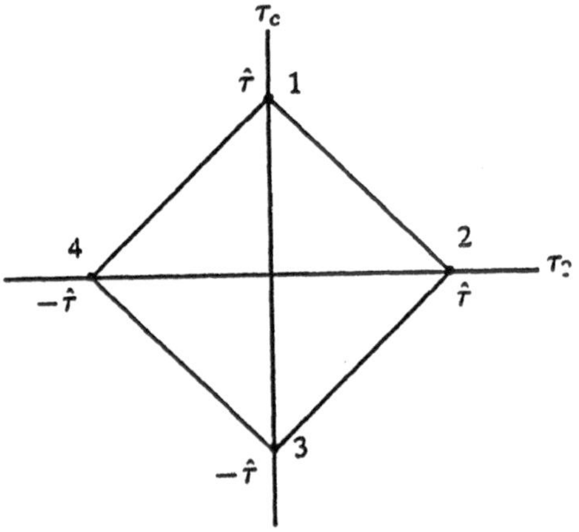

Figure 2: Torque constraints for arm 2.

We assume that the total torque is limited by some magnitude $\hat{\tau}$. This gives the control parallelogram shown in Fig. 3, with the corresponding vertices labelled by the same numbers as in Fig. 2; specifically the points (1) through (4) are given by:

$$(1) \quad u^{21} = \frac{\hat{\tau}}{I'_c}, \qquad u^{22} = -\frac{\hat{\tau}}{I'_c},$$

$$(2) \quad u^{21} = -\frac{\hat{\tau}}{I'_c}, \quad u^{22} = \left(\frac{\hat{\tau}}{I_2}\right) + \left(\frac{\hat{\tau}}{I'_c}\right)$$

$$(3) \quad u^{21} = -\frac{\hat{\tau}}{I'_c}, \quad u^{22} = \frac{\hat{\tau}}{I'_c},$$

$$(4) \quad u^{21} = \frac{\hat{\tau}}{I'_c}, \qquad u^{22} = -\left(\frac{\hat{\tau}}{I_2}\right) - \left(\frac{\hat{\tau}}{I'_c}\right)$$

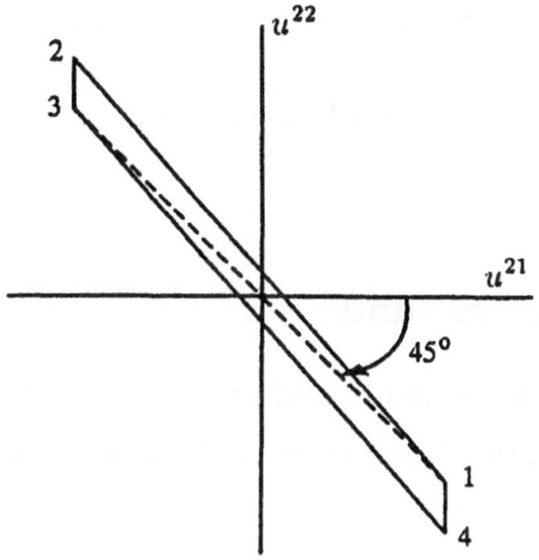

Figure 3: Control constraints for arm 2.

Let

$$\left(\frac{\hat{\tau}}{I_1}\right) \triangleq \hat{u}^1, \qquad \left(\frac{\hat{\tau}}{I_c'}\right) \triangleq \hat{u}^{21}, \qquad \left(\frac{\hat{\tau}}{I_2}\right) = \hat{u}^{22},$$

then the boundary of the constraint set in Fig. 3 becomes

$$u^{22} = -\left(1 + \frac{\hat{u}^{22}}{2\hat{u}^{21}}\right) u^{21} \pm \frac{\hat{u}^{22}}{2}$$

for the upper line 1-2 and lower line 3-4, respectively. Hence the admissible controls are the values in the sets U_i given by

$$\left.\begin{array}{l} U_1 : -\hat{u}^1 \leq u^1 \leq \hat{u}^1 \\ U_2 : -\hat{u}^{21} \leq u^{21} \leq \hat{u}^{21} \\ -\left(1 + \frac{\hat{u}^{22}}{2\hat{u}^{21}}\right) u^{21} - \frac{\hat{u}^{22}}{2} \leq u^{22} \leq -\left(1 + \frac{\hat{u}^{22}}{2\hat{u}^{21}}\right) u^{21} + \frac{\hat{u}^{22}}{2} \end{array}\right\} \qquad (4)$$

According to the work scenario described, the successful completion of the operation task of arm 1 is to place an object in a given small angular neighborhood of B_2 in stipulated terminal time with zero velocity, while the task of arm 2 is to do the same in the neighborhood of B_1. Given four corresponding tolerances $\varepsilon_1, \ldots, \varepsilon_4 > 0$ and the instant $t_f < \infty$ we aim at $x_i(t_f)$, $i = 1, 2$ satisfying the above for small values of the terminal velocities $\dot{x}(t_f)$, $i = 1, 2$, with the values $x_j(t_f)$, $\dot{x}_j(t_f)$, $j = 1, 2$, arbitrary.

This gives the targets of the players

$$T^i = \{(x_1, \ldots, x_4) | x_i \leq \epsilon_i,\ x_{2+i}\},\ i = 1, 2,$$

Letting $\mathcal{A}^i$ be the subset of Δ which the arm i must avoid, we obviously have

$$\mathcal{A}^j = \{(x_1, \ldots, x_4) | x_j \le \epsilon_j,\ x_{2+j} \le \epsilon_{2+j}\},\ j = 2, 1$$

yielding $\mathcal{T}^i = \mathcal{A}^j,\ i \ne j$.

Game Theoretic Background

Before continuing our case, it is useful to outline briefly some theoretic results to be used. For convenience let us write the state equations (3) in the general vectorial format

$$\dot{\bar{x}}_i = \bar{f}_i(\bar{x}, \bar{u}^1, \bar{u}^2) \tag{5}$$

where $\bar{x} = (x_1, \ldots, x_4)^T \epsilon \Delta$. The coordination controls are generally specified by feedback programs, see [9]:

$$\bar{u}^i \triangleq \bar{\mathcal{P}}^i(\bar{x}, u^{ji}), \qquad i, j = 1, 2, \qquad i \ne j,$$

In particular $\bar{u}^i = (u^{ii}, u^{ij})^T$, $u^{ii} = \bar{\mathcal{P}}^{ii}(\bar{x})$, $u^{ij} = \bar{\mathcal{P}}^{ij}(\bar{x}, u^{ji})$. Equation (5) under suitable conditions generates unique solutions, called trajectories in Δ.

The semi-game for the arm σ will require knowledge of all the options of the other arm and thus may be expressed by the contingent vector equation in the format

$$\dot{\bar{x}} \epsilon \{\bar{f}(\bar{x}, \bar{u}^1, \bar{u}^2) | \bar{u}^i \epsilon \bar{\mathcal{P}}^i(\bar{x}, u^{ji}),\ \bar{u}^j \epsilon U^j\} \tag{6}$$

with solutions represented by the trajectories of (5) $k(\bar{x}^0, t), t \ge 0$, forming at each $\bar{x}^0 \epsilon \Delta$ a class $\mathcal{K}(\bar{x}^0)$, subject to required sufficient conditions, see [16].

Given the targets $\mathcal{T}^i$, $i = 1, 2$ and the anti targets (avoidance sets) $\mathcal{A}^i$, $i = 1, 2$ with $\mathcal{A}^i = \mathcal{T}^j, i \ne j$, all in the work envelope Δ, the objective of the semi-game in favor of the arm i, briefly the i-game, reduces to the following.

i-OBJECTIVE

(1) *Stabilization:* A given set $\Delta_0 \subset \Delta$ is positively invariant under (6) if given $\bar{u}^1, \bar{u}^2$ in (4), $\bar{x}^0 \epsilon \Delta$ implies $k(x^0,\ t) \epsilon \Delta_0$, $t \ge 0$.

(2) *Reaching:* There is a time interval $t_c^i < \infty$, possibly stipulated, such that for the above $\bar{u}^1, \bar{u}^2, \bar{x}^0 \epsilon \Delta_0$ implies $k(x^0, t) \epsilon \mathcal{T}^i$, $\forall t > t_c^i$.

(3) *Avoiding collision:* There is a time interval t_A^i, $t_A^i \ge t_c^i$ such that for the above $\bar{u}^1, \bar{u}^2, x^0 \epsilon \Delta_0$ implies $k(x^0, t) \cap \mathcal{A}^i = \phi$, $\forall t \le t_A^i$, while (2) holds.

Definition 1: The i-game is strongly controllable at $\bar{x}^0$ for the i-objective, if simultaneously

(a) there is a control program $\bar{\mathcal{P}}^i(\bar{x}, u^j)$ such that for any $k(\cdot)\epsilon\mathcal{K}(\bar{x}^0)$ the subobjective (3) holds.

(b) given some $\bar{u}^j(t)$, the program in (a) secures the subobjectives (1), (2).

The set Δ^i_D of all such $\bar{x}^0$ forms the region of strong controllability for the i-objective. Any subset of such a region is strongly controllable for the said objective. Now let $\Delta^i_\epsilon(\bar{x}^0)$ be the closure of an open subset of Δ such that $\Delta^i_\epsilon \supset \mathcal{A}^i$ and $\partial\Delta^i_\epsilon \cap \partial\mathcal{A}^i = \phi$. Given $\bar{x}^0$ we term $\Delta^i_A(\bar{x}^0) = \Delta^i_\epsilon(\bar{x}^0) - \mathcal{A}^i$ the safety "slow down" zone about $\mathcal{A}^i$. Moreover let $\Delta_0 \subset \Delta$ be a set on which we want the i-objective to hold. We abbreviate the strong controllability for i-objective to "strong i-controllability".

Theorem 1. A given set Δ_0 is strongly i-controllable, if there is a safety zone $\Delta^i_A(\bar{x}^0)$, two controllers $\bar{\mathcal{P}}^i(\bar{x}, u^j)$, $\bar{\mathcal{P}}^j(\bar{x}, u^i)$, and two C^1-functions $V^i(\cdot) : \Delta_0 \to \mathcal{R}$, $V^i_A(\cdot) : \Delta^i_A \to \mathcal{R}$ such that

i. $a^i \leq V^i(\bar{x}) \leq b^i$,

where

$a^i = \inf V^i(\bar{x}) | \bar{x}\epsilon\partial\mathcal{T}^i$,

$b^i = \inf V^i(\bar{x}) | \bar{x}\epsilon\partial\Delta_0$.

ii. $0 < a^i_A < b^i_A < \infty$

where

$a^i_A = \sup V^i_A(\bar{x}) | \bar{x}\epsilon\partial\mathcal{A}^i$,

$b^i_A = \inf V^i_A(\bar{x}) | \bar{x}\epsilon\partial\Delta^i_A$.

iii. for each $(\bar{u}^i, \bar{u}^j)\epsilon\bar{\mathcal{P}}^{ii} \times \bar{\mathcal{P}}^{ji}$ there is

$c^i = \text{const} > 0$ such that

$$\nabla V^i(\bar{x})^T \cdot \bar{f}(\bar{x}, \bar{u}^1, \bar{u}^2) \leq -c^i, \tag{7}$$

$i, j = 1, 2, i \neq j$.

iv. for each $\bar{u}^i\epsilon\bar{\mathcal{P}}^i(\bar{x}, u^j)$ there is

a constant $c^i_A < \left[\dfrac{b^i_A - a^\sigma_A}{b^i - a^i}\right] c^i$, such that

$$\nabla V^i_A(\bar{x})^T \cdot \bar{f}(\bar{x}, \bar{u}^1, \bar{u}^2) \geq -c^i_A \tag{8}$$

for all $\bar{u}^j\epsilon U^j$, $i, j = 1, 2, i \neq j$.

For the proof see [15 and 16].

When t_c^i is stipulated we impose

$$c_i = \frac{b^i - a^i}{t_c^i}, \qquad c_A^i = \frac{b_A^i - a_A^i}{t_A^i}, \qquad t_A^i > t_c^i \tag{9}$$

When $t_A^i \to \infty$ we secure permanent avoidance of $\mathcal{A}^i$. The control programs may be found from the following corollary. Let

$$\mathcal{L}^i(\bar{x}, \bar{u}^1, \bar{u}^2) = \nabla V^i a(\bar{x})^T \cdot \bar{f}(\bar{x}, \bar{u}^1, \bar{u}^2)$$

$$\mathcal{L}_A^i(\bar{x}, \bar{u}^1, \bar{u}^2) = \nabla V_A^i(\bar{x})^T \cdot \bar{f}(\bar{x}, \bar{u}^1, \bar{u}^2)$$

Corollary 1: Given $\bar{x}^0 \epsilon \Delta_0$ if there are u_*^1, u_*^2 such that

$$\mathcal{L}_A^i(\bar{x}, u_*^1, u_*^2) = \min_{u^i} \max_{u^j} \mathcal{L}^i(\bar{x}, \bar{u}^1, \bar{u}^2) \le \frac{b^i - a^i}{t_A^i} \tag{10}$$

$$\mathcal{L}_A^i(\bar{x}, u_*^1, u_*^2) = \max_{u^i} \min_{u^j} \mathcal{L}^i(\bar{x}, \bar{u}^1, \bar{u}^2) \ge \frac{b_A^i - a_A^i}{t_A^i} \tag{11}$$

then conditions (iii), (iv) are met with $u^i = u_*^i$. The region Δ_D^i, maximal Δ_0, may be found from

$$\Delta_D^i : \frac{V^i(\bar{x}) - a^i}{c^i} < \frac{V_A^i(\bar{x}) - a_A^i}{c_A^i} \tag{12}$$

with c^i, c_A^i given as estimates of $\dot{V}^i(\bar{x})$, $\dot{V}_A^i(\bar{x})$ or from (9).

We are especially interested in the interface between the two semi-games. In particular we want to determine regions in the state space dexterous for the arms, i.e. Δ_D^1, Δ_D^2, which will lead to the state map of options mentioned in the introduction. Introduce the set $C\Delta_D^i = \Delta/\Delta_D^i$, the compliment of Δ_D^i in Δ, called semi-neutral. It is covered by points where the strong i-controllability is contradicted. Then introduce a surface $\sum^i$ subdividing Δ into two disjoint sets: $\Delta^i \supset \Delta_D^i$ called interior and $C\Delta^i = \Delta/\Delta^i$ called exterior with the property that for $\bar{x}^0 \epsilon \sum^i$ there is $\bar{\mathcal{P}}^j(\bar{x}, u^i)$, $j \ne i$, such that no $k(\bar{x}^0, t)$, $t \ge 0$ enters the interior. We call $\sum^i$ nonpermeable for arm i, briefly i-nonpermeable.

The following theorem was proved in [18].

Theorem 2. A surface S subdividing Δ into disjoint Δ^i. $C\Delta^i$ is $\sum^i$, if there are $\bar{\mathcal{P}}^j(\cdot)$ and a C^1-function $V_B^i(\cdot) : D \to \mathcal{R}$, $D(\text{open}) \supset S$, such that for all $\bar{x} \epsilon \Delta^i$

(i) $\quad V_B^i(\bar{x}) < V_B^i(\zeta), \zeta \epsilon S$;

(ii) for each $\bar{u}^j \epsilon \bar{\mathcal{P}}^j(\bar{x}, u^i)$,

$$\nabla V_B^i(\bar{x})^T \cdot \bar{f}(\bar{x}, \bar{u}^1, \bar{u}^2) \geq 0, \forall u^i \epsilon U_i, \tag{13}$$

Corollary 2: Given $\bar{x} \epsilon \sum^i$, if there is an admissible pair u_*^1, u_*^2 such that

$$\mathcal{L}_B(\bar{x}, u_*^1, u_*^2, t) = \max_{u^j} \min_{u^i} \mathcal{L}_B(\bar{x}, \bar{u}^1, \bar{u}^2) \geq 0 \tag{14}$$

then condition (ii) is met with $u_*^j \epsilon \bar{\mathcal{P}}^j(\bar{x}, u_*^i)$, making it possible to deduce $\bar{\mathcal{P}}^j(\cdot)$ from (13).

Obviously $\sum^i \subset C\Delta_D^i$, and there may be many of them. We are interested in $\sum^i$ which is closest to Δ_D^i, thus if $\partial \Delta^i$ is not well defined, we choose the $\sum^i$ closest to the target T^i which is always given and obviously in Δ_D^i. We call such $\sum^i$ the *i*-semi barrier, denoted B^i. In particular, when $\partial \Delta_c^i$ is defined and satisfies Theorem 2 i.e. is some $\sum^i$, then $\partial \Delta_D^i = B^i$.

We may now attempt to interface both strong controllabilities on the work envelope Δ. Let us define the neutral zone $\Delta_N = \Delta/(\Delta_D^1 \cup \Delta_D^2)$, closed if Δ_D^1, Δ_D^2 are open, possibly empty and also possibly embedded in $\Delta_D^1 \cup \Delta_D^2$. Then we define the *barrier* $B = B^1 \cap B^2$ obviously in Δ_ν and separating Δ_D^1, Δ_D^2, if not empty. However B will not, in general, divide Δ into two disjoint sets. In view of the above, any candidate for B is confirmed by using Theorem 2 twice, i.e. by finding two functions $V_B^1(\cdot)$, $V_B^2(\cdot)$ each satisfying conditions (i), (ii) on Δ^1, Δ^2, respectively. It is convenient to choose $V^2(\bar{x}) = \text{const} - V^1(\bar{x})$.

In general we do not have to have Δ_D^1, Δ_D^2 disjoint: $\Delta_D^1 \cap \Delta_D^2 \neq o$. The sets $D^1 = \Delta_D^1/\Delta_D^2$ and $D^2 = \Delta_D^2/\Delta_D^1$ will be called the "winning" sets, while $D^{12} = \Delta_D^1 \cap \Delta_D^2$ is the set of guaranteed dexterity, which is of our primary interest when two manipulators operate in the same workspace.

The Pick-and-Place Controllers

We return now to our pick-and-place scenario in terms of the dynamic turret game. Let us consider the 1-game first. To apply Theorem 1. we set up $V^1 = x_1 + x_3$ and $V_A^1 = x_2 + x_4$ and introduce the sets Δ_1, Δ_2;

$$T^1 \supset \Delta_1 = \{\bar{x} \epsilon \Delta | V^1 \leq C_1\}$$

$$\mathcal{A}^1 \subset \Delta_2 = \{\bar{x} \epsilon \Delta | V_A^1 \leq C_2\}$$

where $C_1 = \varepsilon_1 + \varepsilon_3$, $C_2 = \varepsilon_2 + \varepsilon_4$. We now need to find $C^1, C_A^1 > 0$ such that for each $\bar{x} \epsilon \Delta_D^1$, defined by (12), conditions (iii), (iv) are satisfied. In view of (10), (11), the above conditions hold if

$$\hat{u}^{21} - \hat{u}^1 \leq -\tilde{c}^1, \qquad \tilde{c}^1 = c^1 + \hat{x}_3, \tag{15}$$

and

$$\hat{u}^{22} \leq \tilde{c}_A^1, \qquad \tilde{c}_A^1 = c_A^1 + \hat{x}_4, \tag{16}$$

respectively.

For $\tilde{c}^1$ to exist, this implies that $\hat{u}^{21} - \hat{u}^1 < 0$, that is $\gamma_1 = (\hat{u}^{21}/\hat{u}^1) < 1$. Select the control program $u^1_* = \hat{u}^1$ for all t. Then the largest $\tilde{c}^1$ compatible with (15) is $\tilde{c}^1 = \hat{u}^1 - \hat{u}^{21}$. In turn, the smallest $\tilde{c}_A^1$ compatible with (16) is $\tilde{c}_A^1 = \hat{u}^{22}$. So for $\gamma_1 < 1$, setting $\gamma_2 = \hat{u}^{22}/\hat{u}^1$, we have

$$\begin{aligned} D^1 &= \left\{ x \epsilon \Delta \left| \frac{\hat{u}^1 - \hat{u}^{21}}{(x_1 + x_3) - (\varepsilon_1 + \varepsilon_3)} > \frac{\hat{u}^{22}}{(x_2 + x_4) - (\varepsilon_2 + \varepsilon_4)} \right. \right\} \\ &= \left\{ x \epsilon \Delta \left| (x_2 + x_4) - (\varepsilon_2 + \varepsilon_4) > \frac{\gamma_2}{1 - \gamma_1} [(x_1 + x_3) - (\varepsilon_1 + \varepsilon_3)] \right. \right\} \end{aligned}$$

The above D^1 is a subset of Δ_D^1 for it is seen by integrating the state equations that if player 1 selects u^* he wins from all $\bar{x}^0$ satisfying $(x_2+x_4)-(\varepsilon_2+\varepsilon_4) > \gamma_2[(x_1+x_3)-(\varepsilon_1+\varepsilon_3)]$. The greatest effect player 2 can have on the outcome is when $u^{21} = 0$, $u^{22} = \hat{u}^{22}$ to give the greatest rate of decrease of $x_2 + x_4$ towards $\varepsilon_2 + \varepsilon_4$. So D^1 is an underestimate of Δ_D^1, thus

$$\Delta_D^1 = \{\bar{x} \epsilon \Delta | (x_2 + x_4) - (\varepsilon_2 + \varepsilon_4) > \gamma_2[(x_1 + x_3) - (\varepsilon_1 + \varepsilon_3)\}$$

For $\gamma_1 > 1$, $\Delta_D^1 = \phi$.

We turn now to the 2-game, taking $V^2 = x_2 + x_4$, $V_A^2 = x_1 + x_3$. Introducing the constants $\tilde{c}^2 = c^2 + \hat{x}_4$, $\tilde{c}_A^2 = c_A^2 + \hat{x}_3$, similarly as for the 1-game we obtain

$$-u^{22} \leq -\tilde{c}^2 \tag{17}$$

and

$$\hat{u}^1 - u^{21} \leq \tilde{c}_A^2 \tag{18}$$

to imply conditions (iii) and (iv), respectively. From (17) a maximum value of $\tilde{c}^2$ is given by $\tilde{c}^2 = \hat{u}^{22}/(1+\delta)$ for a selection of $u^{22}_* = \hat{u}^{22}/(1+\delta)$, with $\delta \epsilon [0, \infty]$. This means that acceptable values for u^{21} satisfy

$$0 \leq u^{21} \leq \left(1 - \frac{1}{1+\delta}\right) \hat{u}^{21} = \frac{\delta}{1+\delta} \hat{u}^{21}.$$

Hence we can write, selecting the smallest c_A^2 in (18), given a value of u^{21},

$$\frac{c_A^2}{c^2} = \frac{\hat{u}^1 - \beta\delta\hat{u}^{21}/(1+\delta)}{\hat{u}^{22}/(1+\delta)}$$

$$= \frac{1+\delta(1-\beta\gamma_1)}{\gamma_2} \quad \text{with} \quad \beta\epsilon[0,1].$$

Now, provided $\gamma_1 \leq 1$, that is $\hat{u}^{21} \leq \hat{u}^1$, the smallest value of c_A^2/c_2 is obtained when $\delta = 0$. So $c_A^2/c_2 = 1/\gamma_2$ when $u_*^{22} = \hat{u}^{22}$ and $u_*^{21} = 0$. The maximum winning region for arm 2, considering the result for Δ_D^1, is

$$\Delta_D^2 = \{\bar{x}\epsilon\Delta|\,(x_2 + x_4) - (\varepsilon_2 + \varepsilon_4) < \gamma_2[(x_1 + x_3) - (\varepsilon_1 + \varepsilon_3)]\}.$$

For $\gamma_1 > 1$, it is possible with $\beta = 1$ to select δ sufficiently large, thus defining the selection of u_*^{21} and u_*^{22} to make the above expression for c_A^2/c_2 negative. Equivalently, for $\gamma_1 > 1$ we can select a $\beta\epsilon(0,1)$ such that $\beta\hat{u}^{21} = \hat{u}^1$. Then we require that player 2 play $u_*^{21} = \beta\hat{u}^{21}$, $u_*^{22} = \hat{u}^{22}(1-\beta)$. In this case then, Δ_D^2 must be all of Δ.

It is interesting to observe that the selected winning controls $\hat{u}_*^1$, u_*^{21}, u_*^{22} coincide with the optimal controllers chosen by the classical Pontriagin Principle (necessary conditions) in the original turret game, [10], for the costs:

$$J_i = -k_1 x_j^2(t_f) + k_2 \int_0^{t_e} dt, \quad i = 1,2, j = 2,1.$$

Indeed, the above controls u_*^1, u_*^{21}, u_*^{22} coincide with the upper line 1-2 in Fig. 3 which is the natural setting for the optimal controls in the mentioned classical turret game.

Note that in the case $\gamma_1 > 1$, arm 2 is guaranteed of a win in all of Δ with 2-objective: arm 1 cannot hope to win unless arm 2 does not play one of its winning controls. So the case is trivial and we discuss the interface between the two semi-games only for the case of $\gamma_1 < 1$.

Quite obviously the above obtained Δ_D^1, Δ_D^2 are disjoint, consequently we may expect a barrier surface between them somewhere in the neutral $\Delta_N = \Delta/(\Delta_D^1 \cup \Delta_D^2)$. Indeed it was shown in [18] that the surface $B = \{\bar{x}\epsilon\Delta|(x_2+x_4)-(\varepsilon_2+\varepsilon_4) = \gamma_2[(x_1+x_3)-(\varepsilon_1+\varepsilon_3)]\}$ is nonpenetrable for both arms and that it is unique in Δ_N, satisfying Theorem 2.

Conclusions

The obtained D^1, D^2 separated by a given Barrier B give the dexterous regions of operation for the two arms in our pick-and-place scenario. As $D^{12} \neq \phi$, there is no guaranteed dexterity region when both grippers operate together - which is in agreement with the required manufacturing scenario.

References

[1] Luh, J. Y. S., Zheng, Y. F., *Assumption of input generalized focus for robots with closed kinematic chains*, IEEE J. Robotics, RA1, 1985, 95–103.

[2] Hemami H., *Kinetics of two arm robots*, IEEE J. Robotics, RA2, 1986, 275–228.

[3] Vukobratovic, M., Potkonjak, V., *Dynamics of Manipulation Robots*, Springer, 1982.

[4] Tan, T. J., Bejczy, A. K., Yun, X. *Nonlinear feedback control of multiple robot arms. Proc. Workshop on Space Telerobotics*, J.P.L. Pub. 87–13, 1987, Vol. 3, 179–192.

[5] Hayati, S. A., *Dynamics and control of coordinated multiple manipulators, Proc. Workshop on Space Telerobotics*, J.P.L. Publ. 87–13, 1987 Vol. 3, 193–204.

[6] Koivo, J. J., *Adaptive position-velocity-force control of two manipulators*, Proc. 24th IEEE CDC, Ft. Lauderdale, 1985, 1529–1532.

[7] Seragi, H. *Adaptive control of dual arm robots*, Proc. Workshop on Space Telerobotics, J.P.L. Publ 87–13, 1987, Vol. 3, 159–170.

[8] Luh, J. Y. S., Zheng, Y. F., *Motion coordination and programmable teleseparation between two industrial robots*, Proc. Workshop on Space Telerobotics, J.P.L. Publ. 87–13, 1987, Vol. 2, 325–334.

[9] Skowronski, J. M. *Control Theory of Robotic Systems*, World Scientific Publ., N. Jersey-London-Singapore, 1989.

[10] Heymann, M. Ardema, M. D., Rajan N., *A formulation and analysis of combat games*, NASA Rep. TP 2487, 1987.

[11] Ardema, M. D., Heymann, M., Rajan J., *Combat games*, J. Opt. Th. Appl. Vol 46, 1985, 391–398.

[12] Ardema, M. D., Heymann, M., and Rajan, N. *Analysis of a Combat Problem: The Turret Game*, J. Opt. Th. Appl. Vol. 54, 1987, 23–42.

[13] Ardema, M. D., ed., *Singular Perturbations in Systems and Control, International Centre for Mechanical Sciences Courses and Lectures*, Vol. 280, Springer, 1983.

[14] Ardema M.D., Skowronski J.M., *Coordination Controllers for multi-arm manipulators*, in C.T. Leondes Control & Dynamic Systems. Vol. 14, Acad. Press, 1990.

[15] Getz, W. M., Leitmann G., *Qualitative differential games with two targets*, J. Math. Aml. Applic., Vol. 68, 1979, 421–430.

[16] Filippov. A. F., *Classical solutions of differential equations with multivalued right hand sides*, SIAM J. Control, Vol. 5, 1967, 609–621.

[17] Skowronski, J. M., *Coordination control of independent two robot arms on moving platform*, AIAA Paper No. 89–0584, 1989.

[18] Skowronski, J. M., Storier, R. J., *Two person qualitative differential games with two objectives*, Comp. Math. & Applic., Vol. 18, 1989, 133–150.

Approximation of differential games of pursuit-evasion by discrete-time games

Martino BARDI & Pierpaolo SORAVIA
Dipartimento di Matematica Pura e Applicata, Universita` di Padova
via Belzoni 7, I-35131 Padova, Italy

Abstract. In this paper we consider the classical problem of pursuit and evasion for continuous-time and discrete-time systems. We prove the convergence, as the time step goes to 0, of the upper and lower value functions of the discrete-time game to the upper and lower values of the differential game. This is done assuming a capturability condition either on the differential game, or on the discrete-time game uniformly for small values of the time step. An application is the existence of the value in the sense of Fleming under rather general conditions.

0. Introduction.

We consider a dynamical system controlled by two players

$$(0.1)\qquad \begin{cases} \dot{y} = f(y,a,b) & t > 0\ , \\ y(0) = x & x \in \mathbb{R}^N\ , \end{cases}$$

$a \in A$, $b \in B$, A, B compact subsets of $\mathbb{R}^M$, and its discrete-time counterpart

$$(0.2)\qquad \begin{cases} y_{n+1} = y_n + h\, f(y_n,a_n,b_n)\ , & n \in \mathbb{N}\ , \\ y_o = x & x \in \mathbb{R}^N\ , \end{cases}$$

that is the Euler scheme with step $h > 0$ associated to (0.1). We are also given a closed target $\mathcal{T} \subset \mathbb{R}^N$, and we are interested in the following game: the first player "a" seeks to minimize the time taken by the system to reach $\mathcal{T}$, while the second player "b" wants to maximize it. This leads to a differential game for the system (0.1) and to a discrete-time dynamic game for (0.2). It is clear that a suitable choice of $\mathcal{T}$ gives the classical problem of pursuit and evasion, while the absence of the second player leads to the minimum time problem in optimal control theory.

In this paper we study the convergence, as the time step h goes to 0, of the upper and lower value functions of the discrete-time game to the unique viscosity solutions of a suitable boundary value problem for the upper and lower Isaacs equations of Dynamic Programming (DP), and these solutions are the upper and lower value functions in the sense of Varaiya-Roxin-Elliott-Kalton (VREK) of the differential game [V, R, EK], by our results in [BS1, BS2, BS3]. This is done under the main assumption that the (upper or lower) value function of the differential game be continuous on the boundary of the target. This amounts to assuming that the whole boundary of $\mathcal{T}$ be "usable". We call this property "capturability", because the corresponding one for problems with a single controller is the local controllability of the system around $\mathcal{T}$. Alternatively we can assume a sort of capturability for the discrete-time game for all small h.

Our method is based on the theory of viscosity solutions for Hamilton-Jacobi equations, see Crandall & P.L. Lions [CL], Crandall, Evans & Lions [CEL] and Lions [L]. The applications of this theory to differential games was first made by Barron, Evans & Jensen [BEJ], Evans & Souganidis [ES] and Souganidis [S], see also the references in [BS1, BS2]. In particular we use a method for taking weak limits introduced by Barles & Perthame [BP] and adapted to discretization problems by Bardi & Falcone [BF1], who study the minimum time problem. We remark that our result improves the one in [BF1] even in the case of a single controller in that we need the controllability either of (0.1) or of (0.2), while in [BF1] both conditions are assumed. The improvement is obtained by using the strong comparison theorems for the Isaacs equation we proved in [BS2] and [BS3].

Our result shows the consistency between continuous-time and discrete-time models, and we have two further motivations for studying this problem. On one hand we view the discrete-time game as an approximation of the differential game. We show that the Isaacs DP equation for the discrete-time game can be solved by finding the fixed point of a contraction, and that by means of this solution we can sinthesize optimal controls and strategies in feedback form for the two players. Therefore we view our result as the first step towards the numerical solution of the pursuit-evasion problem. This approach to approximation and computation in optimal control problems follows Capuzzo Dolcetta [CD], Capuzzo Dolcetta & Ishii [CDI], Falcone [Fa] and Bardi & Falcone [BF1, BF2], who studied problems with a single controller. Adiatulina & Tarasyev [AT] extended some of the previous results to infinite horizon differential games, while Alziary [A] has used recently a different approximation scheme for a problem of pursuit-evasion with state constraints.

On the other hand our result gives the existence of the upper and lower values in the sense of Fleming [F1], which are by definition the limits of the values of the discrete game when h tends to 0, and show that they coincide with the viscosity solution of the Isaacs equation and with the VREK values. For finite horizon games the first problem was solved by Fleming [F1, F2], the second by Souganidis [S], and the third follows from the work of Evans & Souganidis [ES] and a uniqueness theorem for viscosity solutions due to Crandall & Lions [CL]. The only result we know for pursuit-evasion games is stated in [F1] under rather strong assumptions, see Remark 3.7.

1. The differential game.

In this section we recall the definitions of value for pursuit-evasion games following [V, R, EK], the relation between the capture time function and the Hamilton-Jacobi-Isaacs equation in the sense of the theory of viscosity solutions and some uniqueness theorems about this equation.

Throughout the paper we will assume that $f : \mathbb{R}^N \times A \times B \to \mathbb{R}$ is continuous and such that

$$|f(x,a,b) - f(y,a,b)| \le L\,|x-y| \qquad \text{for all } a, b, x, y .$$

We will denote by $y_x(.;a,b)$ or simply $y(.)$ the solution of (0.1) corresponding to a choice of a and b in the usual sets of admissible controls $\mathcal{A} := \{ a : \mathbb{R}_+ \to A \text{ measurable} \}$, $\mathcal{B} := \{ b : \mathbb{R}_+ \to B$ measurable $\}$. We are given the closed "target" set $\mathcal{T} \subset \mathbb{R}^N$ and define

$$t_x = t_x(a,b) := \inf \{ t : y(t) \in \mathcal{T} \} \le +\infty ,$$

where $t_x = +\infty$ if $y_x(.)$ never reaches the target. The set of admissible strategies for the first player is

$$\Delta := \{ \alpha : \mathcal{B} \to \mathcal{A} : b(t) = b'(t) \text{ for all } t \le t' \text{ implies } \alpha[b](t) = \alpha[b'](t) \text{ for all } t \le t' \},$$

and we will indicate by Γ the corresponding set of strategies for the second player. The lower capture time is defined by

$$T(x) := \inf_{\alpha \in \Delta} \sup_{b \in \mathcal{B}} t_x(\alpha[b],b)$$

and in the same way $\tilde{T} = \sup_{\beta \in \Gamma} \inf_{a \in \mathcal{A}} t_x(a,\beta[a])$ is the upper capture time. For the moment we will treat only the case of the lower capture time T, but everything we are going to describe can be translated in terms of $\tilde{T}$. We now define the (lower) capturability set, namely the set of starting points of the game such that the first player has a choice for the strategy which forces the system into the target in finite time, no matter which is the control selected by the second, i.e.

$$\mathcal{R} := \{x : T(x) < +\infty\}.$$

We need also to define the Hamiltonian of the lower Isaacs equation:

$$H(x,p) := \min_{b \in B} \max_{a \in A} \{-f(x,a,b) \cdot p - 1 \}, \tag{1.2}$$

and the following change of variables (Kruzkov transform) which is crucial in the sequel

$$\psi(r) = 1 - \exp(-r) . \tag{1.3}$$

Consider the function defined by $v(x) := \psi(T(x))$ if $x \in \mathcal{R}$ and extended by $v(x) := 1$ if $x \in \mathcal{R}^c$. It is important to remark that v is bounded and it is itself the (lower) value function of a differential game. In fact, as it is easily seen

$$v(x) := \inf_{\alpha \in \Delta} \sup_{b \in \mathcal{B}} \int_0^{t_x} e^{-s} ds \qquad \text{for all } x \in \mathbb{R}^N \setminus \mathcal{T}.$$

We now briefly recall the definition of discontinuous viscosity solution of a Hamilton-Jacobi equation as introduced by Ishii [I1]. Let $\Omega \subset \mathbb{R}^N$ be an open set and $F : \Omega \times \mathbb{R} \times \mathbb{R}^N \to \mathbb{R}$ be a continuous function. $u_1, u_2 : \Omega \to \mathbb{R}$ are respectively a viscosity subsolution and supersolution of $F(x,u,Du) = 0$ in Ω if u_1 is upper semicontinuous, u_2 is lower semicontinuous and for all $\varphi \in C^1(\Omega)$ such that $u_1 - \varphi$ attains a local maximum point at y (resp. $u_2 - \varphi$ attains a local minimum point at y) we have $F(y,u_1(y),D\varphi(y)) \leq 0$ (resp. $F(y,u_2(y),D\varphi(y)) \geq 0$). A function $u : \Omega \to \mathbb{R}$ is a viscosity solution of $F(x,u,Du) = 0$ in Ω if the upper and lower semicontinuous envelopes of u

$$u^*(x) := \limsup_{y \to x} u(y) \ , \ u_*(x) := \liminf_{y \to x} u(y).$$

are respectively a subsolution and a supersolution.

The first result based on Dynamic Programming relates the value functions T, v and the Hamiltonian in (1.2). Its proof is got by combining the arguments in [ES] and [I 2].

<u>Proposition 1.1.</u> (i) If $\mathcal{R}$ is open and T is locally bounded, then T is a viscosity solution of

$$(1.4) \qquad H(x,DV) = 0 \qquad \text{in } \mathcal{R} \setminus \mathcal{T}.$$

(ii) v is a viscosity solution of

$$(1.5) \qquad v(x) + H(x,Dv(x)) = 0 \qquad \text{in } \mathbb{R}^N \setminus \mathcal{T}.$$

In the following we will use the notation $\Omega := \mathbb{R}^N \setminus \mathcal{T}$. Note that $\partial\Omega = \partial\mathcal{T}$.

Next we recall three comparison theorems for equation (1.5) taken from [BS2, BS3], which will be important tools in Section 3. For other results on comparison and uniqueness of viscosity solutions of equations like (1.4) and (1.5) with suitable boundary conditions we refer to [CL, CEL, L, BP, I2].

<u>Theorem 1.2.</u> Let $u_1, u_2 : \overline{\Omega} \to \mathbb{R}$ be respectively a viscosity sub and supersolution of (1.5) and assume that they are bounded, continuous at each point of $\partial\Omega$, and such that $u_1 \leq u_2$ on $\partial\Omega$. Then $u_1 \leq u_2$ in Ω.

<u>Proof.</u> See Theorem 2.1 in [BS 2]. ///

Theorem 1.3. Let $u \in C(\overline{\Omega})$ be a viscosity solution of (1.5) such that $u(x) = 0$ if $x \in \partial\Omega$. Then $u = v$ in $\overline{\Omega}$ and thus $\mathcal{R} = \{x: u(x)<1\} \cup \mathcal{T}$, $T(x) = \psi^{-1}(u(x))$ for all $x \in \mathcal{R}$.

Proof. See Proposition 2.3 in [BS 2]. ///

For the next result we need a further assumption regarding the regularity of $\partial\mathcal{T}$.

$$\text{(RB)} \quad \begin{cases} \text{there is a bounded, uniformly continuous function } \eta : \overline{\Omega} \to \mathbb{R}^N \text{ and a} \\ \text{constant } c>0 \text{ such that } B(x+t\eta(x),ct) \subseteq \Omega \text{ for all } x \in \overline{\Omega} \text{ and } 0<t\le c\,; \end{cases}$$

where $B(y,r)$ denotes the open ball of radius r centered at y. Condition (RB) seems rather technical but it essentially means that $\mathcal{T}$ is the closure of an open set and $\partial\mathcal{T}$ is a Lipschitz surface.

Theorem 1.4. Assume (RB) and suppose that $|f(x,a,b)| \le C$ for all $a \in A$, $b \in B$, $x \in \partial\mathcal{T}$. Let $u_1, u_2 : \overline{\Omega} \to \mathbb{R}$ be bounded functions and respectively a viscosity sub and supersolution of (1.5). Assume moreover that u_1 is continuous and nonpositive at each point of $\partial\Omega$ and u_2 is a supersolution of

$$\text{(1.6)} \qquad u(x) = 0 \quad \text{or} \quad u(x) + H(x,Du(x)) = 0 \qquad \text{on } \partial\Omega$$

in the viscosity sense. Then $u_1 \le u_2$ in $\overline{\Omega}$. The same conclusion holds if u_2 is continuous and nonnegative at each point of $\partial\Omega$ and u_1 is a subsolution of (1.6).

Proof. See Theorem 3.1 in [BS 3]. ///

We recall that u_1 is a subsolution of (1.6) in the viscosity sense if for all $\varphi \in C^1(\overline{\Omega})$ such that $u_1 - \varphi$ attains a local maximum point in $\overline{\Omega}$ at $y \in \partial\Omega$ we have that $u(y) \le 0$ or $u(y) + H(y,D\varphi(y)) \le 0$.

2. The discrete-time game.

The discrete version of the pursuit-evasion problem we propose is the Euler scheme (0.2) with step h, $h > 0$, associated to (0.1). The sets of admissible controls for the discrete-time system are the sets of sequences taking values in A and B respectively, which we denote by $A^{\mathbb{N}}$ and $B^{\mathbb{N}}$. The set of nonanticipating strategies for the first player is

$$\Lambda := \{ \alpha : B^{\mathbb{N}} \to A^{\mathbb{N}} : b_j = b'_j \ \forall j \le j' \text{ implies } \alpha[b]_j = \alpha[b']_j \ \forall j \le j' \},$$

while the set Θ of nontacipating strategies for the second player is defined in an obvious symmetric way. We define the analogues of $t_x(a,b)$, $T(x)$ and $\mathcal{R}$ for the discrete-time problem:

$$n_h(x,a,b) := \min\{ j \in \mathbb{N} : y_j \in \mathcal{T} \} \le +\infty,$$

where y_j, $j \in \mathbb{N}$, is the solution of (0.2) corresponding to $a \in A^{\mathbb{N}}$ and $b \in B^{\mathbb{N}}$, and $n_h = +\infty$ if y_j never reaches the target,

$$N_h(x) := \inf_{\alpha \in \Lambda} \sup_{b \in B^{\mathbb{N}}} n_h(x, \alpha[b], b),$$

and

$$\mathcal{R}_h := \{ x : N_h(x) < +\infty \}.$$

N_h is the lower value of the discrete-time game, while the upper value is

$$\tilde{N}_h(x) := \sup_{\beta \in \Theta} \inf_{a \in A^{\mathbb{N}}} n_h(x, a, \beta[a]).$$

It is easy to see that the lower value N_h corresponds to the "minorant game" introduced by Fleming [F1], where at each move at time $t_j = hj$ the second player "b" must commit himself first and the first player "a" has the advantage of knowing the move of "b" before making his choice. The upper value corresponds to the opposite information pattern, and is equivalent to the "majorant game" of Fleming.

As in the continuous version we make use of the change of variable (1.3) and define the cost

$$J_h(x,a,b) := \left[\sum_{j=0}^{n_h(x,a,b)-1} e^{-jh} \right] (1 - e^{-h}) \, \chi_{\mathcal{T}^c}(x) = \begin{cases} 1 - e^{-h n_h(x,a,b)}, & \text{if } n_h < \infty, \\ 1 & \text{otherwise}, \end{cases}$$

where $\chi_{\mathcal{T}^c}(x)$ is the characteristic function of $\mathcal{T}^c$. The lower value function for this problem is given by

$$v_h(x) = \inf_{\alpha \in \Lambda} \sup_{b \in B^{\mathbb{N}}} J_h(x, \alpha[b], b) = \begin{cases} 1 - e^{-h N_h(x)}, & \text{for any } x \in \mathcal{R}_h \\ 1, & \text{for any } x \notin \mathcal{R}_h. \end{cases}$$

<u>Remark 2.0.</u> Note that the inf is indeed a min because it is taken over a discrete set whose only possible limit point is 1, and for the same reason the sup is a max if $x \in \mathcal{R}_h$.

Proposition 2.1 (Discrete Dynamic Programming Principle) The value function v_h satisfies

(2.1) $$v_h(x) = \min_{\alpha \in \Lambda} \sup_{b \in B^{\mathbb{N}}} \gamma^j v_h(y_j) + (1-\gamma)\sum_{i=0}^{j-1} \gamma^i \quad , \text{ where } \gamma := e^{-h},$$

for any $x \in \mathbb{R}^N \setminus \mathcal{T}$ and $0 < j \le N_h(x)$.

Proof. For $x \in \mathcal{R}_h \setminus \mathcal{T}$ the proof is standard. For $x \notin \mathcal{R}_h$ the right hand side of (2.1) is 1 because for all α there exists b such that $y_1 \notin \mathcal{R}_h$, by definition of $\mathcal{R}_h$, and then $v_h(y_j) = 1$ for all j. ///

Theorem 2.2 (Discrete Isaacs Equation) The value function v_h is the unique bounded solution of

(2.2) $$v_h(x) = \sup_{b \in B} \inf_{a \in A} \gamma v_h(x + h f(x,a,b)) + 1 - \gamma \quad \text{in } \Omega = \mathbb{R}^N \setminus \mathcal{T}$$

such that $v_h = 0$ on $\mathcal{T}$.

Proof. Consider the Dynamic Programming Principle (2.1) for $j = 1$ and fix $\bar{\alpha}$ such that

(2.3) $$v_h(x) \ge \gamma v_h(x + h f(x, \bar{\alpha}[b]_0, b_0)) + 1 - \gamma \quad \text{for all } b \in B^{\mathbb{N}}.$$

By the definition of nonanticipating strategy, $\bar{\alpha}[b]_0$ depends only on b_0, and then (2.3) implies that $v_h(x)$ is greater than or equal to the right hand side of (2.2). Now we assume by contradiction that this inequality be strict. Then there exists $\varepsilon > 0$, and for each $b \in B$ there is $a(b)$, such that

$$v_h(x) - \varepsilon \ge \gamma v_h(x + h f(x, a(b), b)) + 1 - \gamma .$$

Thus we can define $\bar{\alpha} \in \Lambda$ by $\bar{\alpha}[b]_j := a(b_j)$ and get

$$v_h(x) - \varepsilon \ge \gamma v_h(x + h f(x, \bar{\alpha}[b]_0, b_0)) + 1 - \gamma \quad \text{for all } b \in B^{\mathbb{N}},$$

which contradicts the Dynamic Programming Principle for $j = 1$.

To prove the uniqueness we consider the map

$$Su(x) := \sup_{b \in B} \inf_{a \in A} \gamma v_h(x + h f(x,a,b)) + 1 - \gamma \quad \text{for } x \in \Omega ,$$

and note that (2.2) says that v_h is a fixed point of S. Thus we conclude by claiming that S is a contraction mapping of constant $\gamma = e^{-h}$ on the set of bounded functions, null on $\mathcal{T}$. The proof of the last claim is got by easy modifications of the proof of Thm. 2.3 in [BF1]. ///

Theorem 2.3 (Synthesis of optimal feedback strategies and controls) Let $u : \mathbb{R}^N \to \mathbb{R}$ be a bounded solution of (2.2), null on $\mathcal{T}$. Then $\mathcal{R}_h = \{ x : u(x) < 1 \}$, and there exist $F : (\mathbb{R}^N \setminus \mathcal{T}) \times B \to A$, $G : \mathcal{R}_h \to B$ such that

$$
(2.4) \qquad
\begin{aligned}
& u(x + h\, f(x, F(x,b),b)) = \min_{a \in A} u(x + h\, f(x,a,b)) \quad \text{for all } b \in B \text{ and } x \in \mathbb{R}^N, \\
& u(x + h\, f(x, F(x, G(x)), G(x))) = \sup_{b \in B} u(x + h\, f(x, F(x,b),b)) \quad \text{for all } x \in \mathcal{R}_h .
\end{aligned}
$$

Any such pair of F and G has the property that the solution z_j of

$$
(2.5) \qquad
\begin{cases}
z_{j+1} = z_j + h\, f(z_j, F(z_j, G(z_j)), G(z_j)), & j \in \mathbb{N}, \\
z_0 = x & x \in \mathcal{R}_h ,
\end{cases}
$$

is a saddle trajectory, i.e. the control $\bar{b}_j := G(z_j)$ and the nonanticipating strategy $\bar{\alpha}[b]_j := F(z_j, b_j)$ are a saddle of the game, that is

$$
v_h(x) = J_h(x, \bar{\alpha}[\bar{b}], \bar{b}), \quad N_h(x) = n_h(x, \bar{\alpha}[\bar{b}], \bar{b}) \quad \text{for all } x \in \mathcal{R}_h .
$$

Proof. By Theorem 2.2 $u(x) = v_h(x)$, and then the argument of Remark 2.0 implies the existence of F and G as above. The Isaacs equation (2.2), and (2.4), (2.5) imply

$$
u(z_j) = \gamma\, u(z_{j+1}) + 1 - \gamma \qquad \text{for all } 0 \le j < n_h(x, \bar{\alpha}[\bar{b}], \bar{b}),
$$

which shows that $z_j \in \mathcal{R}_h$ for all $0 < j \le n_h(x, \bar{\alpha}[\bar{b}], \bar{b})$, and gives easily by iteration, using also the boundary condition $u = 0$ on $\mathcal{T}$,

$$
u(x) = 1 - \gamma^{\,n_h(x, \bar{\alpha}[\bar{b}], \bar{b})},
$$

which is the desired equality. ///

3. Convergence theorems.

In the sequel we need one of the following conditions

(A1) T is continuous at each point of $\partial\mathcal{T}$ and $\partial\mathcal{T}$ satisfies (RB) and $|f(x,a,b)| \leq C$ for all $a \in A$, $b \in B$, $x \in \partial\mathcal{T}$;

(A2) for each $R>0$ here exists a function $\sigma_R : \mathbb{R}^+ \times \mathbb{R}^+ \to \mathbb{R}^+$ nondecreasing in both variables, continuous at (0,0) and such that $\sigma_R(0,0) = 0$, which verifies $v_h(x) \leq \sigma_R(d(x),h)$ for $x \in \overline{\Omega}$, $|x| < R$, $h \in \mathbb{R}^+$; where $d(x) := \mathrm{dist}(x,\partial\mathcal{T})$.

We note that if there is only one controller (A1) is equivalent to the local controllability of the continuous-time system around the whole target, for smooth targets; (A2) is a sort of local controllability assumption on the discrete-time system for all small time-step h. For a true pursuit-evasion game we can call them "capturability" assumptions, or assumptions of "usability" of the whole target, for the continuous-time and the discrete-time systems respectively. We also call $(\tilde{A}i)$, $i = 1, 2$, the analogous condition on $\tilde{T}$ or $\tilde{v}_h = \psi(h\tilde{N}_h)$.

We can now state our main result.

Theorem 3.1. Assume that either (A1) or (A2) holds. Then, as $h \to 0^+$, v_h converges to v uniformly on compact subsets of $\overline{\Omega}$, and hN_h converges to T uniformly on compact subsets of $\mathcal{R}$.

The proof of this theorem is based on the method of "weak limits", following [BF1]. We introduce the two functions

$$\underline{v}(x) = \liminf_{\substack{h\to 0^+ \\ y\to x}} v_h(y) \quad , \quad \overline{v}(x) = \limsup_{\substack{h\to 0^+ \\ y\to x}} v_h(y).$$

It is easily seen that $\underline{v}, \overline{v}$ are respectively lower and upper semicontinuous and defined in $\mathbb{R}^N$. The main step of the proof is the following.

Theorem 3.2. $\overline{v}$ and $\underline{v}$ are respectively viscosity sub and super solutions of the Dirichlet problem:

$$(3.0) \qquad \begin{cases} u(x) + H(x,Du(x)) = 0 & \text{in } \Omega, \\ u=0 \text{ or } u(x) + H(x,Du(x)) = 0 & \text{on } \partial\Omega. \end{cases}$$

Proof. We prove only that $\overline{v}$ is a subsolution, the proof of the other statement being similar. Let therefore $\varphi \in C^2(\overline{\Omega})$ be such that $\overline{v} - \varphi$ attains a strict maximum point in $\overline{\Omega}$ at y, and that $(\overline{v} - \varphi)(y) = 0$. If $v_h^* - \varphi$ attains its maximum in $B(y,1) \cap \overline{\Omega}$ at x_h, then by Lemma A.1 of [BP]

$$(3.1) \qquad x_h \to y \quad \text{and} \quad v_h^*(x_h) \to \overline{v}(y) \quad \text{as } h \to 0^+.$$

We have to prove that $\overline{v}(y) + H(y,D\varphi(y)) \leq 0$ in both cases that $y\in\Omega$ and that $y\in\partial\Omega$ satisfies $\overline{v}(y)>0$. Then we can choose $\varepsilon>0$ such that $B(y,\varepsilon)\subset\Omega$ if $y\in\Omega$ and

(3.2) $\qquad \varphi(x) \geq \varepsilon \quad$ in $B(y,\varepsilon)$ $\quad$ if $y\in\partial\Omega$

(we extend φ as a C^1 function in a neighbourhood of $\partial\Omega$). We choose a sequence $x_h^n \to x_h$ as $n \to +\infty$ such that $v_h(x_h^n)\to v_h^*(x_h)$ and first prove that the set $\{n\in\mathbb{N} : x_h^n\in\partial\Omega\}$ is finite if h is sufficiently small. On the contrary select a subsequence in $B(y,\varepsilon)\cap\partial\Omega$, denoted again by x_h^n, then by (3.2) we get $\varphi(x_h^n) \geq \varepsilon = \varepsilon+v_h(x_h^n)$ and passing to the limit as $n\to+\infty$ we obtain $\varphi(x_h)-v_h^*(x_h) \geq \varepsilon$, which gives a contradiction with (3.1) if h is small. Therefore we can restrict ourselves to consider values of h small enough such that $\{n\in\mathbb{N} : x_h^n\in\partial\Omega\}$ is finite, $x_h\in B(y,\varepsilon/2)$ and $|hf(x,a,b)| \leq \varepsilon/2$ for all a, b.

Fix now $\rho>0$ and let $x\in B(y,\varepsilon/2)\cap\Omega$, then by (2.2) there exists $b_x\in B$ such that

(3.3) $\qquad v_h(x) - \gamma v_h(x+hf(x,a,b_x)) - (1-\gamma) < \rho h \qquad$ for all $a\in A$.

We choose a subsequence of $\{x_h^n : n\in\mathbb{N}\}$, still denoted by x_h^n, such that $x_h^n\in\Omega$ and $b_{x_h^n}\to \overline{b} = \overline{b}(h)\in B$ as $n\to+\infty$. For a fixed $a\in A$ we consider another subsequence, still denoted by x_h^n such that

(3.4) $\qquad v_h(x_h^n + hf(x_h^n,a,b_{x_h^n})) \to w \leq v_h^*(x_h + hf(x_h,a,\overline{b}))$ as $n\to+\infty$.

We compute (3.3) at the points x_h^n, pass to the limit as $n\to+\infty$, use (3.4) and obtain

(3.5) $\qquad v_h^*(x_h) - \gamma v_h^*(x_h+hf(x_h,a,\overline{b})) - (1-\gamma) < \rho h \qquad$ for all $a\in A$.

We now observe that

(3.6) $\qquad v_h^*(x_h) - \varphi(x_h) \geq v_h^*(x_h+hf(x_h,a,\overline{b})) - \varphi(x_h+hf(x_h,a,\overline{b}))$.

In fact this is obvious if $x_h+hf(x_h,a,\overline{b}) \in \overline{\Omega}$, while if $x_h+hf(x_h,a,\overline{b}) \in \text{int}\mathcal{T}$, (3.2) implies that $\varphi(x_h+hf(x_h,a,\overline{b})) \geq \varepsilon = \varepsilon + v_h^*(x_h+hf(x_h,a,\overline{b}))$ and moreover if h is small enough $\varepsilon \geq \varphi(x_h) - v_h^*(x_h)$. Thus we use (3.6) in (3.5) and get

$$(1-\gamma)\, v_h^*(x_h) + \gamma(\varphi(x_h) - \varphi(x_h+hf(x_h,a,\overline{b}))) - (1-\gamma) < \rho h \qquad \text{for all } a\in A\,,$$

and then

$$\min_{b\in B}\max_{a\in A} \{ (1-\gamma)\, v_h^*(x_h) + \gamma(\varphi(x_h) - \varphi(x_h+hf(x_h,a,b))) - (1-\gamma) \} \le \rho h\ .$$

Finally we divide by h, recall that $\gamma = e^{-h}$, use Lemma 3.3 below to pass to the limit as $h\to 0^+$, and get

$$\overline{v}(y) + H(y,D\varphi(y)) \le 0. \qquad ///$$

Lemma 3.3. Let $F : [0,1]\times A\times B\to \mathbb{R}$ and $G : A\times B\to \mathbb{R}$ be functions such that $F(h,a,b) - G(a,b)$ tends to 0 uniformly in a and b as $h\to 0^+$. Then

$$\lim_{h\to 0^+} \left| \inf_{b\in B}\sup_{a\in A} F(h,a,b) - \inf_{b\in B}\sup_{a\in A} G(a,b) \right| = 0\,.$$

The proof of this lemma is straightforward. We can now prove the main result.

Proof of Theorem 3.1. We remark that $\underline{v} \le \overline{v}$ by definition. We are going to use Proposition 1.1, the previous Theorem 3.2 and the uniqueness results in Section 1 to prove the converse inequality in $\overline{\Omega}$ and precisely that $v = \underline{v} = \overline{v}$, which implies the local uniform convergence of v_h to v by the definition of weak limits.

Assume (A1). We use twice Theorem 1.4 to compare $\overline{v}$, v and $\underline{v}$. We get $\overline{v} \le v_*$ and $v^* \le \underline{v}$ and therefore $v = \underline{v} = \overline{v}$ because $v_* \le v^*$ by definition.

Assuming (A2) we obtain $0 \le \underline{v}(x) \le \overline{v}(x) \le \sigma_R(d(x),0)$ for all $x\in\overline{\Omega}$, $|x|\le R$, $R>0$, which gives the continuity of $\underline{v}$ and $\overline{v}$ at the points of the boundary of Ω. We now apply Theorem 1.2 and get $\overline{v} \le \underline{v}$ in $\overline{\Omega}$. Thus $u := \underline{v} = \overline{v}$ is a continuous function in $\overline{\Omega}$, null on $\partial\Omega$ and satisfying (1.5) in the viscosity sense. By Theorem 1.3 we then have $u = v$ in $\overline{\Omega}$. The proof of the last statement is easy and we omit it. ///

Remark 3.4. This result improves the one obtained in [BF1] for the case of a single controller, in that there both conditions (A2) and continuity of T are required to prove the convergence of the approximation scheme, and $\mathcal{T}$ is assumed to be bounded.

An example of application of Theorem 3.1 is the following.

Corollary 3.5. Assume that $\mathcal{T}$ is the closure of an open set with C^2 boundary, and

$$\min_{a\in A}\max_{b\in B} f(x,a,b)\cdot n(x) < 0 \quad \text{for all } x\in\partial\mathcal{T}, \tag{3.7}$$

where $n(x)$ is the exterior normal to $\mathcal{T}$. Then (A1) is verified, and therefore the conclusion of the convergence Theorem 3.1 holds.

Proof. See [BS1, BS2]. ///

Remark 3.6. We can give analogous results for the upper capture time $\tilde{T}$ under either condition ($\tilde{A}$1) or ($\tilde{A}$2). In particular Corollary 3.5 holds for the upper value as well. The Hamiltonian in this case is

$$\tilde{H}(x,p) := \max_{a\in A} \min_{b\in B} \{-f(x,a,b)\cdot p - 1\}.$$

The name of lower and upper capture time is justified because $\tilde{H} \leq H$ and therefore in the hypotheses of Theorems 1.2, 1.4 we can prove that $T \leq \tilde{T}$.

Remark 3.7. The limits as $h \to 0^+$ of hN_h and $h\tilde{N}_h$ are, if they exist, the Fleming lower and upper value respectively, see [F1, F2, S]. Theorem 3.1, Corollary 3.5 and Remark 3.6 give conditions for the existence of these values and their coincidence with the corresponding VREK values. If moreover $\tilde{H} = H$ (Isaacs condition), and we have ($\tilde{A}$1) or ($\tilde{A}$2), for instance if (3.7) holds, then the upper and lower values coincide, and so the game has a value in the Fleming and VREK sense. Therefore we improve the result of Fleming [F1], dealing with the case

$$f(x,a,b) = f_1(x) + f_2(x,a) + f_3(x,b),$$

where f_2, f_3 are linear functions in a and b, requiring a slightly stronger assumption than (3.7) and a bound on the duration of the game.

References.

[AT] R.A. Adiatulina, A.M. Taras'yev: A differential game of unlimited duration, J. Appl. Math Mech. 51 (1987), 415-420.

[A] B. Alziary de Roquefort: Jeux de poursuite et approximation des fonctions valeur, Thèse, Universitè Paris-Dauphine 1990.

[BP] G. Barles, B. Perthame: Discontinuous solutions of deterministic optimal stopping time problems, RAIRO Math. Methods and Num. Anal. 21, 1987, 557-579.

[BF1] M. Bardi, M. Falcone: An approximation scheme for the minimum time function, SIAM J. Control Optim. to appear.

[BF2] M. Bardi, M. Falcone: Discrete approximation of the minimal time function for systems with regular optimal trajectories, in "Proceedings of the Ninth International Conference on Analysis and Optimization of Systems", Antibes 1990, Lecture Notes in Control and Information Sciences, Springer Verlag 1990.

[BS1] M. Bardi, P. Soravia: A PDE framework for differential games of pursuit-evasion type, in "Differential games and applications," T. Basar and P. Bernhard eds., pp. 62-71, Lecture Notes in Control and Information Sciences 119, Springer-Verlag 1989.

[BS2] M. Bardi, P. Soravia: Hamilton-Jacobi equations with singular boundary conditions on a free boundary and applications to differential games, Trans. Amer. Math. Soc. to appear.

[BS3] M. Bardi, P. Soravia: On the minimum time problem for smooth targets, preprint Università di Padova 1989.

[BEJ] E.N. Barron, L.C. Evans, R. Jensen: Viscosity solutions of Isaacs' equations and differential games with Lipschitz controls, J. Differential Equations 53 (1984), 213-233.

[CD] I. Capuzzo Dolcetta: On a discrete approximation of the Hamilton-Jacobi equation of dynamic programming, Appl. Math. Optim. 10 (1983), 367-377.

[CDI] I. Capuzzo Dolcetta, H. Ishii: Approximate solutions of the Bellman equation of deterministic control theory, Appl. Math. Optim. 11 (1984), 161-181.

[CEL] M.C. Crandall, L.C. Evans, P.L. Lions: Some properties of viscosity solutions of Hamilton-Jacobi equations, Trans. Amer. Math. Soc. 282 (1984), 487-502.

[CL] M.C. Crandall, P.L. Lions: Viscosity solutions of Hamilton-Jacobi equations, Trans. Amer. Math. Soc. 277 (1983), 1-42.

[EK] R.J. Elliott, N.J. Kalton: The existence of value in differential games, Mem. Amer. Math. Soc. 126 (1972).

[ES] L.C. Evans, P.E. Souganidis: Differential games and representation formulas for solutions of Hamilton-Jacobi equations, Indiana Univ. Math. J. 33 (1984), 773-797.

[Fa] M. Falcone: A numerical approach to the infinite horizon problem of deterministic control theory, Appl. Math. Optim. 15 (1987), 1-13.

[F1] W.H. Fleming: The convergence problem for differential games, J. Math. Anal. Appl. 3 (1961), 102-116.

[F2] W.H. Fleming: The convergence problem for differential games, II, in Advances in Game Theory, M. Dresher,L.S. Shapley, A.W. Tucker eds., Ann. of Math. Studies 52, pp.195-210, Princeton Univ. Press, Princeton 1964.

[I1] H. Ishii: Perron's method for Hamilton-Jacobi equations, Duke Math. J. 55 (1987), 369-384.

[I2] H. Ishii: A boundary value problem of the Dirichlet type for Hamilton-Jacobi equations, Ann. Sc. Norm. Sup. Pisa (IV) 16 (1989), 105-135.

[L] P.L. Lions: Generalized solutions of Hamilton-Jacobi equations, Pitman, Boston 1982.

[R] E. Roxin: Axiomatic approach in differential games, J. Optim. Th. Appl. 3 (1969), 153-163.

[S] P.E. Souganidis: Max-min representations and product formulas for the viscosity solutions of Hamilton-Jacobi equations with applications to differential games, Nonlinear Anal. T.M.A. 9 (1985), 217-257.

[V] P.P. Varaiya: On the existence of solutions to a differential game, SIAM J. Control 5 (1967), 153-162.

ON DISCRETE GAMES OF INFILTRATION.

John M. Auger,

Faculty of Mathematical Studies, University of Southampton, Southampton SO9 5NH, U.K.

A. The Original Problem.

In his book *"Search Games"* [2], in concluding the section on Search Games in Compact Spaces, Gal proposes the following game to be played in discrete time. Consider n points, ordered one to n, and a $(n+1)$th point which is taken to be the target point. At time 0, the two players, the Infiltrator and the Guard, are both located at point one. Thereafter, at the end of each unit of time, the players may decide to move; the Guard can move to any point within a distance of $v \geq 1$, whereas the Infiltrator can move at most a distance of one. During the next time unit, the players are located at their chosen points, and the Guard searches for his opponent. He finds the Infiltrator with probability $1-\lambda$, $0 \leq \lambda < 1$, if ever the players occupy the same point for some time unit, and with probability zero otherwise. As the game progresses neither player receives any feedback concerning his opponents position unless the Guard finds and as a result captures the Infiltrator. The Infiltrator wins the game if he reaches the target point (in finite time) without getting caught; if he fails to do this, he loses.

This apparently simple problem is still open. For $n = 1,2$ and any $v \geq 1$ the game is trivial; optimal play for the Infiltrator is to move immediately towards the target as quickly as possible, and the value of the game, that is the probability of a win for the Infiltrator, is λ and λ^2 respectively. For $n = 3$, the problem is no longer trivial; for $v > 1$ the value is $\frac{1}{2}\lambda^2(1+\lambda)$, and we shall give its solution at the end of this section. However for $n = 3$ and $v = 1$, and especially for general n and v, the solution is unknown.

The general approach that we suggest in this paper is to consider this infinite game as the limit of a sequence of finite games. To do this, we shall denote by $G_{n,v,t}$ the game described above with the added feature of a time limit t, $t \geq n$, within which the Infiltrator must reach the target. Then we consider the game as originally posed to be the limit of $G_{n,v,t}$ as t tends to infinity, and denote it by $G_{n,v}$.

Work which has been carried out on this game has concentrated on two particular values for the speed v of the Guard, $v = 1$ and $v \geq n$. The games associated with these values of v have been dubbed respectively the fast Guard game and the slow Guard game, the former denoted by $G_{n,1,t}$ and the latter by $G_{n,\infty,t}$ since if v is at least as large as n it might just as well be infinite.

In section B we survey the the progress that has been made in considering a game related to $G_{n,v,t}$ in that it has the same features but in addition there is a safe base, or point zero, at which the Infiltrator is located at time 0, and at which he may remain undetected for as long as he wishes. This companion game we shall denote as $H_{n,v,t}$.

To conclude this section, we now give the solution to $G_{3,v}$, the original infinite game played on three points, when $v > 1$.

Consider the pure Infiltrator strategies W_i, $i = 1,2$. Using W_i the Infiltrator remains at point one until time $i-1$ after which he advances full speed towards the target. These strategies are illustrated in Figure 1(a).

point one	two	three	(target)	time	point one	two	three
W_2W_1				0	G_2G_1		
W_2	W_1			1	G_2	G_1	
	W_2	W_1		2			G_2G_1
		W_2	W_1	3			G_2G_1
			W_2W_1	≥ 4			G_2G_1
	(a)					(b)	

Figure 1.

W_1 and W_2 are known as Wait-and-Run strategies, and these will feature again in the later sections. We let I^* denote the mixed strategy which plays W_1 and W_2 with equal probability.

To help us consider how the Guard should play if he is told that the Infiltrator will certainly play I^*, let us think of W_1 and W_2 as two separate Infiltrators (although, because of the probability distribution involved, their existence is mutually exclusive). Let $\bar{G}$ denote the pure Guard strategy which dictates that he searches point one at time 0, where he may find either W_1, W_2 or both of them, point two at time 1, where again he may find W_1, and point three for all later time units; at times 2 and 3 he may again find W_1 and W_2 respectively, but thereafter both W_1 and W_2, if previously uncaptured, will certainly be safe at the target. Note that as $\bar{G}$ has three attempts at detecting W_1, the payoff, $P(W_1, \bar{G})$, when the players use these pure strategies (that is the probability that the Infiltrator is not captured) is given by $P(W_1, \bar{G}) = \lambda^3$; similarly $P(W_2, \bar{G}) = \lambda^2$. Hence, $E(I^*, \bar{G})$, the expected payoff when the Infiltrator plays the mixed strategy I^* and the Guard plays $\bar{G}$, is given by $E(I^*, \bar{G}) = \frac{1}{2}(\lambda^2 + \lambda^3)$.

A consideration of Figure 1(a) shows that the Guard can do no better than play $\bar{G}$ if his opponent is known to be playing I^*. Equally good are other strategies which give him three attempts at one of W_1 and W_2, and two at the other. Decidedly sub-optimal is the strategy which gives him one attempt at W_1 and four at W_2 as this yields an expectation of $\frac{1}{2}(\lambda + \lambda^4)$ which is strictly greater than $\frac{1}{2}(\lambda^2 + \lambda^3)$ for all $0 < \lambda < 1$. Hence, for any Guard strategy G, pure or mixed, $E(I^*, G) \geq \frac{1}{2}(\lambda^2 + \lambda^3)$.

Now consider the pure Guard strategies G_1 and G_2 illustrated in Figure 1(b), and let G^* denote the mixed strategy which plays them with equal probability. It is clear that whatever route the Infiltrator takes to get to the target, he must cross the path of one of G_1 and G_2 at least three times, and the other at least twice. Hence, for any Infiltrator strategy I, $E(I, G^*) \leq \frac{1}{2}(\lambda^2 + \lambda^3)$. Note that for the Guard to play G_2, and occupy the first point at time 1, and the third point at time 2, requires that his speed v be at least 2. So, we have that, for all $v > 1$, $v(G_{3,v}) = \frac{1}{2}(\lambda^2 + \lambda^3)$, and I^* and G^* are optimal strategies for the Infiltrator and the Guard respectively.

Before we leave this example, let us consider one further point. Let us imagine the game obtained by taking $G_{3,v}$ but changing the payoff. Let the Guard win the game if and only if (in finite time) he catches the Infiltrator; otherwise the Infiltrator wins the game. As before let the payoff be the probability of a win for the Infiltrator.

Observe that I^* still ensures the Infiltrator at least $\frac{1}{2}(\lambda^2 + \lambda^3)$ against any Guard strategy. G^*, on the other hand, is no longer of any use to the Guard, as the Infiltrator will just refuse ever to move to point three and hence always collect a payoff of $\frac{1}{2}(\lambda+\lambda^2)$, which is decidedly unsatisfactory for the Guard. What the Guard requires now is a strategy that forces the Infiltrator to move to point three and not just stay at large for ever.

Take ϵ to be arbitrarily small and positive. Let us consider now $\bar{G}^*$, an ϵ-optimal strategy for the Guard. At time 0 the Guard is located at point one, but then at time 1 he chooses randomly between points one and two (as these are the two possible locations of his opponent at this time); for all later time units, he chooses the three points (one, two, three) according to the probability distribution $(\delta,\ \delta,\ 1-2\delta)$, where $\delta = \epsilon/[\lambda(1-\lambda)(1+\lambda)]$; since ϵ is arbitrarily small we can be sure that $0 < \epsilon < \frac{1}{2}\lambda(1-\lambda)(1+\lambda)$, and so both δ and $1-2\delta$ lie in the interval $(0,1)$, and hence may be taken as probabilities. Note that for the Guard to use this strategy requires that his speed be greater than one, so this strategy is only valid for $v > 1$.

If the Guard is playing strategy $\bar{G}^*$, then whatever the Infiltrator does, the probability that he will still be at large at the end of time 1 (that is, 1 – Prob{Infiltrator is captured during time 0 or 1}) is $1 - [(1-\lambda) + \frac{1}{2}\lambda(1-\lambda)] = \frac{1}{2}\lambda(1+\lambda)$, and thereafter, if he is still at large he stands at least a $(1-2\delta)(1-\lambda)$ chance of being captured whenever he moves to point three. If he chooses to remain at large forever by refusing to move to point three, since $\delta > 0$, he will with probability one eventually be caught by the Guard, resulting in a payoff of 0. Hence, for any Infiltrator strategy I, $E(I,\bar{G}^*) \le \frac{1}{2}\lambda(1+\lambda)[1-(1-2\delta)(1-\lambda)] = \frac{1}{2}(\lambda^2+\lambda^3)+\epsilon$.

So, in this game, for all $v > 1$, the value is again $\frac{1}{2}(\lambda^2+\lambda^3)$; I^* is optimal for the Infiltrator, and $\bar{G}^*$ is ϵ-optimal for the Guard. It is clear that changing the payoff in this way makes only an apparent difference to the structure of the game (observe that $\bar{G}^*$ is also an ϵ-optimal strategy in $G_{3,v}$ itself), and so our original formulation of the infinite game $G_{n,v}$ is even more general than at first appears.

B. Two Related Problems.

In *A One-Dimensional Infiltration Game* [3], Lalley considered a game which is both closely related to Gal's original problem, and of interest in its own right. The difference in what he proposes is in the introduction of a safe base, an extra point in the game, effectively a point zero, at which the Infiltrator is to be located at time 0, and at which he can remain for as long as he wants and remain safe from capture (the probability of the Guard finding the Infiltrator at point zero is zero). When the Infiltrator moves away from the base he is then located at point one, and otherwise the game is played as before. To emphasize the relationship between what we have here and the games $G_{n,v,t}$ which we introduced in section A, we shall denote this game by $H_{n,v,t}$.

Lalley considered the case $v = 1$, that is the slow Guard game. He proved the value $v(H_{n,1,t})$ to be given by

$$v(H_{n,1,t}) = \lambda^{q+1}[\lambda(\frac{r}{t-n}) + (1-\frac{r}{t-n})], \tag{1}$$

where q and r are the unique non-negative integers satisfying

$$n-1 = q(t-n)+r, \quad 0 \le r < t-n. \tag{2}$$

Lalley also found optimal strategies for the players. The Infiltrator uses a mixed strategy whose pure components are the Wait-and-Run strategies. We encountered these in Section A, and let us now give a formal definition. For $1 \leq i \leq w = t - n$, the ith Wait-and-Run strategy is denoted by W_i and defined as the pure strategy which determines that the Infiltrator, if he is not captured en route, will wait at the base (point zero) until the end of the $(i-1)$th time unit whereafter he proceeds full speed ahead towards the target, occupying the $(\tau - i + 1)$th point at time τ, for $i \leq \tau \leq i + n - 1$, and the target thereafter. In the slow Guard game $H_{n,1,t}$, it is optimal for the Infiltrator to play each W_i, $1 \leq i \leq w$, with equal probability.

The Guard has an optimal mixed strategy whose components are of a type different from Wait-and-Run. Lalley describes them as Orderly Fallback, and they call for the Guard to occupy point one until time 1, retreat at full speed for a certain amount of time, wait one time unit, retreat again, wait again, and so on until he reachs point n. The number of bursts this fallback employs is w, the same as the number of Wait-and-Run strategies, and the duration of the jth burst is ξ_j, where $\xi_1, \ldots, \xi_w$ are the random variables obtained by sampling without replacement from an urn containing r balls marked $q+1$ and $w-r$ balls marked q, where q and r are as defined by equation (1) above.

The details given above have been adapted from Lalley's own work, and the reader is referred there for further elaboration. But there is one observation we should make at this point, and it shows the motivation behind including this material here.

In proving the optimality of the strategies quoted above, at no point does Lalley rely on the restriction on the Guard's speed. Ostensibly, the game under consideration is $H_{n,1,t}$, but clearly any strategy contained in the strategy space for either player in this game will also be contained in his strategy space in the game $H_{n,v,t}$ for all $v > 1$.

This observation is further justified by the fact that we have the solution to the fast Guard game $H_{n,\infty,t}$ from a completely independent source. In *An Infiltration Game on k Arcs* [1], Auger considers a generalisation of the fast Guard game $H_{n,\infty,t}$ by supposing the base and the target to be joined by k (≥ 1) non-intersecting paths of various vertex lengths. He finds the value of this game to be

$$\lambda^{z+1}\{\frac{t-1}{w} - z\} + \lambda^{z}\{1 + z - \frac{t-1}{w}\}, \tag{3}$$

where the number of Wait-and-Run strategies over the k arcs (equal to $kt - m$ where m is the number of points not counting the base or the target) is again denoted by w, and

$$z = \lfloor\frac{n-1}{w}\rfloor, \text{ where } \lfloor x\rfloor \text{ is the greatest integer } n \text{ such that } n \leq x. \tag{4}$$

Here the optimal strategy for the Infiltrator again involves randomizing over the Wait-and-Run strategies. For the Guard, in general, optimal play becomes more complex and requires different cases to be considered. Again, the interested reader is referred to the work itself for further details.

The real importance here of Auger's results, is that for $k = 1$, the single arc game, we have an immediate solution to the fast Guard game $H_{n,\infty,t}$. Futhermore, we find that in fact the fast Guard game and the slow Guard game have the same value. To prove this, all that is needed is to show the equivalence of equation (1) and equation (3) when $k = 1$, and although we omit the details here this follows directly from the definitions of q, r and z in equations (2) and (4).

We are left then with what is a surprising result. When we are looking at the set of games $H_{n,v,t}$, $v \in \{1, 2, \ldots\}$, giving the Guard an arbitrarily large speed advantage over the Infiltrator does nothing to improve his chances of winning the game. For all $n \geq 1$, and for all $t \geq n + 1$, $v(H_{n,v,t})$ is independent of v.

C. A Finite Approach to the Infinite Game.

Remove the initial safe, and at least in general, this is no longer the case. A consideration of the games $G_{3,1,5}$ and $G_{3,\infty,5}$ will illustrate this. Recall that these are the slow and fast Guard versions of Gal's original game played on just three points and with a time limit of five (recall further that at time zero both Infiltrator and Guard are located at the first point). We observe that for all $0 < \lambda < 1$, $v(G_{3,1,5}) > v(G_{3,\infty,5})$, thereby highlighting a significant difference in behaviour from the games with safe starting point.

Look first at the slow Guard game $G_{3,1,5}$. Consider the mixed strategy I_1^* for the Infiltrator, using which he chooses at random between the Wait-and-Run strategies W_1, W_2, and W_3 (see Figure 2).

			point		
		one	two	three	(target)
	0	$W_3W_2W_1$			
	1	W_3W_2	W_1		
time	2	W_3	W_2	W_1	
	3		W_3	W_2	W_1
	4			W_3	W_2W_1
	5				$W_3W_2W_1$

Figure 2.

Let us now see how the Guard should play if he knows that the Infiltrator is going to play I_1^*. He seeks to minimise $\frac{1}{3}(\lambda^{p_1} + \lambda^{p_2} + \lambda^{p_3})$, where for $i = 1$, 2, and 3, p_i is the number of attempts he has to capture W_i. The first decision he must make is what to do at time 1. At time 0, he must certainly search point one, but his choice then is whether to remain there, giving him a chance to detect both W_2 and W_3, or to move on to point two and have another attempt at W_1 remembering that because the players have equal speeds if he remains at point one, he can never catch up with W_1 again.

Careful consideration of Figure 2 shows that if he follows W_1 to point two at time 1, the best the Guard can then do is to ensure one of p_1, p_2 and p_3 be equal to 3, and the others be equal to 2, giving an expected payoff to the Infiltrator of $\frac{1}{3}(2\lambda^2 + \lambda^3)$. If, on the other hand, he stays at point one at time 1, p_1 will equal 1, and the best he can do is to let one of p_2 and p_3 be equal to 4, and the other 3, giving an expected payoff of $\frac{1}{3}(\lambda + \lambda^3 + \lambda^4)$.

To ensure that $\frac{1}{3}(2\lambda^2 + \lambda^3) \leq \frac{1}{3}(\lambda + \lambda^3 + \lambda^4)$, is equivalent to ensuring $\lambda \leq \lambda^*$, where $\lambda^* = (\sqrt{5} - 1)/2$ (≈ 0.618). In this case we have shown above that, for any Guard strategy G, $E(I_1^*, G) \geq \frac{1}{3}(2\lambda^2 + \lambda^3)$. We now give a mixed strategy for the Guard that ensures he can keep the Infiltrator down to this expected payoff.

Let G_1^* be the probability distribution $(\frac{1}{3}, \frac{1}{3}, \frac{1}{3})$, over the pure strategies G_3, G_4, and G_5 illustrated in Figure 3(a).

point one (a)	*point* two (a)	*point* three (a)	*time*	*point* one (b)	*point* two (b)	*point* three (b)
$G_5G_4G_3$			0	$G_8G_7G_6$		
	$G_5G_4G_3$		1	G_8G_7	G_6	
	G_5G_4	G_3	2		G_8G_7	G_6
	G_5	G_4G_3	3		G_8	G_7G_6
		$G_5G_4G_3$	4,5			$G_8G_7G_6$

(a) (b)

Figure 3.

By inspection of Figure 3(a), the Infiltrator, who must aim to have reached point three at or before time 4, must risk detection at least three times by one of G_3, G_4, and G_5, and at least twice by the other two. Hence, for any Infiltrator strategy I, $E(I, G_1^*) \leq \frac{1}{3}(2\lambda^2 + \lambda^3)$. Hence we have shown that for $0 \leq \lambda \leq \lambda^*$, $v(G_{3,1,5}) = \frac{1}{3}(2\lambda^2 + \lambda^3)$, and I_1^* and G_1^* are optimal strategies for the Infiltrator and the Guard respectively; in fact, provided $\lambda \neq 0$ or λ^* these are the unique optimal strategies.

Note one point of interest here. Provided $\lambda \leq \lambda^*$, the optimal strategies are independent of λ. This is not so if we have $\lambda > \lambda^*$. In this case, the Infiltrator again uses his Wait-and-Run strategies, W_1, W_2 and W_3, but this time with the non-uniform probability distribution $(1 - 2\theta, \theta, \theta)$, where $\theta = (1 + \lambda)/(2 + 4\lambda + \lambda^2)$; we denote this mixed strategy by I_2^*. The Guard uses the pure strategies G_6,G_7, and G_8 illustrated in Figure 3(b), with the probability distribution $(1 - 2\phi, \phi, \phi)$, where $\phi = \lambda/(2 + 4\lambda + \lambda^2)$; we denote this strategy by G_2^*. The reader is left to check, by reference to Figures 2 and 3(b), that, for $\lambda^* \leq \lambda < 1$, I_2^* and G_2^* are in fact optimal, and the value of the game is given by $v(G_{3,1,5}) = (2\lambda^2 + 2\lambda^3 + 2\lambda^4 + \lambda^5)/(2 + 4\lambda + \lambda^2)$.

We have thus seen, that in the slow Guard game $G_{3,1,5}$, λ^* is the critical value of λ when the Guard is planning his strategy. A smaller probability λ of missing his opponent allows him to pursue W_1 to point two at time 1; a greater λ forces him to concentrate at least with a certain probability on trying to detect both W_2 and W_3 by staying at point one, and in so doing letting W_1 get away. It is suggested that this behaviour will be found to feature in the general game $G_{n,1,t}$, and that in each particular game there will be a sequence of critical values of λ mirroring the strategic choices the Guard must make like the one encountered above. Further, we would suggest that, to play optimally, the Infiltrator need only use pure strategies which are Wait-and-Run.

If we look now at the fast Guard game $G_{3,\infty,5}$, we find that if the Guard is to play optimally, he must exploit his speed advantage. We already have the optimal strategies illustrated for us in Figure 1 (although, in the finite game we now have, the pure strategies are truncated for use only up to the end of time unit 5), with the Infiltrator and the Guard choosing randomly between the pure strategies W_1 and W_2, and G_1 and G_2 respectively.

We observe, as we did before, that for the Guard to play G_2 and occupy the first point at time 1 and the third point at time 2 (which in playing optimally he does with probability $\frac{1}{2}$) relies on the superior speed of the Guard. Hence it is no surprise that the value of this game, $v(G_{3,\infty,5}) = \frac{1}{2}\lambda^2(1+\lambda)$, is strictly less than $v(G_{3,1,5})$ for all $0 < \lambda < 1$, thus proving the positive advantage that the Guard's greater speed gives him.

We can now see that the original game clearly faces us with a greater degree of complexity than was encountered with the amendment suggested in Section B. Introducing the safe starting position served to help the Infiltrator in that any speed advantage the Guard might have had was of no real use to him. The ensuing solution while not easy was at least quite straightforward once it had been found. The solution to the original game is still eluding detection. I suggest that a possible strategy for finding the solution to the infinite game $G_{n,v}$ is to obtain the solution to the finite game $G_{n,v,t}$ and take the limit as t tends to infinity. It can be proved that $lim_{t\to\infty} v(G_{n,v,t})$ does equal $v(G_{n,v})$, the value of the infinite game; hence such an approach, were it possible to follow, would certainly be valid. However, at present, the problem is open and characteristics of optimal behaviour for the Guard in particular, be he fast or slow, are still unclear.

D. References.

[1] Auger J.M.: 1990, "An Infiltration Game on k Arcs", *Naval Research Logistics*, Special issue on Search Theory (awaiting publication).

[2] Gal S.: 1980, *Search Games*, Academic Press, New York.

[3] Lalley S.P.: 1988, "A One-Dimensional Infiltration Game", *Naval Research Logistics* **35**, 441–446.

About the resolution of discrete pursuit games and its applications to naval warfare

Vincent Laporte*, Jean Marie Nicolas*, Pierre Bernhard**

*Thomson Sintra ASM
1, avenue Aristide Briand
94117 Arcueil Cedex (France)

**INRIA Sophia Antipolis
2004,Route des Lucioles
06565 Valbonne Cedex (France)

Introduction

Tactical analysis of naval situations is a crucial and difficult problem. Threat evaluation must be done with more or less delayed and incomplete information: for instance, actual kinematics of each actor are only estimated by his opponent. This aspect of missing or ill-defined information is a difficult task for a ship officer, and much work has been devoted to the design of useful real time tools for decision aids, based on Artificial Intelligence : Expert Systems can be applied, for example, to the analysis of the opponent intents [Rai89].

As, at Thomson Sintra ASM, we are involved in modelling for submarine control and control system evaluation [LBN90], we think that complementary tools can be founded upon Games Theory : more, a time-space discrete representation allows a realistic modelling yielding the application of methods based on discrete game theory. This discretization takes into account the fact that, for naval situations, there is uncertainty in the available informations and that these informations are only obtained at discrete times. For example, if a ship officer wants to know position and speed of his opponent, he has to manoeuver in order to obtain these informations with his sonars, and, during this manoeuver, no available information can be provided by some of the sonars. More, a ship officer always makes approximations about kinematics informations of his opponent : indeed, he assumes values of maximum speed and of turn radius which cannot be the actual values of the opponent ship. That is why discrete games seem to be a very interesting tool because it is possible to play a set of games in "real time" with various assumptions (like kinematics, position, speed, opponent intents, ...), and, from this set, the ship officer can take his own decision.

In this paper, we first propose our mathematical modelling giving a method of resolution of discrete pursuit games, then its implementations on a workstation (SUN), finally the possible improvments of this approach.

1 Mathematical modelling

1.1 Preliminaries

1.1.1 Available lattices

Firstly, we want to find out the geometrical framework of the game. Indeed, we choose a geometrical description of the space and of the available controls so that this space description is always available after the motions given by the controls. In our description, a control can be modelled as both a translation and a rotation, and the geometry of the space is chosen so that, after each control, i.e. for each translation and for each rotation, for each player, space representation and controls are the same as previously. There is a famous modelling of such a space with the help of cristallography : indeed, physicists have described all the spaces allowing sorts of couple "translation-rotation" by the famous description of the "Bravais lattices" (see for example [Kit71]). With the help of these works, we know that, in bidimensionnal spaces, there are only five kinds of lattices, owning either twofold, or fourfold, or sixfold rotation axes (let us note that, in three dimensionnal spaces, there are only 14 kinds of lattices, yielding 230 space groups).

For bringing such space description to games applications, we emphasize the fact that the choice of controls is rather restrictive because rotations can only be a multiple of either $\pi/3$ or $\pi/2$. Yet, in our discrete games and for our naval applications, this rather rough approximation of controls seems to suit well with the uncertainty of the situation description.

In order to obtain a realistic description of a naval 2-D situation, we have chosen an hexagonal lattice (so that motion of each player is described as translation on this lattice, and that there are six possibilities for the final speed direction of each player) —as in [Isa65]—.

1.1.2 Solution of the game

We shall see two different ways of solving the problem : the first one, inspired by Pontryagin's paper [Pon68], is based on the building of capture zones W_n in which the evader is caught within n steps, and the second one consists in solving Isaacs' equation using the dynamic programming method. At first, we shall study the game in which the players play alternatively, with the minimizer playing first, which can be solved with pure strategies. Then we shall solve the game in which both players play simultaneously, which requires mixed strategies.

1.2 Alternating game

1.2.1 Notations

- We call x_t the relative coordinates at instant t, $t \in \mathbf{N}$.
- As usual, we call $u_t \in \mathcal{U}$ and $v_t \in \mathcal{V}$ the controls of the minimizer and of the maximizer at instant t. We suppose that $\mathcal{U}$ and $\mathcal{V}$ are both finite sets. The dynamics of the game are described by :

$$x_{t+1} = h(g(x_t, u_t), v_t)$$

  which gives a proper model for this game where the pursuer (P) plays first and the evader (E) plays second, knowing P's control.
- When we don't need to distinguish the minimizer's controls from the maximizer's ones, we will more simply write the classical equation :

$$x_{t+1} = f(x_t, u_t, v_t)$$

  with a given initial position x_0.
- We call C the Capture set.
- The performance index J we want to optimize is the duration of the game .
- We also assume the following conditions on g and h :

$$\begin{cases} x \in C \Rightarrow \forall u \; g(x,u) = x \\ y \in C \Rightarrow \forall v \; h(y,v) = y \end{cases}$$

  which will allow us to consider the finite duration games as infinite duration ones with a stationary trajectory. Let's notice that :

$$x \in C \Rightarrow \forall u \, \forall v \; f(x,u,v) = x$$

Our purpose is to find the value $w(x)$ on the whole game space. We shall prove that a method based on the building of capture zones W_n in which the evader is caught within n steps is equivalent to the classical dynamic programming method that is used to solve Isaacs' equation, and quite faster.

1.2.2 Calculation of w : link between the Capture Zones and Dynamic Programming

In the theory of zones, we first define $W_0 (\equiv C)$, which is the target, then we build the capture zones W_n by induction :

$$W_{n+1} = \{x \; : \; \exists u \, \forall v \; f(x,u,v) \in W_n\}.$$

Remark 1 *if $x \notin W_0$ and if $\exists u^*$ $g(x,u^*) \in W_0$, then $x \in W_1$.*

Proof : $\forall v$ $f(x,u^*,v) = h(g(x,u^*),v) = g(x,u^*) \in W_0$. □

Remark 2 *One can easily check : $\forall n \in \mathbf{N}$ $W_n \subset W_{n+1}$.*

In the theory of dynamic programming, we initialize by giving a nil value to all the points of C, and infinite values elsewhere :

$$\begin{cases} w_0(x) = 0 & \text{if } x \in C \\ w_0(x) = \infty & \text{if } x \notin C\,, \end{cases}$$

and we iterate the following process :

$$w_{n+1}(x) = \min_u \max_v w_n \circ f(x,u,v) + c(x)\,,$$

where

$$\begin{cases} c(x) = 1 & \text{if } x \notin C \\ c(x) = 0 & \text{if } x \in C\,. \end{cases}$$

Remark 3 *If $x \in C$, $f(x,u,v) = x$ and $c(x) = 0$ so $\forall n$ $w_n(x) = w_0(x) = 0$.*

Remark 4 *If $x \notin C$ and if there exists u^* so that $g(x,u^*) \in C$ — i.e. the Pursuer P plays and brings the Evader E into C before E has played—, then $w_1(x) = 1$.*

Proof : We have : $\forall v$ $h(g(x,u^*),v) = g(x,u^*) = f(x,u^*,v)$.
Since $g(x,u^*) \in C$, we can also write :

$$0 \le \min_u \max_v w_0 \circ f(x,u,v) \le \max_v w_0 \circ f(x,u^*,v) = w_0 \circ g(x,u^*) = 0$$

yielding : $w_1(x) = 1$. □

We want to prove the equivalence of the two approaches, formalized in the next Theorem 1. For this purpose, we need a few preliminary results : lemma 1, 2 and 3.

Lemma 1 *For a given x, $(w_n(x))_{n \in \mathbf{N}}$ is decreasing.*

Proof : If $x \in C$, remark 3 proves that $w_0(x) \ge w_1(x)$.
If $x \notin C$, then $w_0(x) = \infty \ge w_1(x)$. So $\forall x$ $w_0(x) \ge w_1(x)$.
Now, assuming that $\forall y$ $w_n(y) \ge w_{n+1}(y)$, let's take x in the game space. We have :

$$w_{n+2}(x) = \min_u \max_v w_{n+1} \circ f(x,u,v) + c(x) \le \min_u \max_v w_n \circ f(x,u,v) + c(x) = w_{n+1}(x).$$

□

Lemma 2 *for a given n :*

$$\begin{cases} i) & x \in W_0 & \Rightarrow & w_{n+1}(x) = w_0(x) = 0 \\ ii) & x \in W_{n+1} - W_n & \Rightarrow & w_{n+1}(x) = n+1 \\ iii) & x \notin W_{n+1} & \Rightarrow & w_{n+1}(x) = \infty\,. \end{cases}$$

Proof : by induction.
Let us call A_n the assertion of the lemma. A_0 is true. Now let us assume A_n is true and let's prove A_{n+1}.
Let $x \in W_{n+2} - W_{n+1}$:

$$x \in W_{n+2} \Rightarrow \exists u^* \forall v\, f(x,u^*,v) \in W_{n+1}.$$

Let $\mathcal{U}^* = \{u^* : \forall v\, f(x,u^*,v) \in W_{n+1}\}$, and let $u^* \in \mathcal{U}^*$:

$$x \in W_{n+2} - W_{n+1} \Rightarrow x \notin W_{n+1} \Rightarrow \exists v^*\, f(x,u^*,v^*) \notin W_n,$$

thus $\qquad f(x,u^*,v^*) \in W_{n+1} - W_n,$

that implies $\qquad w_{n+1} \circ f(x,u^*,v^*) = n+1 \quad$ (due to assertion A_n).

Furthemore, $\forall y \in W_{n+1}$, $w_{n+1}(y) \leq n+1$ (because of remark 2, lemma 1 and assertion A_n), so:

$$\max_v w_{n+1} \circ f(x,u^*,v) \leq n+1 = w_{n+1} \circ f(x,u^*,v^*).$$

This is true for all $u^* \in \mathcal{U}^*$, yielding :

$$\min_{u \in \mathcal{U}^*} \max_v w_{n+1} \circ f(x,u,v) = n+1.$$

Now let $\overline{u} \notin \mathcal{U}^*$:

$$\overline{u} \notin \mathcal{U}^* \Rightarrow \exists \overline{v}\, f(x,\overline{u},\overline{v}) \notin W_{n+1},$$

then, using assertion A_n, $\qquad w_{n+1} \circ f(x,\overline{u},\overline{v}) = \infty,$

so $\qquad \max_v w_{n+1} \circ f(x,\overline{u},v) = \infty,$

and finally $\qquad \min_{u \notin \mathcal{U}^*} \max_v w_{n+1} \circ f(x,u,v) = \infty.$

In conclusion, we have :

$$\min_u \max_v w_{n+1} \circ f(x,u,v) =$$

$$\min\left\{\min_{u \in \mathcal{U}^*} \max_v w_{n+1} \circ f(x,u,v), \min_{u \notin \mathcal{U}^*} \max_v w_{n+1} \circ f(x,u,v)\right\} =$$

$$\min(n+1,\infty) = n+1,$$

so using $c(x) = 1$ (since $x \notin W_0$) , $w_{n+2}(x) = n+2$ which proves *ii)*. □

i) has already been proven in remark 3 and it is easy to check *iii)*.

Lemma 3 *if $x \in W_n$, then $w_{n+1}(x) = w_n(x)$.*

Proof : by induction.
Let us call B_0 the assertion of the lemma. Assertion B_0 is true. Let's assume B_n is true and let's prove B_{n+1}.
If $x \in W_0, B_{n+1}$ is true. Let's take $x \in W_{n+1}$ and $x \notin W_0$:

$$x \in W_{n+1} \textit{ and } x \notin W_0 \Rightarrow$$

$$w_{n+2}(x) = \min_u \max_v w_{n+1} \circ f(x,u,v) + 1 = \max_v w_{n+1} \circ f(x,u_*,v) + 1 .$$

(Such a u^* exists since we supposed that $\mathcal{U}$ is a finite set). Let :

$$\mathcal{U}_* = \{u_* : \min_u \max_v w_{n+1} \circ f(x,u,v) = \max_v w_{n+1} \circ f(x,u_*,v)\} .$$

Let $u_* \in \mathcal{U}_*$. Using : $\forall x \in W_{n+1}\ w_{n+2}(x) \leq w_{n+1}(x) \leq n+1$, we can deduce :

$$\max_v w_{n+1} \circ f(x,u_*,v) \leq n,$$

then $\forall v f(x,u_*,v) \notin W_{n+1} - W_n,$

moreover $\forall v\, f(x,u_*,v) \in W_{n+1}$

(otherwise because of the lemma, $\max_v w_{n+1} \circ f(x,u_*,v) = \infty$),

that implies, since $W_n \subset W_{n+1}$, $\forall v f(x,u_*,v) \in W_n$,

and using assertion B_n, we have :

$$\forall v \quad w_{n+1} \circ f(x,u_*,v) = w_n \circ f(x,u_*,v).$$

This is true for all $u_* \in \mathcal{U}_*$. We can deduce :

$$w_{n+1}(x) = \min_u \max_v w_n \circ f(x,u,v) + 1$$

$$\leq \min_{u \in \mathcal{U}_*} \max_v w_n \circ f(x,u,v) + 1 = \min_{u \in \mathcal{U}_*} \max_v w_{n+1} \circ f(x,u,v) + 1 = w_{n+2}(x),$$

yielding

$$w_{n+1}(x) \leq w_{n+2}(x).$$

As we already had $w_{n+2}(x) \leq w_{n+1}(x)$ (due to the decreasing of function w), we obtain :

$$x \in W_{n+1} \Rightarrow w_{n+2}(x) = w_{n+1}(x). \qquad \square$$

Now, in conclusion, we can assert the following theorem :

Theorem 1
Let's take x in the game space.

- *If $x \notin \bigcup W_k$: it is not capturable and $w_n(x) = \infty$.*
- *If $x \in \bigcup W_k$, let n be the smallest integer such that $x \in W_n$; then the first k verifying $w_k(x) \neq \infty$ is n and, $\forall p \in \mathbb{N}$, $n = w_n(x) = w_{n+1}(x) = \ldots = w_{n+p}(x)$.*

Proof : it is a direct consequence of lemma 3 and remark 1. □

Then let's call $w(x)$ the limit : $w(x) = \lim_{n\to\infty} w_n(x)$.

$$\begin{cases} w(x) = \infty & \text{if } x \notin \bigcup W_k \\ w(x) = n+1 & \text{if } x \in W_{n+1} - W_n \\ w(x) = 0 & \text{if } x \in W_0 . \end{cases}$$

$w(x)$ verifies the following equation:

$$w(x) = \min_u \max_v w \circ f(x,u,v) + c(x), \tag{1}$$

$$\text{with } \begin{cases} c(x) = 0 & \text{if } x \in C \\ c(x) = 1 & \text{elsewhere.} \end{cases}$$

Equation (1) is exactly Isaacs' stationary equation.

1.2.3 Optimal strategy

Now, we can easily see that we have obtained the solution of the game where the players play alternatively. Since the minimizer plays first and the maximizer second, making a decision that depends on the control which the minimizer has just chosen, the notion of strategy is slightly different from the classical concept in differential games. We therefore call strategies ϕ and ψ functions of the type:

$$\begin{cases} u_t = \phi(x_t) \\ v_t = \psi(x_t, u_t). \end{cases}$$

One recognizes the definition of what is usually called upper strategy. A pair of strategies and an initial point induce a unique trajectory, and we can write the dynamics of the game:

$$\begin{cases} x_{t+1} = f(x_t, \phi, \psi) \\ x_{t=o} = x_0 \end{cases}$$

with no ambiguity. Now let's call ϕ^* and ψ^* the strategies defined as follows ($\mathcal{U}$ and $\mathcal{V}$ are finite):

ϕ^* consists in playing $\phi^*(x_t) = u_t^*$ corresponding to $\min_u \max_v w \circ f(x_t, u, v)$

ψ^*consists in playing $\psi^*(x_t, u_t) = v_t^*$ corresponding to $\max_v w \circ f(x_t, u_t, v)$.

Actually, these definitions are ambiguous since they don't define a unique control at each step of the game, but it has no consequence as far as we are concerned, and anyway, we could get round this problem by, for instance, imposing a rule for the choice. We can now assert:

Theorem 2 *With the previous notations, if the game starts on a finite $w(x_0)$, then the duration is finite for all pairs of strategies (ϕ^*, ψ); the pair (ϕ^*, ψ^*) is optimal and $w(x_0)$ is the value of the game.*

Proof : let's place ourselves in a game, at instant t, and position x_t. Then, by definition: u_t^* and v_t^* verify the following inequality:

$$\forall (u,v),\quad w \circ f(x_t, u_t^*, v) \leq w \circ f(x_t, u_t^*, v_t^*) \leq \max_v w \circ f(x_t, u, v),$$

which, in terms of strategies, means:

$$\forall (u,v),\quad w \circ f(x_t, \phi^*, v) \leq w \circ f(x_t, \phi^*, \psi^*) \leq w \circ f(x_t, u, \psi^*),$$

which allows us to write :

$$\begin{aligned} \forall (u,v) \quad & w \circ f(x_t, \phi^*, v) - w(x_t) + c(x_t) \\ & \leq w \circ f(x_t, \phi^*, \psi^*) - w(x_t) + c(x_t) \; = \; 0 \\ & \qquad \leq w \circ f(x_t, u, \psi^*) - w(x_t) + c(x_t), \end{aligned} \tag{2}$$

which is exactly Isaac's equation. The *minmax* equation in terms of controls is a saddle-point equation in terms of strategies, and ϕ^* and ψ^* are the optimal strategies. Indeed, with obvious notations, with a game starting in x_0, we have, using equations (1) and (2) :

$$\forall (\phi, \psi) : \; w \circ f(x_0, \phi^*, \psi) + c(x_0) \leq w(x_0) \leq w \circ f(x_0, \phi, \psi^*) + c(x_0),$$

which can be written:

$$w(x_1^{\phi^*,\psi}) + c(x_0) \leq w(x_0) \leq w(x_1^{\phi,\psi^*}) + c(x_0).$$

In the same manner, we have

$$w(x_1^{\phi,\psi^*}) \leq c(x_1^{\phi,\psi^*}) + w(x_2^{\phi,\psi^*})$$

and

$$c(x_1^{\phi^*,\psi}) + w(x_2^{\phi^*,\psi}) \leq w(x_1^{\phi^*,\psi}),$$

yielding

$$c(x_0) + c(x_1^{\phi^*,\psi}) + w(x_2^{\phi^*,\psi}) \leq w(x_0) \leq c(x_0) + c(x_1^{\phi,\psi^*}) + w(x_2^{\phi,\psi^*}).$$

And we can deduce by induction the following relation :

$$\begin{aligned} \forall t \quad & c(x_0) + c(x_1^{\phi^*,\psi}) + \ldots + c(x_t^{\phi^*,\psi}) + w(x_{t+1}^{\phi^*,\psi}) \leq w(x_0) \\ & \leq c(x_0) + c(x_1^{\phi,\psi^*}) + \ldots + c(x_t^{\phi,\psi^*}) + w(x_{t+1}^{\phi,\psi^*}) \end{aligned}$$

□

Remark 5 *There are two interesting results here:*

- *the algorithm of "iteration on values" converges in this case, in a very simple manner since it is stationary, and allows us to solve Isaacs' stationary equation.*
- *this classical algorithm, where one iterates calculations on the whole game space, can be replaced by the building of these capture zones which we interpreted before; it is a faster algorithm, since one only has to do erosion and dilatation like operations on zones that have a small size at the beginning ($W_0 \equiv C$) and that don't grow fast.*

As a matter of fact, the "n" of W_n is the non linear counterpart of what Pontryagin called the "estimating function" [Pon68] ; it means that the evader will be caught within at most n steps; the previous paragraph proves that, in this case, "n" is also the classical value of the game.

Yet the problem is that we have only found a solution of the game in which the players play one after the other, and our model of incomplete information leads us to solve the game in which both play at the same time. This kind of game requires mixed strategies but we will see that here again, the "iteration on values" method converges and gives the solution of the game.

1.3 Simultaneous game

1.3.1 Calculation of v

Our method consists in initializing the whole game space with the previous $\min_u \max_v$ algorithm that gave us a value function $w(x)$ and in building a new value function $v(x)$ by iterating the process described in theorem 3. Now U and V are considered as random variables of distributions Y and Z. The sets $\mathcal{U}$ and $\mathcal{V}$ of admissible values for u and v are still supposed to be finite, and if we call p and q the number of elements they contain, Y and Z are elements of the p and q dimensional simplices.

Theorem 3 *If $v_0(x) = w(x)$ and $v_{n+1}(x) = \min_Y \max_Z E_{Y,Z}\, v_n \circ f(x,U,V) + c(x)$, then $(v_n(x))_{n\in \mathbb{N}}$ converges for all x.*

Proof : In this case, the *minmax* is the saddle point of the function $E_{Y,Z}\, v_n \circ f(x,u,v))$; it is smaller than or equal to the $\min_u \max_v$:

$$\forall n \in \mathbb{N} \ \min_Y \max_Z E_{Y,Z}\, v_n \circ f(x,U,V) \leq \min_u \max_v v_n \circ f(x,u,v),$$

so: $$\forall x v_1(x) \leq v_0(x),$$

and, by induction, we find $$\forall x v_{n+1}(x) \leq v_n(x).$$

Using $\forall k \in \mathbb{N}\ v_k(x) \geq 0$, since we always calculate saddle points of positive matrices, we prove that the method converges. □

We obtain a function $v(x)$ verifying the following equation:

$$v(x) = \min_Y \max_Z E_{Y,Z}\, v \circ f(x,U,V) + c(x)\,, \tag{3}$$

with the limit condition: $v(x) = v_0(x) = 0$ if $x \in C$.

1.3.2 Optimal Strategy

Now, let's place ourselves in the context of a game. We shall use mixed strategies : the strategy ϕ for P is a function $x_t \mapsto \phi(x_t) = Y_t$, distribution of the random variable u_t, and the strategy ψ for E is a function $x_t \mapsto \psi(x_t) = Z_t$, distribution of the random variable v_t. A pair (ϕ, ψ) of strategies therefore defines a unique random process, and x_t becomes a

random variable that we shall write X_t, of which we know how it is distributed, for given initial conditions.
We shall call $\Omega(x_0, \phi, \psi)$ the set of events ω induced by an initial condition x_0 and the pair of strategies (ϕ, ψ). We shall write with no ambiguity:

$$E_{\phi,\psi}^{|X_k=x_k} v(X_t)\,,\ (k \leq t)$$

instead of

$$E_{\phi,\psi}^{|X_k=x_k} v(X_t^{\phi,\psi})$$

the conditional mean in the process induced by (ϕ, ψ).
The duration of the game is a random variable $D(\omega)_{\omega\in\Omega(x_0,\phi,\psi)}$ and the performance index we want to optimize is M, the mean of D:

$$M(x_0, \phi, \psi) = \int_{\omega\in\Omega(x_0,\phi,\psi)} D(\omega)d\mu(\omega) = E_{\phi,\psi}^{|X_0=x_0} D\,.$$

Let's call ϕ^* and ψ^* the strategies defined as follows: we place ourselves at time t, with a position x_t ($X_t = x_t$); then, by construction of v,

$$v(x_t) = \min_Y \max_Z E_{Y,Z}\, v \circ f(x_t, U, V) + c(x_t) = E_{Y^*,Z^*}\, v \circ f(x_t, U, V) + c(x_t).$$

We shall define:

$$\begin{cases} \phi^*(x_t) = Y^* \\ \psi^*(x_t) = Z^*. \end{cases}$$

Theorem 4 *With the previous notations, if the game starts on a finite $v(x_0)$, then the probability for the duration to be infinite is nil for all pair of strategies (ϕ^*, ψ); the pair (ϕ^*, ψ^*) is optimal for the performance index $M(x_0, \phi, \psi)$, and the value of the game is $v(x_0)$.*

Proof : let ϕ and ψ be two other strategies; we have the following relations :

$$\begin{cases} E_{\phi^*,\psi^*}^{|X_t=x_t} v \circ f(X_t, \phi^*, \psi^*) &= E_{Y^*,Z^*}\, v \circ f(x_t, U, V) = v(x_t) - c(x_t) \ (\text{using}(3)) \\ E_{\phi,\psi^*}^{|X_t=x_t} v \circ f(X_t, \phi, \psi^*) &= E_{\phi(x_t),Z^*}\, v \circ f(x_t, U, V) \\ E_{\phi^*,\psi}^{|X_t=x_t} v \circ f(X_t, \phi^*, \psi) &= E_{Y^*,\psi(x_t)}\, v \circ f(x_t, U, V). \end{cases}$$

The property of v becomes:

$$\begin{aligned} &E_{\phi^*,\psi}^{|X_t=x_t}(v \circ f(X_t, \phi^*, \psi) + c(X_t)) \\ &\quad \leq E_{\phi^*,\psi^*}^{|X_t=x_t}(v \circ f(X_t, \phi^*, \psi^*) + c(X_t)) = v(x_t) \\ &\quad\quad \leq E_{\phi,\psi^*}^{|X_t=x_t}(v \circ f(X_t, \phi, \psi^*) + c(X_t)). \end{aligned} \tag{4}$$

This equation proves that ϕ^* and ψ^* give the solution of the game. Indeed, let x_0 be an initial condition : we can write, using (4):

$$E_{\phi^*,\psi}^{|X_0=x_0}(v(X_1) + c(X_0)) \leq v(x_0) \leq E_{\phi,\psi^*}^{|X_0=x_0}(v(X_1) + c(X_0)),$$

and, in the same manner,

$$v(x_1) \leq E_{\phi,\psi^*}^{|X_1=x_1}(v(X_2)+c(X_1)),$$

or, in terms of random variables,

$$v(X_1) \leq E_{\phi,\psi^*}^{|X_1}(v(X_2)+c(X_1)),$$

which yields

$$E_{\phi,\psi^*}^{|X_0=x_0} v(X_1) \leq E_{\phi,\psi^*}^{|X_0=x_0}(E_{\phi,\psi^*}^{|X_1}(v(X_2)+c(X_1)))$$

The process is a Markov process, so we can write :

$$E_{\phi,\psi^*}^{|X_1}(v(X_2)+c(X_1)) = E_{\phi,\psi^*}^{|X_1,X_0=x_0}(v(X_2)+c(X_1))$$

and

$$E_{\phi,\psi^*}^{|X_0=x_0}(E_{\phi,\psi^*}^{|X_1}(v(X_2)+c(X_1))) = E_{\phi,\psi^*}^{|X_0=x_0}(E_{\phi,\psi^*}^{|X_1,X_0=x_0}(v(X_2)+c(X_1)))$$

and because of the total probability theorem applied to the probability space $\Omega(x_0,\phi,\psi^*)$,

$$E_{\phi,\psi^*}^{|X_0=x_0}(E_{\phi,\psi^*}^{|X_1,X_0=x_0}(v(X_2)+c(X_1))) = E_{\phi,\psi^*}^{|X_0=x_0}(v(X_2)+c(X_1)),$$

yielding :

$$v(x_0) \leq E_{\phi,\psi^*}^{|X_0=x_0}(v(X_2)+c(X_1)+c(X_0)).$$

We have the same inequality on the other side, and by induction,

$$\forall t \qquad E_{\phi^*,\psi}^{|X_0=x_0}(v(X_{t+1})+c(X_t)+\ldots+c(X_1)+c(X_0)) \leq v(x_0)$$
$$\leq E_{\phi,\psi^*}^{|X_0=x_0}(v(X_{t+1})+c(X_t)+\ldots+c(X_1)+c(X_0)),$$

which proves that we optimized the mean of a performance index which is interpreted as the duration of the game if it is finite, since in this case $c(x_t)$ becomes equal to zero for t greater than a certain t_1.

We shall now prove that if we take an initial point x_0 and if $v(x_0)$ is finite, then the probability for the game to have an infinite duration is zero, assuming that P plays ϕ^*. Indeed,

$$\forall\psi\ \forall t\ , E_{\phi^*,\psi}^{|X_0=x_0}(v(X_{t+1})+c(X_t)+\ldots+c(X_1)+c(X_0)) \leq v(x_0) < \infty.$$

Let's call $\Omega_\infty(x_0,\phi,\psi)$ the set of infinite games starting from x_0 with the strategies ϕ and ψ. Using the fact that $\Omega_\infty(x_0,\phi,\psi)$ is a subset of $\Omega(x_0,\phi,\psi)$, and that for a game of infinite duration,
$c(x_0)+c(x_1)+\ldots+c(x_t)+v(x_{t+1}) \geq t,$
we can write:

$$\infty > v(x_0) \geq E_{\phi^*,\psi}^{|X_0=x_0}(v(X_{t+1})+c(X_t)+\ldots+c(X_1)+c(X_0))$$
$$\geq \int_{\Omega_\infty(x_0,\phi^*,\psi)}(v(X_{t+1})+c(X_t)+\ldots+c(X_1)+c(X_0))(\omega)d\mu(\omega)$$
$$\geq t * \mu(\Omega_\infty(x_0,\phi^*,\psi)),$$

which is possible for all t only if

$$\mu(\Omega_\infty(x_0,\phi^*,\psi)) = 0,$$

which proves the assertion. □

2 Implementation

As we already mentionned, we use a hexagonal (or triangular) lattice. Our programs are written in C language, and we run them on "Sun" workstations. For the alternate game, the method of dynamic programming (theorem 3) is easy to compute. The method of zones (theorem 4) requires more subtle programming; we use C memory allocation "malloc" to put the previously calculated zone in memory, and we calculate the following one using only these data. As we expected, the capture zone program is quite faster than the first one: about 5 to 6 times faster.

For the simultaneous game, we use the previous maps (which are actually matrices) for initializing the process. Then we iterate the resolution of the simplex problem on each point of the game space, using the linear programming algorithm.

3 Improvments

Two important ways of improvments which seem promising for naval applications are extensions to higher dimensionnal spaces and to other kinds of games.

3.1 higher dimensional spaces

The previously chosen lattices are 2-D lattices : the two players are moving on a plane, with either rectangular or hexagonal lattices. This kind of game can be restritive for actual applications which, in the case of naval games, can involve submarines. Yet, a submarine warfare application does not need a true 3-D space : indeed, available controls of a submarine are restricted to variation of the depth and have nothing to do with true 3-D controls as for airfights. By the same analogy with cristallography, we propose a graphite-like lattice for discrete game modelling in order to deal with the motions of submarines. In this case, there is an hexagonal subspace, like the previously described one, and the motions available between these subspaces can be restricted to simple variations of depth. By this way, we can apply the mathematical descriptions we propose in section 1, and algorithms giving charts can easily deduced from the 2-D ones.

If we want to deal with a true 3-D space —i.e. we consider that the space is isotropic— a model can be constructed with a 3-D lattice. A first one, similar to the rectangular lattice of the 2-D space, is a cubic lattice : indeed, in this case, each control can be splitted into a translation and a rotation which belong to the space group of this lattice. This case is rather restrictive as final speed is actually parallel to one of the cristal vector. An interesting but more complex lattice can be a diamond-like one. In this case, there is much possibilities for the final speed; indeed, representation in such a space can become easy only for a specialist of cristallography or space groups...

3.2 Survey games

Naval applications required good models of survey games : indeed, a ship can be viewed as a threat if she is too near. Matching of our model to survey games can easily deduced from the previous tools. Nevertheless, we have firstly to estimate if the minimizer, who wishes a survey situation, can win (it is easy to imagine a situation where the maximizer can always evade or attack). In this case, a chart of the space, with which optimal controls can be deduced, can be computed by the same way as previously.

4 Conclusions

These methods for the resolution of discrete pursuit games provide useful tools for naval applications : indeed, they take into account ill-defined data and provide interesting informations about possible strategies. It seems that a possible application for decision aids can be an interactive system displaying several charts to the ship officer, each chart being based on given assumptions about the opponent. As our modelling allows a "real-time" computation of these charts, the ship officer can introduce all kinds of values deduced from ill defined informations and all kinds of opponent intent and, with the help of these charts, he can deduce his own decision and his own strategy.

Finally, we want to emphasize the fact that, on one hand, the discretization of the space is not so much of a problem as far as undersea pursuits are concerned, since ships are rather big, slow, and since specific problems of underwater acoustics yield imprecision in detection and measurements, and on the other hand, our methods, based on this discretization of the space, do not seem to be so appropriate for missiles launching and airfights, that other models of Pursuit Evasion game match well. So naval warfare has to suggest us new game models for which we have to define a pragmatic approach to find suitable solutions.

References

[Isa65] R. Isaacs. *Differential Games*. Wiley, 1965.

[Kit71] C. Kittel. *Introduction to solid state physics*. Wiley, 1971.

[LBN90] Y. Lagoude, B. Babault, and D. Neveu. Parametric variations and multi-level modelling for submarine command and control system evaluation. In *Undersea Defence Technology 1990*, pages 169–176, 1990.

[Pon68] Pontryagin. Linear differential games i and ii. *Soviet Math. doklady*, 8(3 & 4):769–771 & 910–912, 1968.

[Rai89] A. Raimondo. Analyse d'intentions tactiques. In *Ninth International Workshop on Expert Systems and their Applications : Specialized Conference on Artificial Intelligence and Defense*, pages 183–194, 1989.

ON WASHBURN'S DETERMINISTIC GRAPHICAL GAMES

V.J.Baston and F.A.Bostock
Mathematics Department, University of Southampton, Southampton SO9 5NH, U.K.

1. Introduction.

In [4] Washburn defined a Deterministic Graphical (DG) Game as follows. A DG game is a two-person zero-sum game played on a directed graph with $n > 0$ nodes. Nodes with no successor are called terminal nodes and have a payoff to Player 1 associated with them. The other nodes are called continuing nodes and are labelled to indicate which player chooses the successor node. Play starts at a specific node and continues until a terminal node is reached at which point Player 1 obtains the payoff corresponding to that node; if play never ends, the payoff is by convention zero.

Washburn pointed out that special cases of these games had previously occurred in the literature but, that up to then, the class had remained nameless. He also commented that DG games are special cases of recursive games with the consequence that every DG game has a value and, what is more, attention can be restricted to stationary strategies for the players. In other words, for DG games we need only consider strategies in which the player's decision at a node does not depend on the previous history. (The reader is not required to have a knowledge of recursive or stochastic games but anyone wishing to become acquainted with them can consult [1], [2], [3].) The situation is in fact even simpler as the players have pure stationary optimal strategies. Washburn produced an algorithm which provides a method for solving not only DG games but also discounted DG games when the discount factor β lies between 0 and 1; a discounted DG game is played on exactly the same graph but the payoff for a play that terminates after precisely m moves is now β^m times the payoff associated with the terminal node while the payoff for a non-terminating play remains 0. Notice that, discounted games with $\beta \neq 1$ are not recursive since the payoff depends on the previous history. However, for $\beta \leq 1$, the value of the game is assured from the theory of stochastic games. As Washburn acknowledged, his algorithm does not work for $\beta > 1$. The main purpose of this paper is to give an algorithm for solving discounted DG games when $\beta > 1$; for convenience we shall call such games supercounted games. Although supercounted games have many of the characteristics of stochastic games they do not appear to belong to any of the standard classes of games that are known to have a solution. In the course of our algorithm we will therefore also show that supercounted games do have a value under the understanding that the values $\pm\infty$ are allowable. However a supercounted DG game has many structural features in common with the corresponding DG game and we shall exploit some of them in our algorithm.

2. Supercounted Games.

First note that, if play in a supercounted game reaches a terminal node with positive payoff, then it is clearly to player 1's advantage to have taken as many moves as possible to reach it. In particular his payoff will become arbitrarily large if he can arrange to take an arbitrarily large number of moves before reaching a terminal node with positive payoff. Thus, unlike the corresponding DG game, there is the possibility of an infinite expectation. We will now show that, in the very simple game illustrated in figure 1,

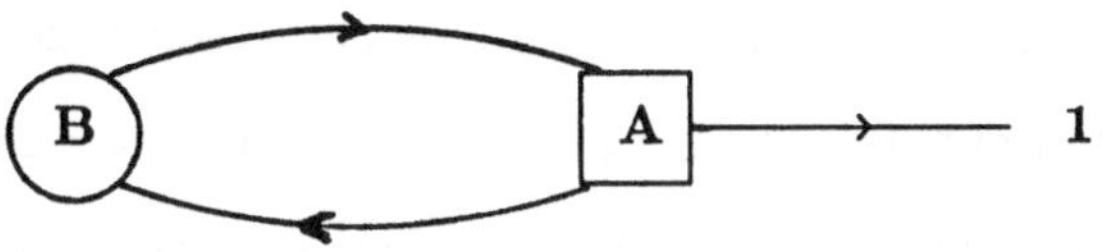

Figure 1.

player 1 can have an expectation of (plus) infinity; node A is a player 1 node while B is a player 2 node. Since there is only one node where a player has a choice, the strategies are effectively given when it is specified what player 1 does at node A.

Let play start at node A and consider the strategy in which Player 1 chooses to go to the terminal node with probability $p = 1 - \beta^{-2}$ whenever he finds himself at A. His expectation is then

$$p\beta + \beta^3(1-p)p + \beta^5(1-p)^2p + \cdots = p\beta + p\beta + p\beta + \cdots = \infty.$$

Notice that, in contrast to DG games, player 1 cannot choose his successor node with probability 1 if he is to act optimally; in other words although he employs a stationary strategy, it is not a pure stationary one. Strictly speaking, to properly define the game, we should specify the strategy spaces of the players and, for definiteness, we assume they are history-remembering ones as in stochastic games. However we feel that a too formal presentation would complicate what is an intuitively natural approach to a simply posed problem so strategy spaces will remain in the background. Nevertheless we will return to the issue later.

Let Γ_β be a supercounted game on a graph G and Γ the corresponding DG game. We use T to denote the set of terminal nodes and, for $i = 1, 2$, V_i to denote the set of nodes in G where the choice of successor is made by player i. For a node $x \in G$, $S(x)$ and $P(x)$ represent the sets of successors and predecessors of x respectively. Further let $v(x)$ be the value of the game Γ when play begins at node x; it is convenient to allow play in both Γ and Γ_β to begin at a terminal node although in these cases the game is trivial. We have already seen that, starting at certain points, it is possible for an infinite expectation to occur in a supercounted game and we now obtain an algorithm which determines such points in Γ_β.

3. Algorithm to find the nodes having an infinite expectation for player 1.

If player 1 is to arrange for play to go through a very large number of steps before arriving at a terminal node with positive payoff, he will obviously need to be able to avoid nodes from which player 2 can force a game to terminate in a finite number of steps, any nodes that do not have positive game values in the undiscounted game and nodes where he must terminate the game. Furthermore it is intuitively reasonable to expect that, if player 1 can avoid such vertices, then he can assure himself of an arbitrarily large payoff. This is the basic idea used in the formulation of the following algorithm. Analogous considerations apply for player 2.

Set $X_1^{(0)} = V_1,\ X_2^{(0)} = V_2,\ X_3^{(0)} = T$,

$$K_1^{(0)} = \{x \in V_1 : v(x) \leq 0\} \cup \{x \in V_1 : \text{ every successor node of } x \text{ is in } T\},$$

$$K_2^{(0)} = \{x \in V_2 : v(x) \leq 0\} \cup \{x \in V_2 : \text{ there is a successor node of } x \text{ in } T\}.$$

Put $X_1^{(1)} = X_1^{(0)} \setminus K_1^{(0)},\ X_2^{(1)} = X_2^{(0)} \setminus K_2^{(0)}$ and $X_3^{(1)} = X_3^{(0)} \cup K_1^{(0)} \cup K_2^{(0)}$. Clearly $x \in X_1^{(1)} \cup X_2^{(1)}$ implies $v(x) > 0$.

Suppose in the general step we have mutually disjoint sets $X_1^{(i)} \subseteq X_1^{(i-1)},\ X_2^{(i)} \subseteq X_2^{(i-1)},\ X_3^{(i)} \supseteq X_3^{(i-1)}$ with $X_1^{(i)} \cup X_2^{(i)} \cup X_3^{(i)} = S_1 \cup S_2 \cup T$. Set

$$K_1^{(i)} = \{x \in X_1^{(i)} : \text{ every successor node of } x \text{ is in } X_3^{(i)}\}$$

$$K_2^{(i)} = \{x \in X_2^{(i)} : \text{ there is a successor node of } x \text{ in } X_3^{(i)}\}.$$

Put $X_1^{(i+1)} = X_1^{(i)} \setminus K_1^{(i)},\ X_2^{(i+1)} = X_2^{(i)} \setminus K_2^{(i)}$, and $X_3^{(i+1)} = X_3^{(i)} \cup K_1^{(i)} \cup K_2^{(i)}$. Clearly $X_1^{(i+1)},\ X_2^{(i+1)}$ and $X_3^{(i+1)}$ have the appropriate properties for the step to be repeated. Thus starting with $i = 1$ repeat the general step until $K_1^{(m)} = \emptyset = K_2^{(m)}$; since G has only a finite number of nodes we must reach such a stage.

We will show that player 1 has an infinite expectation if play starts at a point of $X_1^{(m)} \cup X_2^{(m)}$. First suppose $x \in X_2^{(m)}$; since $K_2^{(m)} = \emptyset$ every successor node of x is in $X_1^{(m)} \cup X_2^{(m)}$. In other words on his move player 2 cannot escape from $X_1^{(m)} \cup X_2^{(m)}$ and in particular he is unable to terminate play because he cannot send play to a terminal node. Now suppose $x \in X_1^{(m)}$; since $K_1^{(m)} = \emptyset$, there is a successor node to x in $X_1^{(m)} \cup X_2^{(m)}$. Thus on his move player 1 can keep play in $X_1^{(m)} \cup X_2^{(m)}$. Hence we can define a function $F : X_1^{(m)} \to X_1^{(m)} \cup X_2^{(m)}$ such that $F(x)$ is a successor to x. Notice that if $x \in X_2^{(m)}$ then player 2 must send play to a member of $X_1^{(m)}$ after a finite number of moves, for, if he could maintain play in $X_2^{(m)}$, $v(x)$ would be zero and we would have a contradiction. Hence it is sufficient to show that every $x \in X_1^{(m)}$ has infinite expectation.

Let play start at $x_0 \in X_1^{(m)}$ and $p = 1/\beta$; since $\beta > 1,\ 0 < p < 1$. Let player 1 send play to $F(x_0)$ and, then whenever play moves to $x \in X_1^{(m)}$, with probability p send play to $F(x)$ and with probability $1-p$ use the strategy which in Γ ensures him an expectation of $v(x)$. Setting $\mu = \min_{x \in X_1^{(m)}} v(x)$ which is positive since $X_1^{(m)}$ is finite, it is easy to see that player 1's expectation is at least

$$(1-p)\beta\mu + (1-p)\beta\mu(p\beta) + (1-p)\beta\mu(p\beta)^2 + \cdots = (1-p)\beta\mu\{1+1+\cdots\} = \infty.$$

Thus player 1 has an infinite expectation at the points of $X_1^{(m)} \cup X_2^{(m)}$.

Let $V^+ = \{z : v(z) > 0 \text{ and } z \notin X_1^{(m)} \cup X_2^{(m)}\}$; the reader is reminded that terminal nodes having positive value are in V^+. Notice that if $z \in V^+$ then $z \in K_1^{(r)} \cup K_2^{(r)}$ for some $r < m$ [by convention $K_1^{(-1)} \cup K_2^{(-1)} = T$] and it is easy to deduce that player 2 can force play to a terminal node in at most m moves. Hence player 1 does not have an infinite expectation at points of V^+. It follows that $X_1^{(m)} \cup X_2^{(m)}$ is precisely the set of points at which player 1 has an infinite expectation.

4. Algorithm to find the value of the supercounted game at points of V^+.

Since $v(z) > 0$ for $z \in V^+$, player 1 can force play in Γ (and so in Γ_β) to a terminal node having positive value in a finite number of moves. Let $x \in V^+ \cap V_1$, then $S(x) \cap (X_1^{(m)} \cup X_2^{(m)}\})$ is empty since $x \in K_1^{(r)}$ for some $r < m$. However $S(x) \cap V^+$ is nonempty as $v(x) > 0$. Thus, on finding himself at x in the game Γ_β player 1 will send play to a member of V^+. If $y \in V^+ \cap V_2$, $S(y) \subseteq V^+ \cup X_1^{(m)} \cup X_2^{(m)}\}$ since $v(y) > 0$; further $S(y) \cap V^+$ is nonempty because $y \in K_2^{(r)}$ for some $r < m$ so, when play is at y, player 2 will send play to a point of V^+. Hence to find the values of nodes in V^+ for the game Γ_β, we can restrict attention to nodes of V^+ and so it is sufficient to consider the corresponding game $\Gamma_\beta(V^+)$ on the subgraph with vertices V^+. Note that $\Gamma_\beta(V^+)$ has the following properties:

(1) player 1 can force play to a terminal node in a finite number of moves,
(2) player 2 can force play to a terminal node in at most m moves,
(3) the node $x \in V^+$ has a value in Γ_β if and only if it has a value in $\Gamma_\beta(V^+)$; furthermore if the values exist they are equal.

If $V^+ \subseteq T$ all the nodes of V^+ have values and there is nothing to do.

(A) Suppose there is a vertex $x \in V^+ \cap V_1$ such that $S(x) \subseteq T$. Clearly, on finding himself at x, player 1 will choose to go to a member of $S(x)$ having largest value, M say. Thus in particular the supercounted game starting at x has value βM and in general we can consider the corresponding game on the subgraph with x replaced by a terminal node having value βM and the arcs joining x to $S(x)$ deleted. Note that this new game inherits the properties (1) and (2) above. Furthermore, by ignoring some dominated strategies for player 1 in $\Gamma_\beta(V^+)$, it is easy to see that we can identify the strategy spaces for the players in the new game with those in $\Gamma_\beta(V^+)$. It follows that a node in the new game has a value if and only if it has a value in $\Gamma_\beta(V^+)$ and that when the values exist, they are equal. Hence it also has property (3).

(B) Suppose (A) does not hold. Then, in play at a node of $V_1 \cap V^+$, player 1 can avoid going to a terminal node. Thus, since player 2 can force play to a terminal node in at most m moves from any point of V^+, there must be a node of $V^+ \cap V_2$ having a terminal node as a successor. Of all such nodes w take y to be one for which $\alpha(w) = \min_{z \in S(w) \cap T} v(z)$ is a minimum. Now, on finding play at y, player 2 will go to a terminal node having value $\alpha(y)$. The reasons for this are that not only can player 1 avoid sending play to a terminal node but that he can also eventually force play to a terminal node; thus, to avoid an arbitrarily large payoff to player 1, player 2 will need to terminate play and, by definition of y, he cannot do better than do so immediately. Hence, in particular, the supercounted game starting at y has value $\beta\alpha(y)$ and we can consider the corresponding game on the graph with y replaced by a terminal node having value $\beta\alpha(y)$ and the arcs joining y to $S(y)$ deleted. Note that this new game also inherits properties (1) and (2). Now player 2's strategy space in the new game can clearly be identified with a subset of his strategy space in $\Gamma_\beta(V^+)$. Furthermore let Y be a strategy for player 2 in $\Gamma_\beta(V^+)$ and $\bar{Y}$ the strategy which differs from Y only in that,when play reaches y, it always sends play to a terminal node having value $\alpha(y)$. It is easy to see that the reasons mentioned above imply that, if player 1 cannot get an expectation of more than w against Y, then he cannot do so against $\bar{Y}$ either. It follows that a node in the new game has a value if and only if it has a value in $\Gamma_\beta(V)$ and that when the values exist, they are equal. Hence the new game also has property (3).

Hence, in either case, we have obtained a corresponding game with fewer arcs. The process can now be repeated until all nodes are terminal nodes. The values of the terminal nodes clearly give the appropriate values of the nodes in Γ_β.

5. Solutions for supercounted games.

Analogous arguments to the ones used in sections 3 and 4 will obviously give the nodes at which player 2 has an infinite expectation and the nodes where he has a positive value. The only nodes remaining for consideration are the points z where $v(z) = 0$; since the players are free to employ their optimal strategies in Γ in the game Γ_β, these nodes have value 0 in Γ_β as well.

We now give an example to show that the players do not necessarily have stationary strategies. Consider the supercounted game played on the graph shown in figure 2; A is a player 1 node and B a player 2 node.

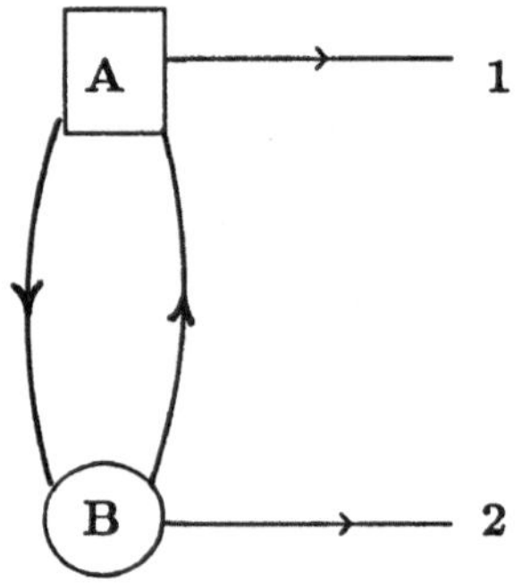

Figure 2.

Clearly for the corresponding undiscounted game $v(A) = 1 = v(B)$ whereas in the supercounted game the values at A and B are $2\beta^2$ and 2β respectively. Notice that, after finding there are no nodes having infinite expectation, our algorithm then replaces B by a terminal node having value 2β deleting the arcs out of B; subsequently it replaces A by a terminal node having value $2\beta^2$ deleting the arcs out of A. Thus at B player 2's action is to move to the terminal node having value 2 while at A player 1's action is to move to B. However moving to B whenever play is at A is not optimal for player 1 since player 2 can achieve a payoff of 0 against it by always moving to A when play is at B. In fact it is easy to see that player 1 does not have a stationary optimal strategy in this game although he does of course have an optimal strategy. Thus, although the algorithms for finding the values of the nodes for the games Γ and Γ_β employ similar ideas (see the following section), the optimal strategies for the players in the two games can have very different forms; in marked contrast to the game Γ, in the game Γ_β the players need not have optimal strategies which are pure or stationary, let alone both. It is perhaps worth mentioning that our choice of history-remembering strategies was more general than necessary; it is not too difficult to check that our arguments would have been equally valid if we had used strategy spaces of the type introduced by Everett [1] for recursive games.

6. DG Discounted Games.

In this section we outline an alternative algorithm to Washburn's for DG discounted games. The algorithm is based on the ideas used in section 4 and employs a dynamic programming backward induction process. Let Γ_β denote a DG game with discount factor $\beta \leq 1$. Since the game is trivial if all the terminal nodes have value 0, assume first of all that there is a terminal node with positive payoff. [If the graph does not contain any terminal nodes with positive payoff, we effectively start at (b) below.] Let T^* be the set of terminal nodes with maximal payoff, p say, and $P(T^*)$ the predecessors of members of T^*. Obtain a new graph as follows.

If $x \in P(T^*) \cap V_1$, delete all arcs out of x, thus converting x into a terminal node, and give it a payoff of p.
If $y \in P(T^*) \cap V_2$, delete any arcs out of y going to a member of T^*; if y does not then have any successors, give it a payoff of p, otherwise do nothing extra.

The rationale for these operations is intuitively obvious. Player 1 cannot send play to a terminal node with value greater than p whatever he does, so he may as well opt for a terminal node with value p when he can and as quickly as possible if $\beta < 1$. On the other hand play will not get to a terminal node with value greater than p whatever player 1 does, so he cannot do worse than avoid a terminal node with value p if he can do so. Even if play subsequently arrives at a terminal node with payoff p, player 2 will be strictly better off if $\beta < 1$.

Having done the above operations for each $a \in A$, we obtain a new DG game played on a subgraph Γ' of Γ which has fewer arcs than Γ. Enumerate the components of Γ' and repeat the process on each component which has a terminal node with positive payoff and so on. Since at each stage we remove at least one arc, the process must end and clearly one of the following must hold.

(a) All nodes of Γ have been assigned a number,
(b) There is a subgraph containing the nodes which have not been assigned a number and all of whose terminal nodes have a payoff which is non-positive.

For each component in (b) which has negative payoffs repeat the above procedure with the roles of the players reversed. Hence we get to the position where

either (α) all nodes of Γ have been assigned a number
or (β) there is a subgraph containing the nodes which have not been assigned a number and all of whose terminal nodes have a payoff of 0.

Although we shall not provide a proof, it is intuitively clear that, at a node of the subgraph in (β), neither of the players can ensure himself of a positive payoff. In other words the value of the game is zero. Thus for a subgraph in (β) assign the value 0 to each of its nodes. The value assigned to a node is then the value of Γ_β starting at that node.

7. Final comments.

Washburn remarked on the possibility of extending DG games by allowing nodes at which random moves are made and concluded that they do not have any feature to exploit that is not already present in recursive games. In the case of supercounted games the extension need not have a value as the following example shows. Consider a supercounted game played on the graph given by figure 3.

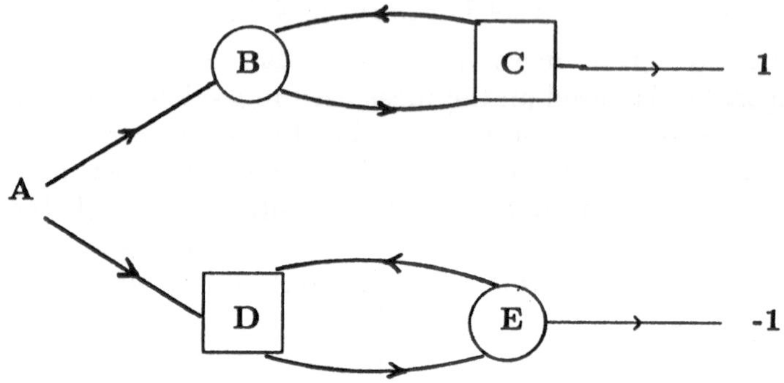

Figure 3.

Nodes C, D are player 1 nodes while B, E are player 2 nodes and A is a random move node where an arc is chosen at random. Nodes B and C have value plus infinity, D and E minus infinity but A does not have a value. Two further extensions of DG games naturally suggest themselves. Firstly the graph could be allowed to have an infinite number of nodes and secondly the terminal nodes could have payoff pairs so that the zero-sum condition is removed. We have obtained some partial results on them and are actively engaged on getting more complete ones.

References.

1. Everett, H., *Recursive games*, in Ann. Math. 39, Princeton University Press, Princeton, 1957, pp. 47-78.
2. Raghavan, T. E. S. and Filar J. A., *Algorithms for stochastic games - a survey*, Zeitschrift für Operations Research (to appear).
3. Vrieze, O. J., *Stochastic games with finite state and action spaces*, CWI Tract 33, Centrum voor Wiskunde en Information, Amsterdam, The Netherlands, 1987.
4. Washburn, A., *Deterministic graphical games*, J. Optim. Theory Appl., (to appear).

GAME THEORY AND H^{∞}-OPTIMAL CONTROL:
The Continuous-Time Case

Tamer Başar
Coordinated Science Laboratory
University of Illinois
1101 W. Springfield Avenue
Urbana, IL 61801/USA

Abstract

We discuss the versatility of differential game theory in the derivation of optimally disturbance attenuating controllers for linear (continuous-) time varying systems under various measurement schemes. These include the perfect state, delayed state, sampled state measurements, and a particular partial state information pattern. In each case we obtain a characterization of the optimal or suboptimal controller, and determine the corresponding minimax attenuation level.

Keywords: H^{∞}-optimal control, differential games, measurement schemes, sampling.

1 Introduction

This is the second part of a two-part paper where the main objective has been to demonstrate the versatility of linear-quadratic dynamic game theory in the design of "optimal" disturbance attenuating controllers for linear systems subject to unknown disturbances. The first part [1] has dealt with the discrete-time problem, in which context optimal controllers have been obtained under various information patterns, including perfect state, delayed state measurements, and a particular partial state information pattern. These controllers were all derived from the saddle-point controller of a linear-quadratic feedback game, using the notion of "strong time consistency," adapted to the given measurement schemes.

In this second part, we direct our attention to the continuous-time disturbance attenuation problem, and derive its solution (again by utilizing the saddle point an appropriate dynamic (differential) game) under different information patterns. Among these are the perfect state (feedback), delayed state, sampled state measurements and a particular partial state information pattern.

The general problem formulation is given in the next section, followed (in Section 3) with the formulation and solution of a related differential game in both finite and infinite horizons. Applications to the disturbance attenuation problem under different information patterns, and also for uncertain nonzero initial state, are discussed in Section 4. The paper ends with the concluding remarks of Section 5.

2 The Basic Controller Design Problem

Let the time interval be $[0, t_f]$, and the plant be described by the vector differential equation

$$\dot{x} = A(t)x + B(t)u(t) + D(t)w(t), \quad x(0) = 0 \tag{2.1}$$

where "t" is the time variable, and the coefficient matrices are taken to be piecewise continuous. The control variable u is square integrable, and is generated by

$$u(t) = \mu(t, x_{[0,t]}), \quad t \in [0, t_f] \tag{2.2}$$

where μ is the controller, belonging to $\mathcal{M}_{CL}$, the class of all "closed-loop" causal mappings under which (2.1) admits a unique solution for every square-integrable disturbance w ($w \in \mathcal{H}_w$). Later, admissible controllers will be further restricted to subclasses (say, generically, $\mathcal{M}$) of $\mathcal{M}_{CL}$, compatible with the measurements available to the controller.

Introduce the quadratic cost function

$$\tilde{L}(u_{[0,t_f]}, w_{[0,t_f]}) = x(t_f)^T Q_f x(t_f) + \int_0^{t_f} [x(t)^T Q(t) x(t) + u(t)^T u(t)]dt, \tag{2.3}$$

where $Q_f \geq 0$, $Q(\cdot) \geq 0$, and the latter is continuous on $[0, t_f]$. The disturbance attenuation problem is to design a controller $\mu^* \in \mathcal{M}$, for a given controller space $\mathcal{M}$, so that the "gain" from the disturbance w, as quantified by $\tilde{L}$, is minimized. Let $T_\mu(w)$ denote the corresponding mapping from w to $z := (x, u)$ under $\mu \in \mathcal{M}$, which was introduced in [1] in abstract terms. Then, the design objective is to minimize $\ll T_\mu \gg$:

$$\inf_{\mu \in \mathcal{M}} \ll T_\mu \gg =: \gamma^* \tag{2.4a}$$

where $\ll \cdot \gg$ denotes the operator norm of T_μ, i.e.,

$$\ll T_\mu \gg := \sup_{w \in \mathcal{H}_w} ||T_\mu(w)||_z / ||w||_w \tag{2.4b}$$

which is defined to be $-\infty$ if $w = 0$. Here, $|| \cdot ||_{(\cdot)}$ denotes the appropriate Hilbert space norm, with the subscript identifying the corresponding space. In particular, $||w||_w^2 := \int_0^{t_f} w^T(t)w(t)dt$, and $||T_\mu(w)||^2 = \tilde{L}(\mu, w)$. Now, let $\mu^* \in \mathcal{M}$ be a controller (assuming that one exists) satisfying the minimax disturbance attenuation bound γ^* in (2.4a). Then, (2.4a) is equivalent to

(i) $||T_{\mu^*}(w)||_z^2 \leq \gamma^{*2}||w||_w^2$, for all $w \in \mathcal{H}_w$

and

(ii) there is no other $\mu \in \mathcal{M}$ (say, $\hat{\mu}$), and a corresponding $\hat{\gamma} < \gamma^*$, such that

$$||T_{\hat{\mu}}(w)||_z^2 \leq \hat{\gamma}^2||w||_w^2, \text{ for all } w \in \mathcal{H}_w.$$

This suggests that μ^* could be obtained as the saddle point of a game with kernel

$$L\gamma(\mu, w) := ||T_\mu(w)||_z^2 - \gamma^2||w||_w^2 \equiv \tilde{L}(\mu, w) - \gamma^2||w||_w^2 \tag{2.5}$$

with a least "allowable" value for γ^2. More precisely, the statements of (i) and (ii) above are equivalent to finding the smallest value of $\gamma > 0$ under which the *upper value* of the associated game with objective function $L_\gamma(\mu, w)$ is bounded, *and* the corresponding controller (if exists) that achieves this bound.

3 A Differential Game And Its Saddle Point

In view of the last remark of Section 2, we now study the solution of the linear-quadratic differential game defined by the state equation (2.1) and cost function (2.5), with the controller belonging to either $\mathcal{M}_{CL}$ or to the restricted class $\mathcal{M}_{FB}$, where the latter is the class of all "memoryless" feedback controllers:

$$u(t) = \mu(t, x(t)), \quad t \in [0, t_f], \tag{3.1}$$

under which the state equation (2.1) admits a unique solution for every fixed $w \in \mathcal{H}_w$. First we present two useful definitions:

Definition 3.1: Saddle point. *A pair ($\mu^* \in \mathcal{M}, w^* \in \mathcal{H}_w$) constitutes a saddle-point solution for the linear-quadratic differential game formulated above, with a controller space $\mathcal{M}$, and with a fixed $\gamma > 0$, if*

$$L_\gamma(\mu^*, w) \le L_\gamma(\mu^*, w^*) \le L_\gamma(\mu, w^*), \quad \forall \mu \in \mathcal{M}, \quad w \in \mathcal{H}_w. \tag{3.2}$$

Definition 3.2: Strong time consistency [5]. *From the original game defined on the time interval $[0, t_f]$ construct a new game on a shorter time interval $[s, t_f]$, by setting $\mu_{[0,s)} = \beta_{[0,s)}$, $w_{[0,s)} = v_{[0,s)}$, where $\beta_{[0,s)}$ and $v_{[0,s)}$ are fixed but arbitrarily chosen, with the only restriction that $\beta_{[0,s)}$ should be compatible with the information structure that determines $\mathcal{M}$. Let ($\mu^* \in \mathcal{M}$, $w^* \in \mathcal{H}_w$) be a saddle-point solution for the original game. Then, $\mu^* \equiv \mu^*_{[0,t_f]}$ is "strongly time consistent", if $\mu^*_{[s,t_f]}$ is a part of the saddle-point solution of the new game (on the interval $[s, t_f]$), regardless of the choices for $\beta_{[0,s)}$, $v_{[0,s)}$, and for every s, $0 < s < t_f$.*

The following theorem now provides a complete characterization of the saddle-point solutions of the linear-quadratic differential game, with initial state x_0 not necessarily equal to *zero*.

Theorem 3.1. *Associated with the linear-quadratic differential game above, introduce the following matrix differential equation:*

$$\dot{Z} + A^T Z + ZA + Q - Z\left(BB^T - \frac{1}{\gamma^2}DD^T\right)Z = 0; \quad Z(t_f) = Q_f \tag{3.3}$$

(i) *Under $\mathcal{M}_{CL}$ or $\mathcal{M}_{FB}$, the differential game admits a* **unique strongly time consistent saddle-point** *policy μ^* for the minimizer if, and only if, the matrix differential equation (3.3) does not have a conjugate point in the interval $(0, t_f]$.*

(*ii*) *The unique strongly time consistent (feedback) saddle-point policy* μ_γ^* *is given by*

$$\mu_\gamma^*(t, x(t)) = -B(t)^T Z_\gamma(t) x(t), \quad t \in [0, t_f], \tag{3.4}$$

where $Z_\gamma(t)$, $t \in [0, t_f]$ *is the unique solution of (3.3). This feedback controller is in equilibrium with the unique open-loop disturbance:*

$$w^*(t) = \frac{1}{\gamma^2} D(t)^T Z_\gamma(t) x^*(t), \quad t \in [0, t_f], \tag{3.5}$$

where $x^*_{[0,t_f]}$ *is the unique saddle-point trajectory, given as the solution of the differential equation*

$$\dot{x}^* = \left(A - \left(BB^T - \frac{1}{\gamma^2} DD^T\right) Z_\gamma\right) x^*; \quad x^*(0) = x_0. \tag{3.6}$$

(*iii*) *The saddle-point value of the game, in terms of the initial state, is*

$$\min_{\mu \in \mathcal{M}_{CL}} \max_{w \in \mathcal{H}_w} L_\gamma(\mu, w) = \max_{w \in \mathcal{H}_w} \min_{\mu \in \mathcal{M}_{CL}} L_\gamma(\mu, w) = x_0^T Z_\gamma(0) x_0. \tag{3.7}$$

(*iv*) *If the disturbance is also allowed to depend on the state, then the above remains intact, save for the fact that (3.5) is no longer the only saddle-point solution for the disturbance. If, for example, we require the disturbance to depend on the current value of the state, then (3.5) is replaced by*

$$w^*(t) = \xi^*(t, x(t)) = \frac{1}{\gamma^2} D(t)^T Z_\gamma(t) x(t), \quad t \in [0, t_f]. \tag{3.5'}$$

(*v*) *If the solution of the matrix Riccati differential equation (3.3) has a conjugate point in the interval* $(0, t_f)$, *then the game does not admit a saddle point under any information pattern for the controller, and in this case there will exist a sequence* $w^{(n)}_{[0,t_f]}$, $n = 1, 2, \ldots$ *in* $\mathcal{H}_w$ *which will drive the value of* L_γ *arbitrarily large (positive).*

(*vi*) *If* $\hat{\mu} \in \mathcal{M}_{CL}$ *is some other saddle-point controller for the game (and there are generally uncountably many of them), then*

$$\hat{\mu}(t, x^*_{[0,t]}) = \mu^*(t, x^*(t)) \equiv -B(t)^T Z(t) x^*(t), \quad t \in [0, t_f]. \tag{3.8}$$

Proof. This result was presented in [6], where (v) was only implicitly given; for a more complete proof of (v) see [8]. Here we provide an alternate proof for the case when all components of the state vector are uniformly (over time) bounded away from zero. However, before going into the details of this proof, we should mention that in the absence of a conjugate point in the interval $[0, t_f]$, the objective function L_γ can equivalently be rewritten as

$$\begin{aligned} L_\gamma(u, w) = x_0^T Z_\gamma(0) x_0 \; &+ \int_0^{t_f} (u + B^T Z_\gamma x)^T (u + B^T Z_\gamma x) dt \\ &- \gamma^2 \int_0^{t_f} (w - \frac{1}{\gamma^2} D^T Z_\gamma x)^T (w - \frac{1}{\gamma^2} D^T Z_\gamma x) dt \end{aligned} \tag{3.9}$$

from which the saddle-point solution (3.4)-(3.5′) readily follows. In view of this, we also note that since the matrix Riccati equation admits a continuously differentiable solution in some neighborhood of the terminal time t_f, one can find a $\bar{t} < t_f$ such that the differential game defined on the interval $[\bar{t}, t_f]$ admits a saddle-point solution, as given in the Theorem.

Now returning to the proof of item (v) of the Theorem, suppose that even though there exists a conjugate point in the interval $(0, t_f)$, say t_s, the game admits a saddle point solution on $[0, t_f]$, say $(\hat{\mu}, \hat{\xi})$, with a corresponding state trajectory $\hat{x}_{[0,t_f]}$. Let $\epsilon > 0$ be sufficiently small, so that $t_s + \epsilon < t_f$. Consider a new game on the shorter interval $[t_s + \epsilon, t_f]$, obtained from the original one by setting

$$\mu_{[0,t_s+\epsilon)} = \hat{\mu}_{[0,t_s+\epsilon)}, \quad w_{[0,t_s+\epsilon)} = \hat{\xi}_{[0,t_s+\epsilon)}(\hat{x}_{[0,t_s+\epsilon)})$$

and with new initial state $\hat{x}(t_s + \epsilon)$. Since the original games does not have a conjugate point on the interval $[t_s + \epsilon, t_f]$, this new (truncated) game admits a saddle point, with value $\hat{x}(t_s+\epsilon)^T Z(t_s+\epsilon)\hat{x}(t_s+\epsilon) + constant$, where the first term becomes arbitrarily large as $\epsilon \downarrow 0$, since t_s is a conjugate point and we have assumed that the state trajectory is bounded away from zero. For the same truncated game we have $(\hat{\mu}_{[t_s+\epsilon,t_f]}, \hat{\xi}_{[t_s+\epsilon,t_f]})$ also a saddle-point pair (by hypothesis) for which the value is bounded above (as $\epsilon \downarrow 0$) since it is part of a saddle point for the entire game (beyond the conjugate point). This leads to a contradiction, by the ordered interchangeability property of multiple saddle points [6], since the saddle-point value has to be unique. ◇

We now study the same game as above, for the infinite-horizon case, where we assume that all coefficient matrices in (2.1) and (2.3) are time-invariant, $Q_f = 0$, and the interval of interest is $[0, \infty)$. Let $\gamma > 0$ be chosen such that the Riccati equation (3.3) does not admit a conjugate point for any $t_f > 0$. Denoting the unique nonnegative solution of (3.3) (with $Q_f = 0$) by $Z_\gamma(t_f; t)$, it is not difficult to see that this solution is in fact monotonically nondecreasing in t_f, for every fixed $t < t_f$; in other words $Z_\gamma(t'_f; t) \geq Z_\gamma(t''_f; t)$ for $t'_f > t''_f$. Then, clearly if

$$Z_\gamma^+ := \lim_{t_f \to \infty} Z_\gamma(t_f; t) \tag{3.10a}$$

exists, it should satisfy the generalized algebraic Riccati equation (GARE)

$$A^T Z + ZA + Q - Z\left(BB^T - \frac{1}{\gamma^2} DD^T\right) Z = 0. \tag{3.10b}$$

This equation may admit more than one nonnegative definite solution, and Z_γ^+ above is in fact the smallest such solution, in the sense that if K is another nonnegative definite solution of the GARE, $Z_\gamma^+ \leq K$ (see [12]). An important (and relevant) question here is whether the limit (as $t_f \to \infty$) of the saddle-point controller (3.4), i.e.,

$$\mu^\infty(x(t)) = -B^T Z_\gamma^+ x(t), \quad t \geq 0, \tag{3.11}$$

constitutes a saddle-point controller for the infinite-horizon LQ differential game? In mathematical terms, whether

$$\sup_{w \in \mathcal{H}_w} L_\gamma(\mu^\infty, w) = x_0^T Z_\gamma^+ x_0. \tag{3.12}$$

A complete answer to this question is provided in the following proposition, which was first presented in [12], with $\mathcal{M} = \mathcal{M}_{FB}$, but can readily be extended to the case $\mathcal{M} = \mathcal{M}_{CL}$ using the result of part (v) of Theorem 3.1.

Proposition 3.1. *Consider the infinite-horizon LQ differential game with (A, C) observable, where $C^T C = Q$.*
Then,

(i) The game has equal upper and lower values if, and only if, the GARE (3.10b) admits a positive definite solution. If Z_γ^+ is the smallest (in the matrix inequality sense) such solution, then the common value is given by (3.12).

(ii) The upper (minimax) value of the game is finite if, and only if, the upper and lower values are equal.

(iii) The controller μ^∞ given by (3.11) attains the upper value (in the sense of (3.12)).

(iv) Whenever the upper value is bounded, the maximizing solution in (3.12) is

$$w^\infty = \xi^\infty(x(t)) = \frac{1}{\gamma^2} D^T Z_\gamma^+ x(t), \quad t \geq 0, \tag{3.13}$$

but it is not necessarily true that

$$\inf_{\mu \in \mathcal{M}_{CL}} L_\gamma(\mu, \xi^\infty) = x_0^T Z_\gamma^+ x_0 ;$$

that is (μ^∞, ξ^∞) may not be in saddle-point equilibrium.

(v) Whenever the upper value is bounded, the two feedback matrices

$$A - \left(BB^T - \frac{1}{\gamma^2} DD^T \right) Z_\gamma^+ \tag{3.14a}$$

and

$$A - BB^T Z_\gamma^+ \tag{3.14b}$$

are stable.

◇

Note that implicit in the statement of part (iii) above is the property that if the upper value is not bounded in the class of feedback controllers, it will remain to be unbounded even if we enlarge the controller space also to include memory controllers (counterpart of this result in the finite horizon is part (ii) of Theorem 3.1.

Before moving on to applications of these two results, we quote one more result, from [6], which will prove useful in the next section.

Theorem 3.2. *For the linear-quadratic differential game of Theorem 3.1, but with open-loop information for the controller, there exists a saddle-point solution if, and only if, the solution $S(\cdot)$ of the following Riccati differential equation does not have a conjugate point in the interval $(0,t_f)$:*

$$\dot{S}+A^TS+SA+Q+\frac{1}{\gamma^2}SDD^TS=0, \quad S(t_f)=Q_f. \tag{3.15}$$

Under this condition, the conjugate-point condition of Theorem 3.1 is satisfied, and the game admits the unique open-loop saddle-point solution

$$\begin{aligned} u^{OL}(t) &= -B(t)^TZ(t)x^*(t), \\ w^{OL}(t) &= \frac{1}{\gamma^2}D(t)^TZ(t)x^*(t), \quad t\in[0,t_f]. \end{aligned} \tag{3.16}$$

where $x^*_{[0,t_f]}$ is the unique solution of (3.6). ◇

4 Applications To The Controller Design Problem

4.1 Perfect State Measurements

As a first application of Theorem 3.1, let us consider the standard disturbance attenuation problem with perfect state measurements. Let Γ be the set of all positive scalars γ for which the conjugate-point condition of Theorem 3.1 is satisfied. (It should be obvious that the condition is satisfied for γ sufficiently large,[1] and hence the set Γ is nonempty.) Let

$$\gamma^* := \inf\{\gamma : \gamma \in \Gamma\}. \tag{4.1}$$

For every $\gamma > \gamma^*$, we know that the associated game admits a saddle-point solution, with the minimizing controller given by (3.4). Unlike the case of the discrete-time problem (see [1], [2]), the limit of μ^*_γ as $\gamma \downarrow \gamma^*$ may not be well-defined, and hence we have to be content with suboptimal solutions. Toward a characterization of one such solution, let $\epsilon > 0$ be sufficiently small, and $\gamma_\epsilon := \gamma^* + \epsilon$. Then, it follows from (3.7), by letting $x_0 \to 0$ that

$$\|T_{\mu^*_{\gamma_\epsilon}}(w)\|_z^2 \le \gamma_\epsilon^2\|w\|_w^2, \quad \forall w\in\mathcal{H}_w.$$

In view of this inequality, we now have the following result, which can also be found in [11].

Theorem 4.1. *For the continuous-time disturbance attenuation problem (with $x_0 = 0$),*

$$\inf_{\mu\in\mathcal{M}_{CL}} \ll T_\mu \gg = \gamma^*$$

which is defined by (3.12). Furthermore, given any $\epsilon > 0$,

$$\ll T_{\mu^*_{\gamma_\epsilon}} \gg \le \gamma_\epsilon := \gamma^* + \epsilon,$$

[1] Because, for $\gamma = \infty$, we have the standard matrix Riccati equation of LQ control, which admits a nonnegative definite solution regardless of the length of the (finite) time interval.

where $\mu^*_{\gamma_\epsilon}$ is defined by (3.4) with $\gamma = \gamma_\epsilon$. ◇

This theorem finds a natural counterpart in the infinite-horizon case, where we now use Proposition 3.1. Let the condition of Proposition 3.1 on observability be satisfied, and Γ^∞ be the set of all positive scalars γ for which the GARE (3.10b) admits a positive definite solution. As the counterpart of (4.1), let

$$\gamma^*_\infty := \inf\{\gamma : \gamma \in \Gamma^\infty\} \tag{4.1'}$$

Then we have the following result, which can also be found in [9]:

Theorem 4.1′. *For the continuous-time infinite-horizon disturbance attenuation problem (with $x_0 = 0$),*

$$\inf_{\mu \in \mathcal{M}_{CL}} \ll T_\mu \gg = \gamma^*_\infty .$$

Given any $\epsilon > 0$,

$$\ll T_{\mu^\infty_{\gamma_\epsilon}} \gg \leq \gamma_\epsilon := \gamma^*_\infty + \epsilon$$

where $\mu^*_{\gamma_\epsilon}$ is defined by (3.11) with $\gamma = \gamma_\epsilon$. Furthermore, for ϵ sufficiently small, $\mu^*_{\gamma_\epsilon}$ leads to an input-output stable system. ◇

4.2 Delayed State Measurements

As a second application of Theorem 3.1, consider the disturbance attenuation problem where the controller has access to state with a delay of τ time units, i.e., permissible controllers are in the form:

$$\begin{aligned} u(t) &= \mu(t, x_{[0,t-\tau]}), \quad t \geq \tau \\ &= \mu(t, x_0), \qquad 0 \leq t < \tau. \end{aligned}$$

We denote this class of controllers by $\mathcal{M}_{\tau D}$. Now let $\gamma > \gamma^*$. Then, for the linear-quadratic differential game covered by Theorem 3.1, the unique strongly time consistent saddle-point controller in $\mathcal{M}_{\tau D}$ is given by (provided that a saddle point exists)

$$\mu^\circ_\gamma(t, x(t-\tau), u_{[t-\tau,t]}) = -B(t)^T Z_\gamma(t) x^*[t, x(t-\tau), u_{[t-\tau,t)}] \tag{4.2a}$$

where for each t, x^* is the solution of the differential equation:

$$\dot{x}^* = \left(A + \frac{1}{\gamma^2} DD^T Z_\gamma\right) x^* + Bu, \quad x^*(t-\tau) = x(t-\tau), \tag{4.2b}$$

i.e.,

$$x^*[t, x(t-\tau), u_{[t-\tau,t)}] = \Phi_\gamma(t, t-\tau) x(t-\tau) + \int_{t-\tau}^t \Phi_\gamma(t,s) B(s) u(s) ds, \tag{4.2c}$$

where Φ_γ is the state transition matrix function associated with $A + \frac{1}{\gamma^2} DD^T Z_\gamma$. We can now eliminate $u(\cdot)$ in (4.2c) by making use of (4.2a), which leads to the following (infinite-dimensional) compensator, where we take $x_0 = 0$:

$$\begin{aligned} \eta(t) &= \Phi_\gamma(t, t-\tau) x(t-\tau) - \int_{t-\tau}^t \Phi_\gamma(t,s) B(s) B(s)^T Z_\gamma(s) \eta(s) ds \\ \eta(s) &= 0 \text{ for } 0 \leq s < \tau. \end{aligned} \tag{4.3}$$

Then, the unique strongly time consistent saddle-point controller (4.2a) can be expressed as

$$\mu_\gamma^\circ(t,\eta(t)) = -B(t)^T Z_\gamma(t)\eta(t), \quad t \in [0,t_f]. \tag{4.4}$$

To determine the range of values of γ for which this will indeed be a saddle-point controller, in equilibrium with (3.5) (or 3.5′), we have to substitute this into L_γ, and require the resulting quadratic function to be strictly concave in its argument $w_{[0,t_f]}$. This involves a quadratic maximization problem in the Hilbert space $\mathcal{H}_w$, for which we do not give the details here. The result, though, is a threshold value $\gamma^\circ \geq \gamma^*$ (with the inequality generally being strict, for $\tau > 0$), so that for all $\gamma > \gamma^\circ$, (4.4) is in saddle-point equilibrium with (3.5). Furthermore, if $\gamma < \gamma^\circ$ (even if $\gamma > \gamma^*$), the linear-quadratic game with delayed state information does not admit a saddle point, which means (in this case) that the upper value is unbounded. (The proof of this result is similar to that of the discrete-time counterpart, given in [1]). A direct implication of this for the disturbance attenuation problem is summarized in the following theorem:

Theorem 4.2. *For the continuous-time disturbance attenuation problem with τ-delayed state information, there exists a positive scalar $\gamma^\circ \geq \gamma^*$ such that*

$$\inf_{\mu \in \mathcal{M}_{\tau D}} \ll T_\mu \gg = \gamma^\circ$$

and for any $\epsilon > 0$,

$$\ll T_{\mu^\circ_{\gamma^\circ_\epsilon}} \gg \leq \gamma^\circ_\epsilon := \gamma^\circ + \epsilon$$

where $\mu^\circ_{\gamma^\circ_\epsilon}$ is given by (4.4), with $\gamma = \gamma^\circ_\epsilon$. ◇

For the infinite-horizon (time-invariant) version, let the condition of Proposition 3.1 be satisfied, γ^*_∞ be as defined by (4.1′), and Z^+_γ again denote the smallest positive definite solution of (3.10b) for $\gamma > \gamma^*_\infty$. Then, (4.4) is replaced by

$$\mu_\gamma^\circ(\eta(t)) = -B^T Z^+_\gamma \eta(t), \quad t \geq 0 \tag{4.4′}$$

where $\eta(\cdot)$ is still given by (the time-invariant version of) (4.3). In this case it is again quite likely that the minimax attenuation level, γ°_∞, is strictly larger than γ^*_∞, for $\tau > 0$, so that (4.4′) with $\gamma = \gamma^\circ_\infty$ will indeed constitute an optimal disturbance attenuating controller for the system.

4.3 Sampled State Measurements

As a third application of Theorem 3.1, we consider the disturbance attenuation problem with sampled state information. Let $t_0, t_1, \ldots, t_{K-1}$ denote K (possibly nonuniform) sampling times, such that $0 = t_0 < t_1 < t_2 \ldots < t_{K-1} < t_f$. Permissible controls are in the form

$$u(t) = \mu(t, x_{[0,k]}), \quad t_k \leq t < t_{k+1}$$

where $x_{[0,k]}$ denotes the sequence $x(t_0), x(t_1), \ldots, x(t_k)$. Let us denote the space of such controllers by $\mathcal{M}_S$. The unique strongly time consistent solution to this differential game readily follows from Theorem 3.1, and is given in the following proposition (see also [3]).

Proposition 4.1. *Let $\gamma > \gamma^*$. For the LQ differential game of Theorem 3.1, but with sampled state information as above,*

(i) There exists a unique strongly time consistent saddle-point controller $\mu_\gamma^S \in \mathcal{M}_S$, if, and only if, the following K matrix Riccati differential equations do not have conjugate points in the given intervals:

$$\dot{S}_k + A^T S_k + S_k A + Q + \frac{1}{\gamma^2} S_k D D^T S_k = 0;$$
$$S_k(t_{k+1}) = Z_\gamma(t_{k+1}); \quad t_k \le t < t_{k+1}, \quad k = K-1, K-2, \ldots, 0. \tag{4.5}$$

(ii) Under the condition above, the strongly time consistent saddle-point controller is

$$\mu_\gamma^S(t, x(t_k)) = -B(t)^T Z_\gamma(t)\phi_\gamma(t, t_k)x(t_k), \quad t_k \le t < t_{k+1} \tag{4.6}$$

where ϕ_γ is the state transition matrix associated with the matrix

$$F_\gamma := A - BB^T Z_\gamma + \frac{1}{\gamma^2} DD^T Z_\gamma. \tag{4.7}$$

(iii) If the condition of (i) is not satisfied, then the upper value of the game is unbounded.

Proof. The result follows from Theorem 3.1, in view of Theorem 3.2, if we note that the original game can be decomposed into a sequence of open-loop games, which have to be solved recursively, and in retrograde time. Because of interchangeability of multiple saddle points, the boundary conditions for the S_k's are determined by the values of $Z_\gamma(\cdot)$ at the sampling times. Note that $Z_\gamma(t)$, $0 \le t \le t_f$, is well defined in this case, because $\gamma > \gamma^*$. ◇

Let Γ^S be the set of all $\gamma > \gamma^*$ which satisfy the condition of Proposition 4.1. Let

$$\gamma^S := \inf\{\gamma : \gamma \in \Gamma^S\}. \tag{4.8}$$

Then clearly (by following the earlier arguments) γ^S is the minimax disturbance attenuation level for the sampled information case. An interesting feature here is that if $\gamma^S > \gamma^*$, then the limit of μ_γ^S as $\gamma \downarrow \gamma^S$ is well defined. Hence,

Theorem 4.3. *For the continuous-time disturbance attenuation problem with sampled state information, let γ^S be as defined above. Then,*

$$\inf_{\mu \in \mathcal{M}_S} \ll T_\mu \gg = \gamma^S.$$

Furthermore, if $\gamma^S > \gamma^*$, the limiting controller

$$\mu_{\gamma^S}^S = \lim_{\epsilon \downarrow 0} \mu_{\gamma^S + \epsilon}^S$$

exists. ◇

For the infinite-horizon time-invariant version, we have the boundary condition on (4.5) replaced by the constant matrix Z_γ^+, and (4.6) simplified to

$$\mu_\gamma^{\infty S}(t, x(t_k)) = -B^T Z_\gamma^+ \exp\{F_\gamma(t - t_k)\}x(t_k), \quad t_k \le t < t_{k+1}. \tag{4.9}$$

where the constant matrix F_γ is given by (4.7). Clearly, the minimax attenuation level, γ_∞^S (as the counterpart of (4.8)), is determined by the conjugate point condition of the set of equations (4.5), and in the case of uniform sampling ($t_{k+1} - t_k =$ constant) one needs to check for the conjugate point of only one of these, since the boundary conditions are all the same. If sampling is not uniform, then γ_∞^S is determined by the conjugate point condition of the Riccati equation defined on the longest sampling interval, and hence still only one equation has to be solved. In all cases though, it is natural to expect $\gamma_\infty^S > \gamma_\infty^*$, so that a well-defined limiting solution exists, given by $\mu_{\gamma_\infty^S}^{\infty S}$.[2] The following example illustrates some of these points.

An Example. Consider the scalar plant

$$\dot{x} = u + w, \quad x(0) = 0, \quad t \ge 0$$

along with the performance index

$$\int_0^{t_f} ([x(t)]^2 + [u(t)]^2)dt.$$

Under perfect state measurements, the minimax attenuation level is determined by the conjugate point condition of the Riccati equation

$$\dot{Z} + 1 - \left(1 - \frac{1}{\gamma^2}\right) Z^2 = 0; \quad Z(t_f) = 0 \tag{$*$}$$

whose unique solution, for $\gamma^2 < 1$, is

$$\begin{aligned} Z_\gamma(t) &= \tfrac{1}{m}\tan[m(t_f - t)], \quad 0 \le t \le t_f \\ m &:= \sqrt{(1-\gamma^2)/\gamma^2} \end{aligned}$$

provided that $0 < mt_f < \frac{\pi}{2}$. Hence, the minimax disturbance attenuation level is

$$\gamma^* = 2t_f/\sqrt{\pi^2 + 4t_f^2}.$$

Given any $\epsilon > 0$, the controller

$$\mu_{\gamma_\epsilon}^*(t, x(t)) = -Z_{\gamma_\epsilon}(t)x(t), \quad 0 \le t \le t_f$$

achieves an attenuation level that is no higher than $\gamma_\epsilon := \gamma^* + \epsilon$. Note that this becomes a "high-gain" controller as $\epsilon \downarrow 0$.

[2] For a further elaboration on this point see [4].

To obtain some explicit (numerical) results for the sampled state measurement scheme, let us take $t_f = \pi$, which leads to $\gamma^* = 2/\sqrt{5}$. Now let there be a single sample point, at $t = (1-\lambda)\pi$, where $0 < \lambda < 1$. Anticipating an attenuation level $\gamma^S > 1$, we first solve (*) to obtain

$$Z_\gamma(t) = \frac{1}{\sigma}\tanh[\sigma(t_f - t)] \equiv \frac{1}{\sigma}\tanh[\sigma(\pi - t)]$$

$$\sigma := \sqrt{(\gamma^2-1)/\gamma^2}$$

which provides the boundary condition to (4.5) at the sampling point $t = (1-\lambda)\pi$. The conjugate point conditions in the intervals $[(1-\lambda)\pi, \pi]$ and $[0, (1-\lambda)\pi]$ dictate, respectively, the following inequalities:

i) $\lambda\pi < \frac{\pi}{2}\gamma$.

ii) $(1-\lambda)\pi + \gamma \arctan\left[\frac{1}{\sqrt{\gamma^2-1}}(\tanh[\lambda\pi\sqrt{\gamma^2-1}/\gamma])\right] < \frac{\pi}{2}\gamma$

for $\gamma > 1$, which indeed turns out to be the case because it can be shown that for $\gamma < 1$, regardless of the value of $\lambda \in (0,1)$, either i) or ii) is violated. At this point we can raise the issue of the "optimal choice" of the sampling time parameter λ, so that γ^S is minimized. Because of the monotonicity property of these conditions, the optimum value of λ is one under which there exists a γ which makes both i) and ii) equalities (simultaneously). Some manipulations bring this condition down to one of existence of a $\lambda \in (0,1)$ to the trigonometric equation

$$\tan\left[\frac{(2\lambda-1)}{\lambda}\cdot\frac{\pi}{2}\right] = \frac{1}{4\lambda^2-1}\tanh\left[\frac{\pi}{2}\sqrt{4\lambda^2-1}\right]$$

which admits the unique solution

$$\lambda \cong 0.6765.$$

This indicates (as to be expected) that uniform sampling does not lead to the best attenuation level in finite horizon problems.

Now, for the infinite horizon version, the minimax attenuation level is $\gamma_\infty^+ = 1$, and for $\gamma > \gamma_\infty^*$ the unique positive solution of the GARE (3.10b) is

$$Z_\gamma^+ = \gamma/\sqrt{\gamma^2-1}, \quad \gamma > 1.$$

Under uniform sampling (which is clearly optimal in the infinite-horizon case), and with a sampling time of t_s, the conjugate-point condition of

$$\dot{S} + 1 + \frac{1}{\gamma^2}S^2 = 0; \quad S(t_s) = Z_\gamma^+$$

(i.e., requiring $S(t)$ to be bounded for $t \in [0, t_s]$), leads to the inequality

$$t_s + \gamma\arctan[1/\sqrt{\gamma^2-1}] < \frac{\pi}{2}\gamma. \tag{**}$$

Hence, for a given t_s, the minimax attenuation level is the smallest positive value of γ that solves (**) as an equality. Numerical experimentation has in fact shown that the solution to (**) is actually unique, so that we do not have to search for the smallest positive root. Furthermore, this unique solution converges to γ^* (from above) as $t_s \downarrow 0$. We give below the unique solution of (**) for three different choices of t_s:

$$t_s = \pi \Rightarrow \gamma_\infty^S \cong 2.653; \quad t_s = \frac{\pi}{2} \Rightarrow \gamma_\infty^S \cong 1.682; \quad t_s = \frac{\pi}{4} \Rightarrow \gamma_\infty^S \cong 1.241.$$

Note that γ_∞^S is a monotonically increasing function of t_s, which is in fact a general property of the solution (in infinite-horizon problems) as already mentioned. Note also that since $\gamma_\infty^S > 1 = \gamma_\infty^+$ in each case, the corresponding optimal controllers are asymptotically well defined as $\gamma \downarrow \gamma_\infty^S$, with the limits given (using (4.9)) by

$$\mu_\gamma^{\infty S}(t, x(t_k)) = -\exp\left\{-\sqrt{\frac{\gamma-1}{\gamma+1}}(t-t_k)\right\} x(t_k)\,, \quad t_k \leq t < t_k + t_s,$$

where $\gamma = \gamma_\infty^S$.

4.4 A Partial State Measurement Scheme

As a problem with a particular partial state measurement scheme, we consider the continuous-time counterpart of the discrete-time problem considered in subsection 4.3 of [1]. We let the first j components of x (denoted y) be measured perfectly, and the remaining components (denoted e) not measured. Compatible with this, the system matrices are partitioned as

$$A =: \begin{pmatrix} A^{yy} & A^{ye} \\ A^{ey} & A^{ee} \end{pmatrix} =: \begin{pmatrix} A^y \\ A^e \end{pmatrix}; \quad B =: \begin{pmatrix} B^y \\ B^e \end{pmatrix}; \quad D =: \begin{pmatrix} D^y \\ 0 \end{pmatrix}. \tag{4.10}$$

The special structure of D reflects the structural assumption that only the measurable components of the state are directly affected by the disturbance. Now, a combination of Theorem 3.1, and the reasoning used in subsection 4.3 of [1] leads to the result that an ϵ-optimal controller for this problem is given by

$$\mu^\square(t, y(t), \eta(t)) = -B(t)^T Z_{\gamma_\epsilon}(t) \begin{pmatrix} y(t) \\ \eta(t) \end{pmatrix}, \quad t \geq 0 \tag{4.11a}$$

where η is generated by the j-dimensional compensator:

$$\dot{\eta} = (A^e - B^e B^T Z_{\gamma_\epsilon}) \begin{pmatrix} y(t) \\ \eta(t) \end{pmatrix}; \quad \eta(0) = 0 \tag{4.11b}$$

with $\gamma_\epsilon := \gamma^* + \epsilon$, and $\epsilon > 0$ being arbitrarily small.

Extension of this result to the most general partial state measurement scheme is considerably involved, and will not be presented here; see [13], [14] for some results on this extension, which use a "completing the squares" argument.

4.5 Unknown Nonzero Initial State

If the initial state of the system is completely unknown (instead of being *zero*), one possible approach is to treat it also as a disturbance. One way of accommodating this generalization in the formulation of Section 2 would be to replace (2.4b) by ([10], [14])

$$\ll T_\mu \gg := \sup_{w \in \mathcal{H}_w} \|T_\mu(w)\|_z \big/ \{\|w\|_w^2 + x_0^T Q_0 x_0\}^{1/2} \tag{4.12}$$

where $x(0) = x_0$, and Q_0 is an appropriately chosen nonnegative definite weighting matrix. Now, using Theorem 3.1, and the argument that led to Theorem 4.1, we arrive at the relationship

$$\max_{w,x_0}\{\|T_{\mu_\gamma^*}(w)\|_z^2 - \gamma^2(\|w\|_w^2 + x_0^T Q_0 x_0)\} = \max_{x_0}\{x_0^T(Z_\gamma(0) - \gamma^2 Q_0)x_0\} \tag{4.13}$$

where we again look for the smallest value of $\gamma > 0$ for which the right-hand side is bounded (actually *zero*). In view of this, the set Γ defined in Section 4.1 is now the set of all positive scalars γ for which the conjugate-point condition of Theorem 3.1 is satisfied *and* the matrix inequality

$$\gamma^2 Q_0 - Z_\gamma(0) \geq 0 \quad (\text{nonnegative} - \text{definite}) \tag{4.14}$$

holds. With this new set Γ, the optimum performance γ^* is again as defined in (4.1), from which the natural counterpart of Theorem 4.1 follows. We should note that it is quite possible (depending on the choice of Q_0) that (4.14) will not impose any additional constraint on γ^*, in particular if $Z_\gamma(\cdot)$ has finite escape inside (rather than on the boundary of) the time interval (i.e., if the conjugate point is of the *even* type [8]).

For the infinite-time horizon, one can analogously introduce the counterpart of Γ^∞, and thus of (4.1′), thereby arriving at the counterpart of Theorem 4.1′ for the problem with unknown initial state. The same extension also applies to the sampled data information case.

5 Concluding Remarks

Perhaps the most natural extension of the results presented in this paper would be to the class of infinite-dimensional state space models in a Hilbert space setting. Interpreting (2.1) as a linear operator equation, and (2.3) as an inner product in the appropriate (Hilbert) space, Theorem 3.1 remains intact with the "transposes" translated into "adjoints" of the corresponding linear operators, and the matrix differential equation (3.3) replaced by the corresponding operator-valued Riccati equation (see [7]). Again the minimax disturbance attenuation level is that value of γ beyond which this operator-valued Riccati equation ceases to have a bounded solution. Different applications discussed in Section 4 again have natural counterparts here, at least at the conceptual abstract level, but of course one has to work out the details of the derivation for specific models, such as distributed parameter systems or systems described by delay-differential equations.

Another fruitful research direction would be to extend these results to nonlinear dynamics and/or nonquadratic performance indices. We already have the counterparts of

Theorem 3.1 and 3.2 for such systems (see [6]), so that optimal controllers under other (than feedback) information patterns can be obtained as particular representations of strongly time consistent feedback saddle-point controllers. Precise delineation of the underlying conditions again require further research.

Acknowledgement. This work was supported in part by the Joint Services Electronics Program under Grant N00014-K-90-J-1270.

References

[1] T. Başar. Game theory and H^∞-optimal control: The discrete-time case. *Proceedings of the 1990 International Conference on New Trends in Communication, Control and Signal Processing*, Ankara, Turkey, July 1990.

[2] T. Başar. A dynamic games approach to controller design: Disturbance rejection in discrete time. In *Proceedings of the 29th IEEE Conf. Decision and Control*, pages 407–414, Tampa, FL, December 13-15, 1989.

[3] T. Başar. Generalized Riccati equations in dynamic games. In S. Bittanti A. Laub, and J.C. Willems, editors, *Riccati Equation in Control Systems and Signals.* Springer-Verlag, December 1990.

[4] T. Başar. Optimum H^∞ designs under sampled state measurements. Submitted for publication, 1990.

[5] T. Başar. Time consistency and robustness of equilibria in noncooperative dynamic games. In F. Van der Ploeg and A. de Zeeuw, editors, *Dynamic Policy Games in Economics*, pages 9–54. North Holland, 1989.

[6] T. Başar and G. J. Olsder. *Dynamic Noncooperative Game Theory.* Academic Press, London/New York, 1982.

[7] A. Bensoussan. Saddle points of convex concave functionals. In H. W. Kuhn and G. P. Szegö , editors, *Differential Games and Related Topics*, pages 177–200. North-Holland, Amsterdam, 1971.

[8] P. Bernhard. Linear-quadratic two-person zero-sum differential games: Necessary and sufficient conditions. *Journal of Optimization Theory & Applications*, 27: 51–69, 1979.

[9] J. Doyle, K. Glover, P.P. Khargonekar, and B. Francis. State-space solutions to standard H_2 and H_∞ control problems. *IEEE Transactions on Automatic Control*, AC-34(8):831–847, 1989.

[10] P.P. Khargonekar. State-space H_∞ optimal control theory. Preprint, November 1989.

[11] D. J. N. Limebeer, B. D. O. Anderson, P. Khargonekar, and M. Green. A game theoretic approach to H_∞ control for time varying systems. *Proceedings of the MTNS*, Amsterdam, 1989.

[12] E. F. Mageirou and Y. C. Ho. Decentralized stabilization via game theoretic methods. *Automatica*, 13:393–399, 1977.

[13] K. Uchida and M. Fujita. On the central controller: characterizations via differential games and LEQG control problems. *Systems & Control Letters*, 13:9-13, 1989.

[14] K. Uchida and M. Fujita. Controllers attenuating disturbances and initial uncertainties for time-varying systems. Presented at this Symposium, August 1990.

CONTROLLERS ATTENUATING DISTURBANCES AND INITIAL-UNCERTAINTIES FOR TIME-VARYING SYSTEMS

Kenko Uchida† and Masayuki Fujita‡

†Dept. of Electrical Engineering, Waseda University
Okubo 3-4-1, Shinjuku, Tokyo 169, Japan
‡Dept. of Electrical and Computer Engineering, Kanazawa University
Kodatsuno 2-40-20, Kanazawa 920, Japan

Abstract : A new type of H^{∞} control problem, which considers a mixed attenuation of disturbances and initial-uncertainties for time-varying systems, is formulated and solved; the approach is based only on completing the square.

1. Introduction

The recent progresses in the H^{∞} control theory have found new state space formulas for H^{∞} controls based on indefinite Riccati equations (for state feedback cases, [3][7]; for output feedback cases, [1][2][4][6]). These results suggest possibilities of finite horizon H^{∞} controls for time-varying systems, which have been established by Tadmor [8] using the maximum principle and by Limebeer et al. [5] using the LQ differential game theory and the plant inverse method [1].

This paper directs its attention to finite horizon output feedback controls for time-varying systems. H^{∞} controls are originally such that they optimize input-output characteristics (transfer functions) on some criteria in the frequency domain, and they are not concerned with zero-input responses; in fact, initial states of the systems are assumed to be zero in the above state space solutions to infinite or finite horizon H^{∞} control problems. In the infinite horizon case for time-invariant systems, the zero-input response, even if it appears, may be no matter since it decays as time tends to infinity. In the finite horizon case for time-varying systems, however, the zero-input response, if it appears, may take a large part of the total response and may not be negligible. Moreover, in the framework of output feedback controls, it is natural to assume that the initial state is unknown. In view of these, we pose a new finite horizon H^{∞} output feedback control problem considering a mixed attenuation of disturbances and initial-uncertainties for time-varying systems, and we solve the problem by using two indefinite Riccati differential equations.

The key point of our approach to the problem is to use a special argument of completing the square [9] for the output feedback case (with no use of the plant inverse methods used in [5]), which makes it possible to treat the mixed attenuation of disturbances and initial-uncertainties and makes our approach rather simple compared with those of [5][8] to standard H^∞ control problems.

Notations $L^2(t_0,t_1;R^k)$ is the space of square integrable functions f such that $f:[t_0,t_1] \to R^k$. $\|\cdot\|_2$ is the L^2-norm in $L^2(t_0,t_1;R^k)$. $\|\cdot\|$ is the Euclidean norm in R^k. I is the identity matrix with appropriate dimension. $(\cdot)'$ denotes the transpose of vector or matrix. $\rho(X)$ is the spectral radius of matrix X.

2. Formulation and the result

We consider the linear time-varying system which is defined on the time interval $[t_0,t_1]$ and described by

$$\frac{d}{dt}x(t) = A(t)x(t) + B(t)u(t) + D(t)v(t), \qquad x(t_0) = x_0 \tag{1a}$$

$$y(t) = C(t)x(t) + w(t) \tag{1b}$$

$$z(t) = F(t)x(t) \tag{1c}$$

where $x(t)\in R^n$ is the state and $x_0\in R^n$ is the initial state; $u(t)\in R^r$ is the control input; $y(t)\in R^m$ is the observed output; $g(t):=(z(t)'\ u(t)')'\in R^{q+r}$ is the controlled output; $h(t):=(v(t)'\ w(t)')'\in R^{p+m}$ is the disturbance. $A(t)$, $B(t)$, $C(t)$, $D(t)$ and $F(t)$ are matrices of appropriate dimensions whose elements are continuous functions of time; each elements of $v(t)$ and $w(t)$ are square integrable functions of time.

For the system (1), every admissible control is given by an admissible feedback operator Φ to be the form

$$u(t) = [\Phi\, y](t) \tag{2}$$

An admissible feedback operator is a linear causal operator from $L^2(t_0,t_1;R^m)$ to $L^2(t_0,t_1;R^r)$ described by a linear time-varying system

$$u(t) = J(t)x_e(t) + K(t)y(t)$$

$$\frac{d}{dt}x_e(t) = G(t)x_e(t) + H(t)y(t), \qquad x_e(t_0) = 0$$

where $J(t)$, $K(t)$, $G(t)$ and $H(t)$ are matrices of appropriate dimensions whose elements are continuous functions of time.

We are concerned with admissible controllers which attenuate disturbances and initial-uncertainties in the way that, for a positive definite matrix Γ, $g=(z' \ u')'$ satisfies

$$\| g \|_2^2 < \| h \|_2^2 + x_0{}'\Gamma x_0 \tag{3}$$

for all $h=(v' \ w')'$ in $L^2(t_0,t_1;R^{p+m})$ and all x_0 in R^n such that $(u \ w \ x_0)\neq 0$. We call such a controller the **disturbance and initial-uncertainty attenuation (DIA) controller.** Our problems are first to judge whether there exists a DIA controller and second to characterize all DIA controllers.

In order to state the results, let us introduce the following conditions:

(A_1) There exists a solution $M(t)$ defined on $[t_0,t_1]$ to the Riccati differential equation

$$\begin{aligned} -\frac{d}{dt}M(t) &= M(t)A(t) + A(t)'M(t) + F(t)'F(t) \\ &\quad - M(t)(B(t)B(t)' - D(t)D(t)')M(t), \qquad M(t_1) = 0; \end{aligned} \tag{4}$$

(A_2) There exists a solution $P(t)$ defined on $[t_0,t_1]$ to the Riccati differential equation

$$\begin{aligned} \frac{d}{dt}P(t) &= A(t)P(t) + P(t)A(t)' + D(t)D(t)' \\ &\quad - P(t)(C(t)'C(t) - F(t)'F(t))P(t), \qquad P(t_0) = \Gamma^{-1}; \end{aligned} \tag{5}$$

(A_3) $\rho(P(t)M(t)) < 1$ for all t in $[t_0,t_1]$.

Theorem

(I) There exists a DIA controller if and only if the conditions (A_1), (A_2) and (A_3) are satisfied.

(II) If the conditions (A_1), (A_2) and (A_3) are satisfied, the set of all DIA controllers is parameterized by

$$u(t) = u^o(t) + [\Psi(y - y^o)](t) \tag{6}$$

where $u^o(t)$ and $y^o(t)$ are given as

$$u^o(t) = -B(t)'S(t)x_e(t), \quad y^o(t) = C(t)(I + P(t)S(t))x_e(t)$$

$$\frac{d}{dt}x_e(t) = (A(t) - B(t)B(t)'S(t) - P(t)C(t)'C(t) + P(t)F(t)'F(t))x_e(t)$$
$$+ P(t)C(t)'y(t) + B(t)[\Psi(y - y^o)](t), \quad x_e(t_0) = 0$$

$S(t)$ is given as $S(t)=M(t)(I-P(t)M(t))^{-1}$, and Ψ is an arbitrary admissible feedback operator from $L^2(t_0,t_1;R^m)$ to $L^2(t_0,t_1;R^r)$ such that

$$\| \Psi f \|_2^2 < \| f \|_2^2 \tag{7}$$

for all nonzero f in $L^2(t_0,t_1;R^m)$.

3. Proof of the part (II): Sufficiency and parametrization

Before turning to the proof we observe positive definiteness of the solution to the indefinite Riccati differential equation (5).

Lemma

The solution $P(t)$ to the Riccati differential equation (5) is positive definite for all t in the interval that the solution exists.

Proof of Lemma Let $\phi(t,s)$ be the transition matrix associated with the system matrix $\{A(t) - (1/2)P(t)(C(t)'C(t) - F(t)'F(t))\}$. Then, the solution $P(t)$ can be expressed as

$$P(t) = \phi(t,t_0)\Gamma^{-1}\phi(t,t_0)' + \int_{t_0}^{t}\phi(t,s)D(s)D(s)'\phi(t,s)'\,ds,$$

which implies that $P(t)$ is positive definite, since Γ^{-1} is positive definite.

Now we prove the part (II) of the theorem; the main idea of argument is the same behind the authors' previous work [9] and the parametrization technique used here is hinted by that used by Limebeer et al. [5] in the full information case. The proof will be divided into three parts. For notational simplicity, we will often suppress the time dependence of vectors and matrices.

1) From (4) and (5), we can show that $S(t) = M(t)(I - P(t)M(t))^{-1}$ satisfies

$$-\frac{d}{dt}S = S(A + PF'F) + (A + PF'F)'S + F'F - S(BB' - PC'CP)S \tag{8}$$

and $S(t_1) = 0$. It follows from Lemma and (5) that $P(t)^{-1}$ exists and satisfies

$$-\frac{d}{dt}P^{-1} = P^{-1}A + A'P^{-1} + P^{-1}DD'P^{-1} - C'C + F'F \tag{9}$$

and $P(t_0)^{-1} = \Gamma$. For an admissible control $u = \Phi y$, consider the functional

$$V(t) := x_e(t)'S(t)x_e(t) + (x(t) - x_e(t))'P(t)^{-1}(x(t) - x_e(t)) \tag{10}$$

where $x(t)$ is given by (1a) and $x_e(t)$ is given by

$$\frac{d}{dt}x_e = Ax_e + Bu + PC'(y - Cx_e) + PF'Fx_e, \quad x_e(t_0) = 0 \tag{11}$$

Differentiating both sides of (10) with respect to t, inserting (1), (8), (9) and (11) into the right hand side, and integrating both sides with respect to t over $[t_0, t_1]$, we obtain the identity

$$\begin{aligned} V(t_1) - V(t_0) &= -\| z \|_2^2 - \| u \|_2^2 + \| v \|_2^2 + \| w \|_2^2 \\ &\quad + \| u - u_{\min} \|_2^2 - \| v - v_{\max} \|_2^2 - \| w - w_{\max} \|_2^2 \end{aligned}$$

where

$$u_{\min}(t) := -B(t)'S(t)x_e(t) \tag{12a}$$

$$v_{\max}(t) := D(t)'P(t)^{-1}(x(t) - x_e(t)) \tag{12b}$$

$$w_{\max}(t) := -C(t)(x(t) - x_e(t)) + C(t)P(t)S(t)x_e(t). \tag{12c}$$

Since $g = (z' \; u')'$, $h = (v' \; w')'$, $S(t_1) = 0$, $P(t_0)^{-1} = \Gamma$ and $x_e(t_0) = 0$, it follows from the above identity and the definition (10) that

$$\begin{aligned} &\| g \|_2^2 - \| h \|_2^2 - x_0'\Gamma x_0 \\ &= \| u - u_{\min} \|_2^2 - \| v - v_{\max} \|_2^2 - \| w - w_{\max} \|_2^2 \\ &- (x(t_1) - x_e(t_1))'P(t_1)^{-1}(x(t_1) - x_e(t_1)). \end{aligned} \tag{13}$$

2) Next we show that any admissible control $u = \Phi y$ has a corresponding expression of the form

$$u(t) = u_{\min}(t) + [\Psi(y - y_{\max})](t) \tag{14}$$

where Ψ is an admissible feedback operator, $u_{\min}$ is given by (12a), and

$$y_{\max}(t) := C(t)x(t) + w_{\max}(t) = C(t)(I + P(t)S(t))x_e(t). \tag{15}$$

Substituting $u = \Phi y$ into (11), using (15), and rearranging terms yield

$$\frac{d}{dt}x_e = (A - PC'C + PF'F)x_e + B\Phi(y - y_{\max}) + B\Phi C(I + PS)x_e$$
$$+ PC'(y - y_{\max}) + PC'C(I + PS)x_e,$$

and moreover using (12a) and (15) yields

$$u(t) = u_{\min}(t) + [\Phi(y - y_{\max})](t) + [\Phi C(I + PS)x_e](t) + B(t)'S(t)x_e(t).$$

These assure the existence of the admissible feedback operator Ψ in the expression (14). Thus, without loss of generality, we can take any admissible control to be the form (14).

3) From (1b) and (15), we see

$$y(t) - y_{\max}(t) = w(t) - w_{\max}(t). \tag{16}$$

Substituting (14) with (16) into (13), we have

$$\begin{aligned} &\| g \|_2^2 - \| h \|_2^2 - x_0{}'\Gamma x_0 \\ &= \| \Psi(w - w_{\max}) \|_2^2 - \| w - w_{\max} \|_2^2 - \| v - v_{\max} \|_2^2 \\ &- (x(t_1) - x_e(t_1))'P(t_1)^{-1}(x(t_1) - x_e(t_1)). \end{aligned} \tag{17}$$

If Ψ satisfies (7), the right hand side of (17) is non-positive, since $P(t_1)^{-1}$ is positive definite; in this case, the right hand side of (17) is equal to zero (i.e., $v = v_{\max}$, $w = w_{\max}$ and $x(t_1) = x_e(t_1)$), only if $(v\ w\ x_0) = 0$. Thus we obtain the attenuation property (3).

Conversely, if (3) holds for a given admissible control (14) with an admissible operator Ψ, it follows from (17) that

$$\begin{aligned} &\| \Psi(w - w_{\max}) \|_2^2 - \| w - w_{\max} \|_2^2 \\ &< \| v - v_{\max} \|_2^2 + (x(t_1) - x_e(t_1))'P(t_1)^{-1}(x(t_1) - x_e(t_1)) \end{aligned}$$

for all $(v\ w)$ in $L^2(t_0,t_1;R^{p+m})$ and all x_0 in R^n such that $(v\ w\ x_0) \neq 0$; if we choose $(v\ w)$ such that $v = v_{\max}$ and $w = f + w_{\max}$ for nonzero f in $L^2(t_0,t_1;R^m)$, we have

$$\| \Psi f \|_2^2 - \| f \|_2^2 < (x(t_1) - x_e(t_1))'P(t_1)^{-1}(x(t_1) - x_e(t_1)) \tag{18}$$

for all nonzero f in $L^2(t_0,t_1;R^m)$ and all x_0 in R^n. Here x_e and $x - x_e$ obey

$$\frac{d}{dt}x_e = (A - BB'S + PC'CPS + PF'F)x_e + B\Psi f + PC'f$$
$$\frac{d}{dt}(x - x_e) = (A + DD'P^{-1})(x - x_e) - (PC'CPS + PF'F)x_e - PC'f,$$

which implies that, for each f, we can choose x_0 such that $x_e(t_0)=0$ and $x(t_1)-x_e(t_1)=0$. Thus, by choosing x_0 in such a way, the right hand side of (18) can be taken to be zero for each nonzero f, and the inequality (7) follows.

Finally, if we define $u^o(t):=u_{min}(t)$ and $y^o(t):=y_{max}(t)$ for any admissible control (14) satisfying (7), we complete the proof.

4. Proof of the part (I): Necessity

We prove here the part (I) of the theorem; only necessity will be proved, since sufficiency has already been established in the section 3.

Before turning to the proof, let us derive two identities. Using the solution $P(t)$ to the Riccati differential equation (5), define the functional

$$U(t) := (x(t) - x_e(t))'P(t)^{-1}(x(t) - x_e(t)) \tag{19}$$

where $x(t)$ is given by (1a) and $x_e(t)$ is given by (11). Differentiating both sides of (19) with respect to t, inserting (1), (9) and (11) into the right hand side, and integrating both sides with respect to t over $[T_1, T_2]$, we obtain the identity

$$\begin{aligned}\int_{T_1}^{T_2}(\| g \|^2 - \| h \|^2)dt &= - U(T_2) + U(T_1)\\ &+ \int_{T_1}^{T_2}(\| Fx_e \|^2 + \| u \|^2 - \| v - v_{max} \|^2 - \| w - w_{o,max} \|^2)dt\end{aligned} \tag{20}$$

where v_{max} is given by (12b) and $w_{o,max}$ is defined by

$$w_{o,max}(t) := - C(t)(x(t) - x_e(t)). \tag{21}$$

Consider again the functional (10). Repeating the same completing the square argument as before and finally integrating over $[T_3, T_4]$, we obtain the identity

$$\begin{aligned}\int_{T_3}^{T_4}(\| g \|^2 - \| h \|^2)dt &= - V(T_4) + V(T_3)\\ &+ \int_{T_3}^{T_4}(\| u - u_{min} \|^2 - \| v - v_{max} \|^2 - \| w - w_{max} \|^2)dt\end{aligned} \tag{22}$$

where u_{min}, v_{max} and w_{max} are given by (12).

Let us start the proof of necessity. We first show that, if there exists a DIA control, the condition (A_2) must be satisfied. Suppose that the condition (A_2) does not hold; then, we can find the smallest time t^* in (t_0,t_1) such that (5) has a solution $P(t)$ on $[t_0,t^*)$. Note that there exists a nonzero vector α such that $\lim_{T\to t^*} P(t)^{-1}\alpha = 0$. Now, for any fixed admissible control u, choosing nonzero $(v\ w\ x_0)$ such that

$$(v(t)\ \ w(t)) = \begin{cases} (\ 0 \quad\ 0\ ), & T \le t \le t_1 \\ (v_{\max}(t)\ \ w_{o,max}(t)), & t_0 \le t < T \end{cases} \quad (T < t^*)$$

and x_0 guarantees $x(T)-x_e(T)=\alpha$, where $\alpha\neq 0$ assures $(v\ w\ x_0)\neq 0$, using the identity (20) with $T_1=t_0$ and $T_2=T\to t^*$, we have

$$\begin{aligned} &\| g \|_2^2 - \| h \|_2^2 - x_0{}'\Gamma x_0 \\ &\ge \int_{t_0}^{t^*}(\| g \|^2 - \| h \|^2)\,dt - x_0{}'\Gamma x_0 = \int_{t_0}^{t^*}(\| Fx_e \|^2 + \| u \|^2)\,dt \ge 0, \end{aligned}$$

which contradicts the existence of a DIA control. Thus, (A_2) must hold.

Next we show that, if there exists a DIA control, the conditions (A_1) and (A_3) must be satisfied. For this purpose, suppose that the Riccati equation (8) with the final condition $S(t_1) = 0$ does not have a solution on the whole interval $[t_0, t_1]$; then, we can find the largest time t_* in (t_0,t_1) such that (8) with $S(t_1) = 0$ has a solution $S(t)$ on $(t_*,t_1]$. Note that the solution $S(t)$ is positive semi-definite, which follows from $S(t_1) = 0$ and the almost same argument as in the proof of Lemma, and that there exists a nonzero vector β such that $\lim_{T\to t_*} \beta' S(T)\beta = \infty$. Now, for any fixed admissible control u, choosing nonzero $(v\ w\ x_0)$ such that

$$(v(t)\ \ w(t)) = \begin{cases} (v_{\max}(t)\ \ w_{\max}(t)), & T < t \le t_1 \\ (\ 0 \quad\ 0\ ), & t_0 \le t \le T \end{cases} \quad (T > t_*)$$

and x_0 guarantees $x(t_1) - x_e(t_1) = 0$ and $x_e(T) = \beta$, where $\beta \neq 0$ assures $(v\ w\ x_0) \neq 0$, substituting these into the identity (22) with $T_3 = T$ and $T_4 = t_1$, we have

$$\begin{aligned} &\| g \|_2^2 - \| h \|_2^2 - x_0{}'\Gamma x_0 \\ &\ge \int_T^{t_1}(\| g \|^2 - \| h \|^2)\,dt - x_0{}'\Gamma x_0 \\ &= \int_T^{t_1}\| u - u_{\min}\|^2\,dt + (x(T)-\beta)'P(T)^{-1}(x(T)-\beta) + \beta' S(T)\beta - x_0{}'\Gamma x_0. \quad (23) \end{aligned}$$

Although the last term of the right hand side of the equality in (23) is non-positive (but bounded), the right hand side of the equality in (23) is non-negative and so the left hand side of the inequality in (23), because we can take $\beta' S(T)\beta$ arbitrary large by letting $T \rightarrow t_*$. This contradicts the existence of a DIA control. Hence, the Riccati equation (8) with $S(t_1) = 0$ must have a solution $S(t)$ on the whole interval $[t_0,t_1]$. Then, letting $M := S(I+PS)^{-1}$, which exists since $P(t)$ is positive definite and $S(t)$ is positive semi-definite, we can show that M satisfies (4), and we have (A_1). Moreover $PM = PS(I+PS)^{-1}$ assures (A_3).

5. Concluding remarks

a) In our formulation, the (standard) H^∞controllers ([5][8]) are such that they are admissible controllers attaining $\|g\|_2^2 < \|h\|_2^2$ for all nonzero $h = (v',w')'$ in $L^2(t_0,t_1;R^{p+m})$ and "$x_0 = 0$". Limebeer et al. [5] showed that there exist the H^∞ controllers if and only if (A_1), $({A_2}^0)$ and (A_3) are satisfied, where $({A_2}^0)$ is the following statement:

$({A_2}^0)$ There exists a solution $P(t)$ defined on $[t_0,t_1]$ to the Riccati differential equation (5) with $P(t_0) = 0$ (instead of $P(t_0) = \Gamma^{-1}$),

and also gave a parametrization of all the H^∞ controllers. By comparing the criterion for the DIA controllers with that for the H^∞ controllers and considering the special case that $x_0 = 0$ in the DIA control problem, we can see that (A_1), (A_2) and (A_3) form a sufficient condition for existence of the H^∞ controllers (in fact, (A_2) is a sufficient condition for $({A_2}^0)$) and our parametrization (6) of the DIA controllers gives a sub-class of all the H^∞ controllers. (Addendum: In the recent paper [10], a generalized H^∞ control problem, which cover both the H^∞ control problem and the DIA control problem, has been formulated and solved by using two indefinite Riccati differential equations with positive semi-definite, possibly singular, terminal conditions.)

b) Another extreme case of the DIA control problem is the initial-uncertainty attenuation problem whose solutions are admissible controllers attaining $\|g\|_2^2 < x_0'\Gamma x_0$ for all nonzero x_0 in R^n and "h=0". It is a matter of course that (A_1), (A_2) and (A_3) are sufficient for solving this problem and the DIA controllers given by (6) are solutions to this problem.

c) If we take $\Psi = 0$ in our parametrization (6), we have a "central controller" u^o. It is noted that the central controller u^o is dual to those used by Limebeer et al. [5] and Tadmor [8] in their parametrizations of the

H^{∞}controllers, provided that the difference of the initial conditions ($P(t_0) = \Gamma^{-1}$ and $P(t_0) = 0$) for (5) is neglected.

d) This paper has dealt with special time-varying systems satisfying the "orthogonality assumptions" (see [1][8]). It will be possible to eliminate the assumptions by generalizing the two Riccati differential equations as in [5].

Acknowledgment : The authors wish to thank Prof. D.J.N. Limebeer for providing the preprint [5] which stimulated them to complete this paper. Also we thank Mitsuo Kubo and Hiroaki Kagaya for typing this manuscript.

References

[1] J. C. Doyle, K. Glover, P. P. Khargonekar and B. A. Francis, State-space solutions to standard H_2 and H_{∞}control problems, IEEE Trans. Automatic Control, 34 (1989), 831-847.

[2] K. Glover and J. C. Doyle, State-space formulae for all stabilizing controllers that satisfy an H_{∞}-norm bound and relations to risk sensitivity, Systems & Control Letters, 11(1988), 167-172.

[3] P. P. Khargonekar, I. R. Petersen and M. A. Rotea, H^{∞}optimal control with state feedback, IEEE Trans. Automatic Control, AC-33 (1988), 783-786.

[4] H. Kimura and R. Kawatani, Synthesis of H^{∞} controllers based on conjugation, Proc. IEEE CDC, Austin, 1988.

[5] D. J. N. Limebeer, B. D. O. Anderson, P. P. Khargonekar and M. Green, A game theoretic approach to H^{∞} control for time varying systems, Proc. MTNS-89, Amsterdam, 1989.

[6] D. J. N. Limebeer, E. M. Kasenally, E. Jaimouka and M. G. Safonov, A characterization of all solutions to the four block general distance problem, Proc. IEEE CDC, Austin, 1988.

[7] I.R. Petersen, Disturbance attenuation and H^{∞}optimization: a design method based on the algebraic Riccati equation, IEEE Trans. Automatic Control, AC-32 (1987), 427-429.

[8] G. Tadmor, H^{∞} in the time domain: the standard four block problem, To appear in Mathematics of Control, Systems and Signal Processing.

[9] K. Uchida and M. Fujita, On the central controller: characterizations via differential games and LEQG control problems, Systems & Control Letters, 13 (1989), 9-13.

[10] K. Uchida and M. Fujita, Finite horizon H^{∞}control problems with terminal penalties, Report of Sci. and Eng. Res. Lab., Waseda University, no.90-1, 3 April, 1990.

Robust Stabilization of Time Delay Systems Based on Riccati Type Operator Equation Arising in Differential Games

Akira KOJIMA†, Kenko UCHIDA†, and Etsujiro SHIMEMURA†

†Department of Electrical Engineering, Waseda University
3-4-1 Ohkubo, Shinjuku-ku, Tokyo 169, JAPAN

Abstract

A robust stabilization problem is discussed for time delay systems. By introducing an indefinite Riccati type operator equation for time delay systems, we propose a design method of robust stabilizing control law against the parametric uncertainties, which occur in lumped parameter element and in all types of delay element. In the derivation of the control law, the key point is that we take a twofold approach; one is an internal approach based on Lyapunov type operator equation and another is an external approach based on small-gain theorem. This twofold approach first enable us to deal the uncertainties of time delay systems in a unified way.

1. Introduction

Time delay phenomena often occur in the transmission of signal or material between different part of a system. They arise either as a result of inherent delays in the components of the system or as a deliberate introduction of time delay into the system for control purpose. In recent years, the control theory for time delay systems has been deeply studied, and among the contributions, the linear quadratic optimal control problem has been emphasized together with the stability properties induced from the Lyapunov type operator equation (Datko 1970, Delfour et al. 1975, Vinter et al. 1981).

In this paper, we consider a robust stabilization problem for uncertain time delay systems. We propose a design method of robust stabilizing control law by introducing an indefinite Riccati type operator equation for time delay systems. The indefinite Riccati type operator equation, which arises in linear quadratic differential games, plays a key role for the design of robust stabilizing control law.

The nominal plant we focus on is a linear time-invariant system described in a form of a functional differential equation;

$$\Sigma: \quad \dot{x}(t) = A_0 x(t) + \int_{-h}^{0} A_{01}(\beta) x(t+\beta)\, d\beta + A_1 x(t-h) + Bu(t), \qquad (1)$$

where $u(t) \in R^r$ is the control and $x(t) \in R^n$ is a part of the state, which is available for control. The parameter perturbations are assumed to occur in lumped parameter element (A_0) and in all types of delay element ($A_{01}(\cdot), A_1$). Our objective in this paper is to design a stabilizing control law such that the resulting closed loop system has robust stability against the perturbations which occur in ($A_0, A_{01}(\cdot), A_1$).

In the above formulation, the nominal plant is suitably described in a form of an evolution equation on appropriately introduced function space ; $M_2 := R^n \times L_2([-h, 0]; R^n)$ (e.g. Delfour et al. 1975). However, it should be noticed that the scheme of designing a robust control law raises mathematical problems, which require new challenges. First, the parameter perturbation which occurs in point delay element (A_1) indicates unbounded perturbation to the nominal plant described in a form of an evolution equation. This fact implies that the parameter perturbation of point delay element is not suitably evaluated from the viewpoint of operator norm. Secondly, even the relation between the parameter perturbations of ($A_0, A_{01}(\cdot)$) and their representation with a bounded operator is not still clarified. Thus, it is also required to investigate this relation in terms of operator norm. Furthermore, these problems mean that the conventional approach on M_2-space does not so fit as to deal the perturbations of ($A_0, A_{01}(\cdot), A_1$) in a unified way.

To attack the robust stabilization problem together with these problems, we take a twofold approach; one is from the internal viewpoint based on Lyapunov type operator equation and another is from the external viewpoint based on small-gain theorem (Desoer et al. 1975). For time delay systems, the equivalence between the internal stability and the external stability (L_2-input/output stability) is shown by Yamamoto et al. (1989).

The paper is organized as follows. First, we consider a preliminary robust stabilization problem against the restricted parameter perturbations, which are assumed to occur only in both lumped parameter element (A_0) and in distributed delay element ($A_{01}(\cdot)$). From the internal approach, we derive a robust stabilizing control law based on an indefinite Riccati type operator equation.

Secondly, we show that the resulting closed loop system also has robust stability against the perturbation of point delay element (A_1). To analyze the robust stability from the external point of view, we introduce an augmented system for the resulting closed loop system, which has exogenous input and regulated output, and define a disturbance attenuation problem. After clarifying the robustness of the resulting closed loop system against parametric and unparametric perturbations, we evaluate the robust stability against the perturbation of point delay element.

Based on this twofold analysis, we lastly derive a design method of robust stabilizing control law for time delay systems. The parameter perturbations which occur in lumped parameter element and in all types of delay element ($A_0, A_{01}(\cdot), A_1$) are totally evaluated in terms of a newly introduced generalized norm.

2. Systems and Preliminaries

Consider a class of uncertain time delay systems described as follows,

$$\tilde{\Sigma}: \quad \dot{x}(t) = (A_0 + \Delta A_0)x(t) + \int_{-h}^{0} (A_{01}(\beta) + \Delta A_{01}(\beta))x(t+\beta)\,d\beta$$
$$+(A_1 + \Delta A_1)x(t-h) + Bu(t) \qquad (2)$$

$$\begin{array}{rcll} \Delta A_0 & := & M\delta D_0, & \delta D_0 \in R^{k\times n} \\ \Delta A_{01}(\cdot) & := & M\delta D_{01}(\cdot), & \delta D_{01} \in L_2([-h,0];R^{k\times n}) \\ \Delta A_1 & := & M\delta D_1, & \delta D_1 \in R^{k\times n} \end{array}$$

where $(\delta D_0, \delta D_{01}(\cdot), \delta D_1)$ are the modeling uncertainties, and $M \in R^{n\times k}$ is a fixed known real matrix which characterizes the structure of the uncertainties. In this formulation, the triplet $(\delta D_0, \delta D_{01}(\cdot), \delta D_1)$ stands for the uncertainties in lumped parameter element (ΔA_0), in distributed delay element $(\Delta A_{01}(\cdot))$ and in point delay element (ΔA_1) respectively. Our objective in this paper is to design a robust stabilizing control law such that the resulting closed loop system has robust stability against these uncertainties (δD_0, $\delta D_{01}(\cdot), \delta D_1$), which occur in lumped parameter element and in all types of delay element.

By introducing a Hilbert space

$$M_2([-h,0];R^n) := R^n \times L_2([-h.0];R^n) \qquad (3)$$

endowed with the inner product

$$\langle x, y\rangle := x^{0\prime}y^0 + \int_{-h}^{0} x^{1\prime}(\beta)\,y^1(\beta)\,d\beta, \qquad (4)$$

$$x = \begin{bmatrix} x^0 \\ x^1 \end{bmatrix} \in M_2, \quad y = \begin{bmatrix} y^0 \\ y^1 \end{bmatrix} \in M_2$$

the time delay system (1) can be written in a form of an evolution equation

$$\dot{\underline{x}} = \mathcal{A}\underline{x}(t) + \Pi Bu(t), \qquad \underline{x}(0) \in M_2([-h,0];R^n) \qquad (5)$$

where

$$\begin{array}{rcll} \underline{x}(t) & = & \begin{bmatrix} x(t) \\ x_t \end{bmatrix}, & x_t(\beta) = x(t+\beta), \quad -h \le \beta \le 0 \\ \Pi v & = & \begin{bmatrix} v \\ 0 \end{bmatrix}, & v \in R^n \end{array}$$

$\mathcal{A}$ is an infinitesimal generator defined as follows,

$$D(\mathcal{A}) = \left\{\phi = \begin{bmatrix} \phi^0 \\ \phi^1 \end{bmatrix} : \phi^1 \in W_2^{(1)}([-h,0];R^n),\ \phi^0 = \phi^1(0)\right\}$$

$$\mathcal{A}\phi = \begin{bmatrix} A_0\phi^0 + \int_{-h}^{0} A_{01}(\beta)\phi^1(\beta)\,d\beta + A_1\phi^1(-h) \\ \frac{d\phi(\beta)}{d\beta} \end{bmatrix}$$

where $W_2^{(1)}$ denotes the Sobolev space of R^n-valued, absolutely continuous functions with square integrable derivatives on $[-h, 0]$.

For time delay systems, following properties on Lyapunov type operator equation are derived by Datko (1970) and Delfour et al. (1975).

Lemma 1 (Datko 1970, Delfour et al. 1975) *The time delay system (5) is internally stable if and only if there exists a positive definite solution $\mathcal{P} \in \mathcal{L}(M_2)$ such that*

$$\mathcal{P}\mathcal{A} + \mathcal{A}^*\mathcal{P} = -\mathcal{Q} \tag{6}$$

where $\mathcal{Q} \in \mathcal{L}(M_2)$ is a given positive definite operator.

Furthermore the solution $\mathcal{P} > 0$ of (6) is characterized by its matrix of operators;

$$\mathcal{P} := \begin{bmatrix} P^{00} & P^{01} \\ P^{10} & P^{11} \end{bmatrix}, \quad \begin{array}{ll} P^{00} \in \mathcal{L}(R^n), & P^{01} \in \mathcal{L}(L_2, R^n) \\ P^{10} \in \mathcal{L}(R^n, L_2), & P^{11} \in \mathcal{L}(L_2) \end{array} \tag{7}$$

$$\begin{array}{lcl} P^{00}\phi^0 & := & P_0\phi^0 \\ P^{01}\phi^1 & := & \int_{-h}^0 P_1(\beta)\phi^1(\beta)\,d\beta \\ (P^{10}\phi^0)(\alpha) & := & P_1'(\alpha)\phi^0 \\ (P^{11}\phi^1)(\alpha) & := & \int_{-h}^0 P_2(\alpha,\beta)\phi^1(\beta)\,d\beta, \qquad -h \le \alpha \le 0 \end{array}$$

We call the triplet $\{P_0, P_1, P_2\}$ an integral kernel of the operator $\mathcal{P}$.

3. Robust Stabilization against Restricted Perturbations

In this chapter, we consider a restricted robust stabilization problem, in which the parametric uncertainties do not concern with point delay element (A_1).

Restricted Problem : *Design a robust stabilizing control law such that the uncertain time delay system $\tilde{\Sigma}$ is stabilized against the following uncertainties;*

$$\left\| \delta D_0\, \delta D_0' + \int_{-h}^0 \delta D_{01}(\beta)\, \delta D_{01}'(\beta)\, d\beta \right\|_2^{\frac{1}{2}} < 1, \qquad \|\delta D_1\|_2 = 0$$

The uncertainties stated in the Restricted Problem, which do not occur in (A_1) but in $(A_0, A_{01}(\cdot))$, imply bounded perturbation to the nominal plant Σ described in a form of an evolution equation. Therefore, in the approach to the Restricted Problem, we can deal the uncertain system $\tilde{\Sigma}$ in a form of an evolution equation;

$$\hat{\Sigma}: \quad \dot{\underline{x}}(t) = (\mathcal{A} + \Pi M \delta\mathcal{D})\underline{x}(t) + \Pi B u(t), \tag{8}$$

in M_2-space, where $\mathcal{A}$ is an infinitesimal generator defined in (5), and $\delta\mathcal{D} \in \mathcal{L}(M_2([-h,0]; R^n), R^k)$ is a bounded operator defined in a form of matrix of operators;

$$\delta\mathcal{D} = \begin{bmatrix} \delta\mathcal{D}^0 & \delta\mathcal{D}^{01} \end{bmatrix}, \quad \begin{array}{ll} \delta\mathcal{D}^0\phi^0 := \delta D_0\phi^0, & \phi^0 \in R^n \\ \delta\mathcal{D}^{01}\phi^1 := \int_{-h}^0 \delta D_{01}(\beta)\phi^1(\beta)\,d\beta, & \phi^1 \in L_2([-h,0]; R^n) \end{array}$$

which represents the parametric uncertainties $(\delta D_0, \delta D_{01}(\cdot))$. Moreover, since $\Pi M \delta\mathcal{D}$ is bounded, the operator $\mathcal{A} + \Pi M \mathcal{D}$ is also defined as an infinitesimal generator in $D(\mathcal{A})$ (Kato 1966).

We first derive a relation between the operator $\delta\mathcal{D}$ and its practical representation $(\delta D_0, \delta D_{01}(\cdot))$ in terms of operator norm.

Lemma 2 *The operator inequality $\delta\mathcal{D}^*\delta\mathcal{D} < k{\cdot}\mathcal{I}$ $(k > 0)$ holds if and only if $(\delta D_0, \delta D_{01}(\cdot))$ satisfies*

$$\left\| \delta D_0\, \delta D_0' + \int_{-h}^{0} \delta D_{01}(\beta)\, \delta D_{01}'(\beta)\, d\beta \right\|_2 < k$$

Based on Lemma 2 and the properties of Lyapunov type operator equation stated in Lemma 1, we obtain the following theorem on the restricted robust stabilization problem.

Theorem 1 *Suppose there exists a positive definite solution $\mathcal{P} \in \mathcal{L}(M_2)$ to the indefinite Riccati type operator equation;*

$$\mathcal{P}\mathcal{A} + \mathcal{A}^*\mathcal{P} - \mathcal{P}\Pi B R^{-1} B'\Pi^*\mathcal{P} + \mathcal{P}\Pi M M'\Pi^*\mathcal{P} + \mathcal{I} + \mathcal{Q} = 0 \tag{9}$$

where $\mathcal{Q} \in \mathcal{L}(M_2)$, $R \in R^{r\times r}$ are given positive definite operator and matrix respectively. Then the control law

$$u(t) = -\mathcal{K}\underline{x}(t), \qquad \mathcal{K} := R^{-1}B'\Pi^*\mathcal{P} \tag{10}$$

stabilizes the system $\hat{\Sigma}$ against the restricted parametric uncertainties;

$$\delta\mathcal{D}^*\delta\mathcal{D} < \mathcal{I}, \qquad \left(\text{i.e. } \left\| \delta D_0\, \delta D_0' + \int_{-h}^{0} \delta D_{01}(\beta)\, \delta D_{01}'(\beta)\, d\beta \right\|_2^{\frac{1}{2}} < 1 \right) \tag{11}$$

(Proof)

Rewriting equation (9) with the positive definite solution $\mathcal{P} \in \mathcal{L}(M_2)$, we have a following Lyapunov type operator equation;

$$\begin{aligned} \mathcal{P}\mathcal{A}_c + \mathcal{A}_c^*\mathcal{P} &= -\tilde{\mathcal{Q}}, \qquad (12)\\ \mathcal{A}_c &:= (\mathcal{A} + \Pi M \delta\mathcal{D}) - \Pi B \mathcal{K} \\ \tilde{\mathcal{Q}} &:= \mathcal{Q} + \mathcal{P}\Pi B R^{-1} B'\Pi^*\mathcal{P} + (\mathcal{I} - \delta\mathcal{D}^*\delta\mathcal{D}) \\ &\quad + \{\mathcal{P}\Pi M - \delta\mathcal{D}^*\}\{\mathcal{P}\Pi M - \delta\mathcal{D}^*\}^* \end{aligned}$$

where $\mathcal{A}_c$ represents a resulting closed loop system from $\hat{\Sigma}$ and (10), and $\tilde{\mathcal{Q}}$ is a positive definite operator since $\mathcal{I} - \delta\mathcal{D}^*\delta\mathcal{D} > 0$.

Hence, by Lemma 1, the closed loop system $\mathcal{A}_c$ is internally stable if there exists a positive definite solution $\mathcal{P} \in \mathcal{L}(M_2)$ to equation (9). Q.E.D.

Furthermore, by defining an integral kernel $\{P_0, P_1, P_2\}$ to the solution $\mathcal{P} \in \mathcal{L}(M_2)$, the stabilizing control law (10) is written down as follows.

$$\begin{aligned} u(t) &= -K_0 x(t) - \int_{-h}^{0} K_1(\beta) x(t+\beta)\, d\beta, \qquad (13)\\ &\quad K_0 := R^{-1}B'P_0, \\ &\quad K_1(\beta) := R^{-1}B'P_1(\beta), \quad -h \le \beta \le 0 \end{aligned}$$

The closed loop system constructed from $\hat{\Sigma}$ and (10) is also shown to be internally stable by introducing a quadratic Lyapunov function

$$\langle \underline{x}(t), \mathcal{P}\underline{x}(t) \rangle, \qquad \mathcal{P} > 0$$

in M_2-space. Hence, the control law stated in Theorem 1 is a solution of quadratic stabilization problem developed for time delay systems (Barmish 1985, Petersen et al. 1986). As far as we focus on the Restricted Problem, a principle of quadratic stabilizability in M_2-space enables us to solve the problem in a unified way. However, it should be noticed that, for time delay systems, these internal approaches are never applicable for the analysis of point delay element (A_1).

4. Main Problem with Disturbance Attenuation

In chapter 4, we first discuss the robustness of the closed loop system constructed from $\hat{\Sigma}$ and (10). We take an external approach based on small-gain theorem to evaluate the robust stability against the uncertainty of point delay element (A_1). And lastly, we solve a main robust stabilization problem defined for the uncertain time delay system $\tilde{\Sigma}$, which involves parametric uncertainties ($\delta D_0, \delta D_{01}(\cdot), \delta D_1$) simultaneously.

Main Problem : *Design a robust stabilizing control law such that the uncertain time delay system $\tilde{\Sigma}$ is stabilized against the following uncertainties;*

$$\|\delta D_0\|_2 + \left\| \int_{-h}^{0} \delta D_{01}(\beta)\, \delta D_{01}'(\beta)\, d\beta \right\|_2^{\frac{1}{2}} + \|\delta D_1\|_2 < 1$$

To analyze the robust stability of the closed loop system constructed from $\hat{\Sigma}$ and (10), we introduce an augmented system;

$$\hat{\Sigma}_A : \quad \begin{aligned} \dot{\underline{x}}(t) &= (\mathcal{A} + \Pi M \delta \mathcal{D})\underline{x}(t) + \Pi B u(t) + \Pi M w(t) \\ z(t) &= \Pi^* \underline{x}(t) \\ u(t) &= -\mathcal{K}\underline{x}(t) \qquad \mathcal{K}\text{: defined in (10)} \end{aligned} \tag{14}$$

where $w \in L_2([0,\infty); R^k)$ is the disturbance and $z \in L_2([0,\infty); R^n)$ is the regulated output. Our objective here is to investigate the relation between the attenuation effect from w to z and the internal parametric uncertainty $\delta\mathcal{D}$ (i.e. $\delta D_0, \delta D_{01}(\cdot)$) of $\hat{\Sigma}_A$. We derive a following theorem on the trade-off between internal uncertainties and disturbance attenuation.

Theorem 2 *Suppose there exists a positive definite solution $\mathcal{P} \in \mathcal{L}(M_2)$ to the equation (9). Then the augmented system $\hat{\Sigma}_A$ is internally stable against the uncertainties;*

$$\delta\mathcal{D}^* \delta\mathcal{D} < \frac{1}{\gamma^2} \cdot \mathcal{I}, \quad \left(\text{i.e. } \left\| \delta D_0 \delta D_0' + \int_{-h}^{0} \delta D_{01}(\beta)\, \delta D_{01}'(\beta)\, d\beta \right\|_2 < \frac{1}{\gamma^2} \right), \quad \gamma > 1 \tag{15}$$

and attenuates the disturbance as follows.

$$\|z\|_{L_2} < \frac{\gamma}{\gamma - 1} \cdot \|w\|_{L_2}, \qquad w \neq 0 \tag{16}$$

(Proof)

Since $\delta\mathcal{D}^*\delta\mathcal{D} < \frac{1}{\gamma^2}\cdot\mathcal{I} < \mathcal{I}$ and $\mathcal{Q} > 0$, the augmented system $\hat{\Sigma}_A$ is shown to be internally stable by Theorem 1.

To verify the inequality (16), we first discuss restricted disturbances;

$$w \in C^1([0,\infty);R^k), \quad \int_0^\infty |w(t)|^2\,dt < \infty \tag{17}$$

and denote the disturbances $w \in C^1L_2([0,\infty);R^k)$. The function space $C^1L_2([0,\infty);R^k)$ is dense in $L_2([0,\infty];R^k)$, and it is known that, if $\underline{x}(0) \in D(\mathcal{A})$ and $w \in C^1L_2([0,\infty);R^k)$, $\underline{x}(t)$ $(t > 0)$ is everywhere differentiable.

We obtain the following equality by rewriting equation (9).

$$\mathcal{P}\mathcal{A}_c + \mathcal{A}_c^*\mathcal{P} + \frac{1}{\delta}\cdot\mathcal{P}\Pi MM'\Pi^*\mathcal{P} + \frac{1}{\delta}\cdot\Pi\,\Pi^* + \hat{\mathcal{Q}} = 0 \tag{18}$$

$$\begin{aligned}
\mathcal{A}_c &:= (\mathcal{A} + \Pi M\delta\mathcal{D}) - \Pi B\mathcal{K} \\
\hat{\mathcal{Q}} &:= \mathcal{Q} + \mathcal{P}\Pi BR^{-1}B'\Pi^*\mathcal{P} + \gamma\left(\frac{1}{\gamma^2}\cdot\mathcal{I} - \delta\mathcal{D}^*\delta\mathcal{D}\right) \\
&\quad + \left(\frac{1}{\sqrt{\gamma}}\cdot\mathcal{P}\Pi M - \sqrt{\gamma}\cdot\delta\mathcal{D}^*\right)\left(\frac{1}{\sqrt{\gamma}}\cdot\mathcal{P}\Pi M - \sqrt{\gamma}\cdot\delta\mathcal{D}^*\right)^* \\
&\quad + \frac{1}{\delta}\cdot(\mathcal{I} - \Pi\,\Pi^*) > 0 \\
\delta &:= \frac{\gamma}{\gamma - 1}
\end{aligned}$$

Consider the functional

$$V(t) := \langle\underline{x}(t), \mathcal{P}\underline{x}(t)\rangle \tag{19}$$

in M_2-space, then differentiating both sides with respect to t, inserting (14) and (18), we have

$$\dot{V}(t) = -\left|\frac{1}{\sqrt{\delta}}\cdot M'\Pi\mathcal{P}\underline{x}(t) - \sqrt{\delta}\cdot w(t)\right|^2 - \langle\underline{x}(t), \hat{\mathcal{Q}}\underline{x}(t)\rangle + \delta\cdot|w(t)|^2 - \frac{1}{\delta}\cdot|z(t)|^2 \tag{20}$$

Integrating both sides with respect to t over the interval $[0,\infty)$, we finally obtain

$$\begin{aligned}
\langle\underline{x}(\infty), \mathcal{P}\underline{x}(\infty)\rangle - \langle\underline{x}(0), \mathcal{P}\underline{x}(0)\rangle &= -\int_0^\infty \left|\frac{1}{\sqrt{\delta}}\cdot M'\Pi\mathcal{P}\underline{x}(t) - \sqrt{\delta}\cdot w(t)\right|^2 dt \\
&\quad - \int_0^\infty \langle\underline{x}(t), \hat{\mathcal{Q}}\underline{x}(t)\rangle\,dt + \delta\cdot\|w\|_{L_2}^2 - \frac{1}{\delta}\cdot\|z\|_{L_2}^2
\end{aligned} \tag{21}$$

Since $\hat{\Sigma}_A$ is shown to be internally stable, $\underline{x}(\infty) = 0$, and without loss of generality in the input/output analysis, we can assume the initial state as $\underline{x}(0) = 0$.

Hence it follows from equality (21) that the inequality (16) holds for the restricted disturbances $w \in C^1L_2([0,\infty);R^k), w \neq 0$.

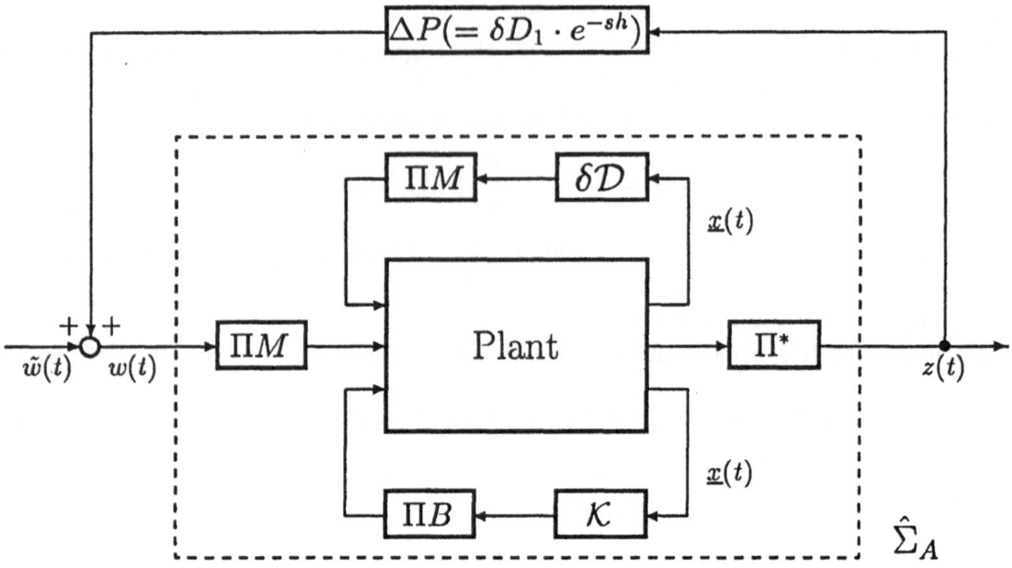

Figure Augmented System with Perturbations

For given $w \in L_2([0,\infty);R^k)$, there exists a sequence $\{w^n\}$, $w^n \in C^1L_2([0,\infty);R^k)$ which converges to $w \in L_2([0,\infty);R^k)$. Hence it then follows from the fact that $\underline{x}(t)$, $(t > 0)$ depends continuously on $\{w^n\}$ that (16) holds for all disturbances $w \in L_2([0,\infty);R^k)$, $w \neq 0$. Q.E.D.

Considering the pure delay perturbation

$$\Delta P(s) = \delta D_1 \cdot e^{-sh}, \quad \text{i.e.} \quad w(t) = \delta D_1 z(t-h) + \tilde{w}(t) \tag{22}$$

to the augmented system $\hat{\Sigma}_A$ as shown in Figure, we can evaluate the robust stability based on small-gain theorem (Desoer et al. 1975).

$$\|\Delta P\|_\infty = \|\delta D_1\|_2 < \left(\frac{\gamma}{\gamma-1}\right)^{-1} = 1 - \frac{1}{\gamma}, \qquad \Delta P \in H^\infty \tag{23}$$

The other hand, the augmented system with this perturbation

$$\hat{\Sigma}_{AP}: \quad \begin{array}{l} \dot{\underline{x}}(t) = (\mathcal{A} + \Pi M \delta\mathcal{D})\underline{x}(t) + \Pi B u(t) + \Pi M w(t) \\ z(t) = \Pi^* \underline{x}(t) \\ u(t) = -\mathcal{K}\underline{x}(t), \\ w(t) = \delta D_1 z(t-h) + \tilde{w}(t) \end{array} \tag{24}$$

can be written down in the following way.

$$\begin{aligned}\dot{x}(t) &= (A_0 + M\delta D_0)x(t) + \int_{-h}^{0}(A_{01}(\beta) + M\delta D_{01}(\beta))x(t+\beta)\,d\beta \\ &\quad +(A_1 + M\delta D_1)x(t-h) + Bu(t) + M\tilde{w}(t) \\ z(t) &= x(t) \\ u(t) &= -K_0x(t) - \int_{-h}^{0} K_1(\beta)x(t+\beta)\,d\beta\end{aligned} \tag{25}$$

Hence the augmented system with perturbation $\hat{\Sigma}_{AP}$ represents the stabilized uncertain system $\tilde{\Sigma}$, which involves uncertainties $(\delta D_0, \delta D_{01}(\cdot), \delta D_1)$ simultaneously.

Together with this external analysis, we finally obtain the main result on robust stabilization problem defined for time delay systems.

Theorem 3 (Main Result) *Suppose there exists a positive definite solution $\mathcal{P} \in \mathcal{L}(M_2)$ to the indefinite Riccati type operator equation*

$$\mathcal{P}\mathcal{A} + \mathcal{A}^*\mathcal{P} - \mathcal{P}\Pi BR^{-1}B'\Pi^*\mathcal{P} + \mathcal{P}\Pi MM'\Pi^*\mathcal{P} + \mathcal{I} + \mathcal{Q} = 0 \tag{26}$$

where $\mathcal{Q} \in \mathcal{L}(M_2)$, $R \in R^{r\times r}$ are given positive definite operator and matrix respectively. Then the control law

$$\begin{aligned}u(t) &= -K_0x(t) - \int_{-h}^{0} K_1(\beta)x(t+\beta)\,d\beta \\ &\quad K_0 := R^{-1}B'P_0, \\ &\quad K_1(\beta) := R^{-1}B'P_1(\beta), \quad -h \le \beta \le 0 \\ &\quad \{P_0, P_1, P_2\};\ \textit{integral kernel of the solution } \mathcal{P}\end{aligned} \tag{27}$$

stabilizes the uncertain system $\tilde{\Sigma}$ against the parametric uncertainties;

$$\|\delta D_0\|_2 + \left\|\int_{-h}^{0} \delta D_{01}(\beta)\,\delta D_{01}'(\beta)\,d\beta\right\|_2^{\frac{1}{2}} + \|\delta D_1\|_2 < 1 \tag{28}$$

(Proof)

For given $(\delta D_0, \delta D_{01}(\cdot), \delta D_1)$, there exists $\gamma > 1$ such that

$$\|\delta D_0\|_2 + \left\|\int_{-h}^{0} \delta\, D_{01}(\beta)\delta D_{01}'(\beta)\,d\beta\right\|_2^{\frac{1}{2}} < \frac{1}{\gamma} \tag{29}$$

$$\|\delta D_1\|_2 < 1 - \frac{1}{\gamma} \tag{30}$$

Furthermore, the following inequality holds sufficiently under the condition (29).

$$\left\|\delta D_0\,\delta D_0' + \int_{-h}^{0} \delta D_{01}(\beta)\,\delta D_{01}'(\beta)\,d\beta\right\|_2 < \frac{1}{\gamma^2}, \quad \text{i.e. } \delta\mathcal{D}^*\delta\mathcal{D} < \frac{1}{\gamma^2}\cdot\mathcal{I} \tag{31}$$

Hence, by Theorem 2 and the condition (23), the resulting closed loop system from $\tilde{\Sigma}$ and (27) is shown to be internally stable. Q.E.D.

5. Conclusion

In this paper, we have solved a robust stabilization problem defined for time delay systems. The key point is that we take a twofold approach to solve the problem; one is the internal approach based on Lyapunov type operator equation and another is the external approach based on small-gain theorem.

The internal approach enables us to introduce an indefinite Riccati type operator equation for time delay systems and solve a restricted robust stabilization problem, in which the uncertainties do not concern with point delay element. However, as the internal approach is not applicable for the analysis of point delay element, we secondly employ the external approach to evaluate the robust stability against the uncertainty of point delay element. In this sense, the twofold approach developed in this paper is indispensable for time delay systems to deal all types of uncertainties in a unified way.

In the practical application of this method, we need an integral kernel representation $\{P_0, P_1, P_2\}$ of the positive definite solution $\mathcal{P} \in \mathcal{L}(M_2)$. Averaging method (Banks et al. 1978, Gibson 1983) are easily applied for the numerical calculation and the computational work needed to obtain the solution can be said quite moderate.

References

Banks, H.T. and Burns, J.A. 1978, *Hereditary control problems: numerical methods based on averaging approximations*, SIAM J. Control and Optimization, 16, pp.169-208

Barmish, B.R. 1985, *Necessary and sufficient conditions for quadratic stabilizability of an uncertain linear system* , J. Optimiz. Theory Appl., 46, pp.399-408.

Datko, R. 1970, *Extending a theorem of A.M.Lyapunov to Hilbert spaces*, J. Math. Anal. Appl., 32, pp.610-616.

Delfour, M.C., McCalla, C. and Mitter, S.K. 1975, *Stability and the infinite-time quadratic cost problem for linear hereditary differential systems*, SIAM J. Control, 13, pp.48-88.

Desoer, C.A. and Vidyasagar, M. 1975, *Feedback Systems: Input/Output Properties*, Academic Press.

Gibson, J.S. 1983, *Linear-quadratic optimal control of hereditary differential systems: infinite dimensional Riccati equations and numerical approximations*, SIAM J. Control and Optimization, 21, pp.95-139

Kato, T. 1966, *Perturbation theory for linear operators*, Springer-Verlag, Berlin

Petersen, I.R. and Hollot, C.V. 1986, *A Riccati equation approach to the stabilization of uncertain linear systems*, Automatica, 22, pp.397-411.

Vinter, R.B. and Kwong, R.H. 1981, *The infinite time quadratic control problem for linear systems with state and control delays : An evolution equation approach*, SIAM J. Control and Optimization, 19, pp.139-153.

Yamamoto, Y. and Hara, S. 1989, *Robust stability condition for infinite-dimensional systems with internal exponential stability*, to appear in Proc. MTNS-89, Amsterdam.

QUADRATIC DIFFERENTIAL GAMES WITH OVERTAKING OPTIMALITY

Akira Ichikawa
Department of Electrical Engineering
Shizuoka University, Hamamatsu 432, Japan

1. Introduction.

It is known that values and optimal pairs of infinite-time quadratic differential games for linear time-invariant systems are given by non-negative solutions of algebraic Riccati equations [11]-[13]. Riccati equations of the same type also play an important role in the recent H_∞ theory [7] and in the stabilization of uncertain systems [14], [15].

The problem of tracking periodic signals for a time-invariant linear system has been considered by Artstein and Leizarowitz [1]. It is an infinite-time problem and the notion of overtaking optimality is used. We have generalized their results to time-varying systems [9]. Closely related problems are considered in Da Prato and Ichikawa [5], [6].

In this paper we consider infinite-time quadratic differential games for time-varying systems and the differential game version of the tracking problems. In section 2 we study Riccati equations over an infinite horizon and give sufficient conditions for the existence of bounded solutions. We show that bounded stable solutions of Riccati equations give values and optimal pairs of quadratic games. In section 3 we consider the tracking problems and characterize optimal pairs in the overtaking sense. We also consider a special case where ideal inputs and a response to the given signal exist. We then introduce a more natural payoff function and reduce the problem to an infinite-time quadratic game considered in section 2.

2. Quadratic games and Riccati equations.

Consider the differential game defined by

$$x' = A(t)x + B_1(t)u + B_2(t)v, \quad x \in R^n,\ u \in R^{m_1},\ v \in R^{m_2}$$
$$x(t_0) = x_0, \tag{1}$$
$$y(t) = C(t)x(t), \quad y \in R^p$$

$$J_T(u,v)=\int_{t_0}^{T}[|y|^2+|u|^2-|v|^2]dt, \tag{2}$$

where A, B_i, C are bounded piecewise continuous matrices of appropriate dimensions and u tries to minimize (2) while v tries to maximize it. The Riccati equation associated with this game is

$$Q'+A^TQ+QA+C^TC-Q(B_1B_1{}^T-B_2B_2{}^T)Q=0, \tag{3}$$
$$Q(T)=0. \tag{4}$$

A pair (u,v) is called admissible if it is feedback and yields a unique solution to (1) and u, v as functions of t are in $L_2(t_0,T)$ (square integrable). It is well-known that if there exists a symmetric solution $Q_T(t)$ to (3), (4) it is unique and nonnegative. The feedback pair

$$\bar{u}=-B_1{}^TQ_Tx$$
$$\bar{v}=B_2{}^TQ_Tx$$

is a saddle point (optimal pair) in the sense

$$J_T(\bar{u},v)\leqq J_T(\bar{u},\bar{v})\leqq J_T(u,\bar{v})$$

for any admissible $(\bar{u},v)$ and $(u,\bar{v})$ and

$$J_T(\bar{u},\bar{v})=x_0{}^TQ_T(t_0)x_0.$$

Consider the special case of (1), (2) with v=0:

$$x'=Ax+B_1u,$$
$$x(t_0)=x_0,$$

$$J_T(u)=\int_{t_0}^{T}[|Cx|^2+|u|^2]dt.$$

There exists a unique nonnegative solution $P_T(t)$ to the Riccati equation

$$P'+A^TP+PA+C^TC-PB_1B_1{}^TP=0,$$
$$P(T)=0.$$

The feedback law

$$u_1=-B_1{}^TP_T(t)x$$

is optimal. Since we have inequalities

$$0\leqq x_0{}^TP_T(t_0)x_0=J_T(u_1)\leqq J_T(\bar{u})$$
$$=J_T(\bar{u},0)\leqq J_T(\bar{u},\bar{v})=x_0{}^TQ_T(t_0)x_0$$

we obtain

$$0\leqq P_T(t_0)\leqq Q_T(t_0).$$

Lemma 1. For each t, $Q_T(t)$, if it exists, is monotone increasing in T.

Proof. Consider the Riccati equation (3) with $Q(T)=F\geqq 0$. Then the solution is nonnegative and monotone increasing in F. The assertion follows from this easily [6].

For the infinite horizon problem we replace (2) by

$$J_\infty(u,v)=\int_{t_0}^{\infty}[|Cx|^2+|u|^2-|v|^2]dt. \tag{5}$$

We call a pair (u,v) admissible if it is a feedback pair for which (1) has a unique solution x such that $x(t)\to 0$ as $t\to\infty$ and u, $v\in L_2(t_0,\infty)$.

Theorem 1. Assume the following:

(H1): The Riccati equation (3) has a bounded solution Q on $[t_0,\infty)$ and $A-(B_1B_1^T-B_2B_2^T)Q$ is (uniformly exponentially) stable.

Then the feedback pair

$$\bar{u}=-B_1^TQx$$
$$\bar{v}=B_2^TQx$$

is optimal for (5) in the sense

$$J_\infty(\bar{u},v)\leqq J_\infty(\bar{u},\bar{v})\leqq J_\infty(u,\bar{v})$$

for any u, v with $(\bar{u},v)$, $(u,\bar{v})$ admissible.

Proof. Similar to Theorem 3.2 [6]. See the proof of Theorem 4.

Now our primary concern is to find conditions which assure (H1). We call Q a stable solution of (3) if it satisfies (H1). First we study some properties of bounded solutions of (3).

Lemma 2. Let Q be a bounded nonnegative solution of (3). If (C,A) is detectable [6], [9], then $A_1=A-B_1B_1^TQ$ is stable.

Proof. We can rewrite (3) to get

$$Q'+A_1^TQ+QA_1+C^TC+Q(B_1B_1^T+B_2B_2^T)Q=0.$$

From this we obtain Cx_1, $B_1^TQx_1\in L_2(t_0,\infty)$, where x_1 is the solution of

$$x'=A_1x,\quad x(t_0)=x_0. \tag{6}$$

Since (C,A) is detectable, A-JC is stable for some bounded J. Now we rewrite (6) to get

$$x'=(A-JC)x+(JC-B_1B_1^TQ)x,\quad x(t_0)=x_0.$$

By Datko's result [4] A_1 is stable.

Lemma 3. Suppose $B_1B_1^T-B_2B_2^T\geqq 0$. Then a bounded stable solution of the Riccati equation (3) is maximal (hence unique if it exists).

Proof. Similar to Proposition 3.2 [6].

Since we know that $Q_T(t)$ is increasing in T we give a condition for its uniform boundedness and hence for the existence of its limit. We assume the following as in H -problems [10], [7]:

$u_K = Kx$ is stabilizing i.e., $A_K = A + B_1 K$ is stable, for some bounded K and $|G| < 1$, where $G: L_2(t_0, \infty; R^{m2}) \rightarrow L_2(t_0, \infty; R^p)$ is defined by

(H2)
$$(Gv)(t) = C_K(t) \int_{t_0}^{t} S(t,r)B_2(r)v(r)dr$$

and $C_K = (C^T C + K^T K)^{1/2}$ and $S(t,r)$ is the fundamental solution for A_K.

Now consider
$$x' = A_K x + B_2 v, \quad x(t_0) = x_0,$$

$$J_{KT}(v) = \int_{t_0}^{T} [|C_K x|^2 - |v|^2] dt,$$

where v maximizes $J_{KT}(v)$. Under (H2) there exists a unique nonnegative solution P_{KT} to the Riccati equation
$$P' + A_K^T P + P A_K + C_K^T C_K + P B_2 B_2^T P = 0, \tag{7}$$
$$P(T) = 0. \tag{8}$$
The feedback law $v_K = B_2^T P_{KT} x$ is optimal and $J_{KT}(v_K) = x_0^T P_{KT}(t_0) x_0$. Furthermore we have

Lemma 4. Assume (H2). Then $P_{KT}(t) \uparrow P_K(t)$ which is a bounded nonnegative solution to (7) and $A_K + B_2 B_2^T P_K$ is stable.

Remark 1. One can show that the existence of P_K as in Lemma 4 implies the condition (H2) [10].

Using Lemma 4 we can establish a bounded solution to (3).

Theorem 2. Suppose (H2) holds. Then $Q_T(t)$ is uniformly bounded and $Q_T(t) \uparrow Q(t)$ which is a minimal bounded nonnegative solution of (3).

Proof. By the optimality of $(\bar{u}, \bar{v})$ we have
$$x_0^T Q_T(t_0) x_0 = J_T(\bar{u}, \bar{v}) \leq J_T(u_K, \bar{v})$$
$$= J_{KT}(\bar{v}) \leq J_{KT}(v_K) = x_0^T P_{KT}(t_0) x_0 \leq x_0^T P_K(t_0) x_0.$$
Hence Q_T is uniformly bounded. The rest follows as Theorem 3.1 [6].

Next we shall show the existence of a stable $Q(t)$. For this purpose we rewrite (1), (2) as

$$x' = A_K x + B_1 w + B_2 v, \quad w = u - Kx$$
$$x(t_0) = x_0, \tag{9}$$
$$J_T(w,v) = \int_{t_0}^{T} [\,|C_K x|^2 + 2w^T Kx + |w|^2 - |v|^2\,]dt.$$

First we consider

$$J_{T,a}(w,v) = \int_{t_0}^{T} [\,|C_K x|^2 + 2w^T Kx + (1+a)|w|^2 - |v|^2\,]dt.$$

Then under (H2) the functional $J_{T,a}(w,v)$ is stricly convex-concave for all $0<T\leqq\infty$ and $J_{T,a}(w,v)\to\infty$ $(-\infty)$ as $|w|\to\infty$ ($|v|\to\infty$ respectively). Hence there exists a unique optimal pair (w_a,v_a) in $L_2(t_0,T)$ or in $L_2(t_0,\infty)$. If T is finite we have [2], [8]

$$w_a = -(1/1+a)B_1^T Q_{T,a} x$$
$$v_a = B_2^T Q_{T,a} x$$

where x is the solution of (9) corresponding to (w_a,v_a) and $Q_{T,a}$ is the solution of the Riccati equation

$$Q' + A_K^T Q + Q A_K + C_K^T C_K - (QB_1 + K^T)(1+a)^{-1}(B_1^T Q + K) + QB_2 B_2^T Q = 0, \tag{10}$$
$$Q(T) = 0. \tag{11}$$

If $T=\infty$ following [6], [7] we have

$$w_a = -(1/1+a)B_1^T Q_a x$$
$$v_a = B_2^T Q_a x$$

where x is the solution of (9) corresponding to (w_a,v_a) and Q_a is the bounded nonnegative solution of the Riccati equation (10). Moreover for each t, $Q_{T,a}(t)\uparrow Q_a(t)$ as $T\to\infty$ and Q_a is stable. But $Q_{T,a}$ and Q_a are nonnegative and monotone decreasing as $a\to 0$. Hence there exist limits $\bar{Q}_T$ and $\bar{Q}$. $\bar{Q}_T$ satisfies

$$Q' + A_K^T Q + Q A_K + C_K^T C_K - (QB_1 + K^T)(B_1^T Q + K) + QB_2 B_2^T Q = 0, \tag{12}$$
$$Q(T) = 0, \tag{13}$$

while $\bar{Q}$ is a bounded solution of (12). The Riccati equations (12) and (3) are the same and $\bar{Q}$ is a solution of (3).

Theorem 3. Suppose (H2) holds. Then $\bar{Q}$ is stable.

Proof. Since w_a and v_a, as functions of time, are in $L_2(t_0,\infty)$ the responce x_a is also in $L_2(t_0,\infty)$. The functions w_a and v_a are bounded in $L_2(t_0,\infty)$ and there exist a subsequence $(\tilde{w}_a,\tilde{v}_a)$ which is weakly convergent. Let $(\tilde{w},\tilde{v})$ be the limit and let $\tilde{x}$ be its responce. Then $\tilde{x}_a(t)\to\tilde{x}(t)$

as $a \to 0$ for each t. Hence $\tilde{x}$ is in $L_2(t_0,\infty)$. But $\tilde{x}$ is also the responce to the feedback law

$$\bar{w}=-B_1{}^TQx$$
$$\bar{v}=B_2{}^TQx.$$

Hence $\bar{Q}$ is stable.

If the system is periodic or time-invariant we have the following:

Corollary 1. If A, B_i and C are θ-periodic, then the bounded solutions of (3) in Theorems 2, 3 are also θ-periodic. If the matrices are constant, then (3) becomes an algebraic Riccati equation.

Remark 2. The result above establishes the necessary condition for the existence of γ-suboptimal controller in H -type problems [10], [7].

We have also a condition for nonexistence of a bounded solution to (3). Suppose there exists a bounded L such that the control problem

$$x' = A_L x + B_1 u, \quad x(t_0)=x_0, \quad A_L=A+B_2L$$

$$J_L(u)=\int_{t_0}^{\infty} [\,|Cx|^2-|Lx|^2+|u|^2]dt$$

has no finite solution ($J_L(u)=+\infty$) for some x_0. Then there is no bounded nonnegative solution to (3).

3. Tracking problem.

Consider

$$x' = Ax+B_1u+B_2v,$$
$$x(t_0)=x_0, \qquad (1)$$
$$y(t)=Cx(t),$$

$$J_T(u,v)=\int_{t_0}^{T} [\,|y-h|^2+|u|^2-|v|^2]dt \qquad (14)$$

where h is a bounded piecewise continuous function. As is known we have in general $J_T(u,0)\uparrow\infty$ as $T\uparrow\infty$. Hence the infinite horizon problem does not make sense [1]. But we can formulate a well-posed problem by introducing overtaking optimality [1], [9]. A feedback pair (u,v) is said to be admissible if it yields a bounded solution to (1) and u, v as functions of t are locally L_2. A pair $(\bar{u},\bar{v})$ is overtaking optimal if for any admissible $(\bar{u},v)$ and $(u,\bar{v})$ there exist $T(u)$, $T(v)$ such that

$$J_T(\bar{u},v)\leqq J_T(\bar{u},\bar{v}) \text{ for all } T\geqq T(v)$$
$$J_T(\bar{u},\bar{v})\leqq J_T(u,\bar{v}) \text{ for all } T\geqq T(u).$$

Suppose there exists a bounded stable solution to the Riccati equation (3). Then the following equation has a unique bounded solution on $[t_0,\infty)$ [6].

$$r' +[A-(B_1B_1^T-B_2B_2^T)Q]^Tr-C^Th=0.$$

Theorem 4. Suppose there exists a bounded stable solution Q to the Riccati equation (3). Then the feedback pair

$$\bar{u}=-B_1^T(Qx+r)$$
$$\bar{v}=B_2^T(Qx+r)$$

is overtaking optimal. For any admissible $(\bar{u},v)$ (respectively $(u,\bar{v})$) with $v\neq\bar{v}$,$(u\neq\bar{u})$ there exist $\varepsilon>0$ and $T(v)$ $(T(u))$ such that

$$J_T(\bar{u},v)\leqq J_T(\bar{u},\bar{v})-\varepsilon \text{ for all } T\geqq T(v)$$
$$J_T(\bar{u},\bar{v})+\varepsilon \leqq J_T(u,\bar{v}) \text{ for all } T\geqq T(u).$$

Proof. For any u, v in $L_{2loc}(t_0,\infty)$ we have

$$x(T)^TQ_T(T)x(T)+2r(T)^Tx(T)+J_T(u,v)=x_0^TQ_T(t_0)x_0+2r(t_0)^Tx_0$$
$$+\int_{t_0}^{T}[|u+B_1^T(Qx+r)|^2-|v-B_2^T(Qx+r)|^2+|h|^2-|B_1^Tr|^2+|B_2^Tr|^2]dt.$$

The assertion follows as in the single control case [9].

As in [3], [16] we assume that ideal bounded inputs u^*, v^* and response x^* exist such that

$$x^{*\prime} =Ax^*+B_1u^*+B_2v^*,$$
$$y^*(t)=Cx^*(t)=h(t).$$

Then a more natural choice of the payoff function is

$$J_T(u,v)=\int_{t_0}^{T}[|y-h|^2+|u-u^*|^2-|v-v^*|^2]dt \qquad (15)$$

Letting $\hat{x}=x-x^*$, $\hat{u}=u-u^*$ and $\hat{v}=v-v^*$ we can rewrite (1) and (15) as

$$\hat{x}' =A\hat{x}+B_1\hat{u}+B_2\hat{v}$$
$$J_T(\hat{u},\hat{v})=\int_{t_0}^{T}[|C\hat{x}|^2+|\hat{u}|^2-|\hat{v}|^2]dt$$

Hence under the condition of Theorem 4 the pair

$$\bar{u}=-B_1^TQ(x-x^*)+u^*$$
$$\bar{v}=B_2^TQ(x-x^*)+v^*$$

is overtaking optimal. It is also optimal for the infinite horizon problem (13) with $T=\infty$. Note that $(y-h)(t)\to 0$ as $t\to\infty$.

Example 1. Consider

$$x' = x+2u+v, \quad x(0)=x_0, \tag{16}$$

$$J_T(u,v)=\int_0^T [\,|x-h|^2+|u-u^*|^2-|v|^2]dt$$

where $h(t)$ and u^* are periodic functions of period 2 given by

$$h(t)=\begin{cases} t, & 0\leq t\leq 1 \\ 2-t, & 1\leq t\leq 2 \end{cases}, \qquad u^*(t)=\begin{cases} 1/2, & 0\leq t\leq 1 \\ -1/2, & 1<t<2 \end{cases}$$

and $x^*=h$, $v^*=0$. In fact in this case x^*, u^*, v^* satisfy (16) with $x_0=0$ and $x^*=h$. They are obtained by Laplace transform [3]. The Riccati equation here is algebraic and $3q^2-2q-1=0$. It has a unique positive solution $q=1$. The optimal pair is

$$\bar{u}=-2(x-h)+u^*$$

$$\bar{v}=x-h.$$

This example satisfies the condition of Lemma 3 and (H2) with $K>4/3$.

References.

[1] Z. Artstein and A. Leizarowitz, Tracking periodic signals with the overtaking criterion, IEEE Trans. Automat. Contr., AC-30 (1985), 1123-1126.

[2] A. Bensoussan, Saddle points of convex concave functionals with application to linear quadratic differential games, Differential Games and Related Topics, H.W. Kuhn and G.P. Szego, eds., American Elsevier, New York, 1971, 177-199.

[3] F. Bouslama and A. Ichikawa, Remarks on tracking problems, 12th SICE Symposium on Dynamical System Theory, Toyohashi, Japan, 1989.

[4] G. Da Prato and A. Ichikawa, Liapunov equations for time-varying linear systems, Systems Control Letters 9 (1987), 165-172.

[5] G. Da Prato and A. Ichikawa, Quadratic control for linear periodic systems, Appl. Math. Optim., 18 (1988), 39-66.

[6] G. Da Prato and A. Ichikawa, Quadratic control for linear time-varying systems, SIAM J. Control Optimiz., 28 (1990), 359-381.

[7] J.C. Doyle, K. Glover, P.P. Khargonekar and B.A. Francis, State-space solutions to standard H_2 and H_∞ control problems, IEEE Trans. Automat. Contr., AC-34 (1989), 831-847.

[8] A. Ichikawa, Linear quadratic differential games in a Hilbert space, SIAM J. Control Optimiz., 14 (1976), 120-136.

[9] A. Ichikawa, Optimal overtaking control for linear time-varying systems, Int. J. Control (1990), to appear.

[10] A. Ichikawa and H. Katayama, Differential games and H_∞-type problems for time-varying systems, to appear.

[11] D.H. Jacobson, On values and strategies for infinite-time linear quadratic games, IEEE Trans. Automat. Contr., AC-22 (1977), 490-491.

[12] E.F. Mageirou, Values and strategies for infinite time linear quadratic games, IEEE Trans. Automat. Contr., AC-21 (1976), 547-550.

[13] M. Pachter, Some properties of the value matrix in infinite-time linear-quadratic differential games, IEEE Trans. Automat. Contr., AC-23 (1978), 746-748.

[14] I.R. Petersen, Linear quadratic differential games with cheap control, Systems Control Letters, 8 (1986), 181-188.

[15] I.R. Petersen, Some new results on algebraic Riccati equations arising in linear quadratic differential games and the stabilization of uncertain linear systems, Systems Control Letters, 10 (1988), 341-348.

[16] M. Suzuki and K. Shimizu, Synthesis of optimal servosystems by dynamic compensator, 12th SICE Symposium on Dynamical System Theory, Toyohashi, Japan, 1989.

TRACKING STRATEGIES FOR ASYMPTOTICALLY ADMISSIBLE TARGET PATHS IN ECONOMIC MODELS

Jacob C. Engwerda
Tilburg University,
Hogeschoollaan 225, 5037 GC Tilburg
The Netherlands

Abstract

In this paper we discuss the pros and cons of several optimal control policies that can be used to track desired target paths which are asymptotically admissible.
To that end we first properly define what we mean by an asymptotically admissible target path, and give a characterization of these paths in case the considered system is described by a linear, finite dimensional, time-varying difference equation.

I Introduction

In the past much research has been done on the subject of designing, for a dynamic economic model, a policy yielding some predescribed desired behaviour. In literature it is usually known as the theory of optimal stabilization see e.g. [1,5,14,17,18, 19]. In this theory, however, mostly no attention is paid to the question whether a prescribed target path can be tracked at all. Only indirectly by considering an infinite planning horizon, which then usually gives rise to the imposition of rather severe additional conditions on the system, attention has been paid to this problem. Contrarily, much attention has been addressed to the problem whether it is possible to track (exactly) a given set of economic target variables along any time path by means of an appropriate choice of the policy instruments. This so-called Target Path Controllability (TPC) problem was studied e.g. in [2,3,4,6,12,13,15,16,17,20]. So, on the one hand much work has been done to obtain conditions on the system which guarantee tracking of any desired target path, whereas on the other hand this tracktability property of a desired trajectory is almost neglected in the design of optimal control strategies.

In this paper we view the TPC-problem as a problem of optimal stabilization. That is, we first give a characterization of all target paths which can be ultimately tracked and next pay attention to the question how the goal can be optimally achieved.

The optimality criteria we deal with are derived from the quadratic cost criterion as considered e.g. in [14]. This criterion expresses that positive and negative deviations of target variables from desired levels are weighted equally and that they are increasingly costly. We consider in section 2 two special cases of this criterion. The first one is obtained by taking the planning horizon equal to one. The advantage of the corresponding optimal controller, the so-called Minimum Variance (MV) controller, is that it is a rather simple one and that it immediately takes account of changes in the system.

In the second criterion, we consider, the planning horizon is extended to infinity. This criterion can be viewed as an approximation of a criterion with a large planning horizon. The advantage of the resulting controller is that it always stabilizes the closed-loop (CL) system. Since most economic models are just approximations of some observed process it is in fact a must that any controller that is used to regulate the process has this stabilization property.

As the MV controller not always has this stabilization property, we study in section 3 its robustness property in more detail. Since it strongly depends on the chosen weighting matrix in the criterion, and there is often much freedom in choosing this matrix, we discuss weight matrices yielding a deadbeat and minimal norm MV controller, respectively.

This paper extends and summarizes to some extent results reported in [8,9].

II Preliminaries and Optimal Controllers

The base system analyzed in this paper is described by the following linear, finite dimensional, time-varying difference equation:

$$\Sigma_{yd}: \quad \begin{aligned} x(k+1) &= A(k)x(k) + B(k)u(k) + G(k)d(k); \; x(k_0) = x_0 \\ y(k) &= C(k)x(k), \end{aligned}$$

where $x(k)$ is an n-dimensional vector consisting of endogenous variables (state variables) observed at time k; $u(k)$ is an m-dimensional vector of policy instruments (control variables); $d(k)$ is an s-dimensional vector of deterministic (non-controllable) variables, assumed to be known at time k; $y(k)$ is an r-dimensional vector of target variables, and x_0 is the initial state of the system. We assume that the system parameters (i.e. the matrices A(k), B(k), G(k) and C(k)) are bounded in time. Throughout the paper we use $|A| := (\sum_{i,j} a_{ij}^2)^{\frac{1}{2}}$, for a matrix A, to denote its Euclidean norm, A^T to denote its transpose and A^+ for its Moore-Penrose inverse (see [10, section 12.8]).

Definition 1

A reference trajectory $y^*[0,.]$ is called asymptotically admissible for x_0 at time $k = 0$, if there exists a control sequence $u[0,.]$ such that $|y(k) - y^*(k)| \to 0$ for $k \to \infty$. In that case we call u(.) a successful control.

From [9] we have the following characterization for asymptotically admissible target paths.

Theorem 2

A reference trajectory $y^*[0,.]$ is asymptotically admissible for x_0 at time $k = 0$, if and only if (iff), there exists $u^*[0,.]$ and $v[0,.]$ such that:

$$\begin{aligned} y^*(k) &= C(k)x^*(k) + v(k) \\ x^*(k+1) &= A(k)x^*(k) + B(k)u^*(k) + G(k)d(k), \; x^*(0) = x_0. \end{aligned} \tag{1}$$

where $v(.) \to 0$. □

Next consider the minimum variance cost functional

$$J : |y(k+1) - y^*(k+1)|^2_{Q(k+1)} \tag{2}$$

where $y^*(.)$ is a prespecified desired trajectory for y(.), Q is a semi-positive definite weight matrix and the abbreviation $|x|_Q^2$ is used to denote x^TQx. Note that since matrix Q is semi-positive definite, we can factorize it as S^TS. This decomposition will be often used. Moreover, we introduce the notation P(k,Q) to denote the matrix $I - B(k)(S(k+1)C(k+1)B(k))^+S(k+1)C(k+1)$. We proceed with stating the optimal controller (the so-called Minimum Variance (MV) controller) that results if we minimize this cost functional J w.r.t. Σ_{yd}. A proof of this result is given in [6].

Theorem 3

An optimal controller for the minimization problem $\min_{u(k)} J$ w.r.t. Σ_{yd} is given by

$$u(k) = -(S(k+1)C(k+1)B(k))^+S(k+1)(C(k+1)A(k)x(k) + C(k+1)G(k)d(k) - y^*(k+1)) \tag{3}$$

Moreover, the resulting command error, $y(k+1) - y^*(k+1)$, equals

$$C(k+1)P(k,Q)A(k)x(k) + C(k+1)P(k,Q)G(k)d(k) +$$

$$+ (C(k+1)B(k)(S(k+1)C(k+1)B(k))^+S(k+1) - I)y^*(k+1) \qquad (4)$$

□

Note that by manipulating the Q(.) matrix, different controllers are obtained. The set consisting of all controlers of this type is denoted by $\mathcal{U}_{MV} := \{u(k) \mid \exists Q > 0$ such that $u(k)$ satisfies (3)$\}$. From [8] we have the following result.

Theorem 4

If Q is such that the system $x(k+1) = (PA)(k)x(k)$ is asymptotically stable, then every asymptotically admissible target path is tracked by the corresponding MV-controller (3). □

From this result, it is obvious why the question to determine the subset of stabilizing controllers in the set $\mathcal{U}_{MV}$ is an interesting one. In section 3 we go into more detail on this question.

Next, we consider an optimal controller minimizing an infinite planning horizon. The advantage of this controller is that under rather mild conditions it stabilizes the closed-loop system. So, it is robust for small perturbations. Its disadvantage is, that it requires much information concerning the future development of the system parameters, a property which is only rarily satisfied by economic systems. For that reason we do not elaborate on this controller here.

Theorem 5

Assume that (A(.), B(.)) is uniformly periodically smoothly exponentially stabilizable and (A(.), C(.)) is periodically smoothly exponentially detectable (see [6] for a formal definition).
Moreover let $y^*(k)$ be an asymptotically admissible target path, decomposed as (1).
Then the controller

$$u(k) = u^*(k) - F(k)(x(k) - x^*(k))$$

where $F(k) = (I + B^T(k)K(k+1)B(k))^{-1}B^T(k)K(k+1)A(k)$

and K(k) is given by

$$K(k) = \lim_{N\to\infty} K_N(k)$$

where $K_N(t) = A^T(t)\{K_N(t+1) - K_N(t+1)B(t)(I + B^T(t)K_N(t+1)B(t))^{-1}B(t)K_N(t+1)\}$
$A(t) + C^T(t)C(t)$

$K_N(N) = C^T(N)C(N)$, is successful.

Moreover, this controller stabilizes the closed-loop system and whenever $v(k) \in \text{Ker } C(k)\ \forall k$, minimizes the cost criterion

$$\lim_{N\to\infty} \sum_{k=0}^{N-1} \{|y(k) - y^*(k)|^2 + |u(k) - u^*(k)|^2\} + |y(N) - y^*(N)|^2$$

subject to Σ_{yd}. □

Proof

Consider the minimization problem

$$\min_{u[0,.]} \lim_{N\to\infty} \sum_{k=0}^{N-1} \{|y(k) - y^*(k)|^2 + |u(k) - u^*(k)|^2\} + |y(N) - y^*(N)|^2$$

subject to Σ_{yd} and (1).

Let $x'(k) := x(k) - x^*(k)$ and $u'(k) := u(k) - u^*(k)$.

With this notation, the problem is equivalent with

$$\min_{u'[0,.]} \lim_{N\to\infty} \sum_{k=0}^{N-1} \{|C(k)x'(k) - v(k)|^2 + |u'(k)|^2\} + |C(N)x'(N) - v(N)|^2$$

subject to $x'(k+1) = A(k)x'(k) + B(k)u'(k)$; $x'(0) = 0$.

Since $C(k)x'(k)$ and $v(k)$ are perpendicular, the problem reduces to:

$$\min_{u'[0,.]} \lim_{N\to\infty} \sum_{k=0}^{N-1} \{|x'(k)|^2_{C^T(k)C(k)} + |u'(k)|^2\} + |x'(N)|_{C^T(N)C(N)} + \sum_{k=0}^{N} |v(k)|^2$$

subject to $x'(k+1) = A(k)x'(k) + B(k)u'(k)$; $x'(0) = 0$.

This minimization problem is a special case of the more general EQL-problem studied in [6, section 5]. Therefore application of these results yield the stated result. □

III The MV-controller

In section 2 we noted that the weight matrix in the MV-controller can be used to manipulate the stability of the closed-loop system matrix PA.
To prove our first result, on this subject, we need the following lemma.

Lemma 6

$$\min_F |A + BFC| \text{ is obtained by } F = -B^+AC^+$$

Proof
Rewrite $A + BFC$ as $(I - BB^+)A + BB^+A(I - C^+C) + (BB^+AC^+C + BFC)$. Since both $B^T(I - BB^+) = 0$ and $(I - C^+C)C^T = 0$, we have that

$$|A + BFC|^2 = |(I - BB^+)A|^2 + |BB^+A(I - C^+C)|^2 + |B(B^+AC^+ + F)C|^2.$$

Consequently $\min_F |A + BFC|^2$ is obtained by $F = -B^+AC^+$. □

In fact this lemma states that, whenever static output injection is used to control the system, $u(k) = -B^+AC^+y(k)$ yields a closed-loop system matrix which has a minimal Euclidean norm.
From this lemma it is clear that if the system is full-state observable (i.e. matrix C is invertible), then the choice $S = C^{-1}$ in (3) yields an MV controller that additionally minimizes the Euclidean norm of the closed-loop system matrix PA.
A second interesting case to investigate is, which S-matrices yield PA = 0 or, more generally for time-invariant systems, $(PA)^n = 0$. To study this problem we note first that without loss of generality we can assume that matrix B is injective. This, since whenever B is not injective there exist an unitary matrix U and injective matrix $\bar{B}$ such that $BU = (\bar{B}\ 0)$. Elementary calculations then show that matrix PA equals $I - \bar{B}(SC\bar{B})^+SC$, which proves this assertion. We have the following two results.

Theorem 7
Let B be injective.
Then there exists an invertible matrix S such that PA = 0 iff
i) $\operatorname{Im} A \subset \operatorname{Im} B$; ii) CB is injective.

Proof
"⇐" Since $\operatorname{Im} A \subset \operatorname{Im} B$, there exists an matrix X such that $A = BX$.

Consequently $PA = B(I - (SCB)^{+}SCB)X$. As both S and CB are injective it is clear now that $PA = 0$.

"$\Longrightarrow$" From $PA = 0$, we deduce that $A = B(SCB)^{+}SCA$. This immediately yields i). Using this result, we know that there exists a surjective matrix X such that $A = BX$. So $PA = 0$ can be rewritten as $B(I - (SCB)^{+}SCB)X = 0$. Since B is injective and X surjective we consequently obtain that $I - (SCB)^{+}SCB = 0$. Or, stated differently, $(SCB)^{+}SCB = I$. This last equality implies however that SCB is injective. As S is invertible this finally yields ii). □

Theorem 8

Let B be injective and (A,B) controllable. Then there exists a matrix S such that $(PA)^{n} = 0$ if one of the following two conditions is satisfied

i) C is injective

ii) C is surjective, CB is invertible and $\text{Im } C^{T} = \text{Im } \bar{S}^{T}\bar{S}B$, where $\bar{S}$ is such that $\bar{S}A\bar{S}^{-1}$ is in Luenberger's phase canonical form (see [11]).

Proof

i) Due to our assumptions we have that SCB is injective. Consequently $PA = (I - B(B^{T}C^{T}S^{T}SCB)^{-1}B^{T}C^{T}S^{T}SC)A$. From [7] we have that, since (A,B) is controllable, there exists a positive definite matrix $\bar{Q}$ such that $(I - B(B^{T}\bar{Q}B)^{-1}B^{T}\bar{Q})A$ is nilpotent. Now, let $W^{T}DW$ and $U^{T}(\Sigma\ 0)V$ be singular value decompositions of $\bar{Q}$ and C, respectively.

Next choose $S = \begin{bmatrix} D^{\frac{1}{2}}W\ U^{T}\Sigma^{-1} & 0 \\ 0 & I \end{bmatrix}V$. Then it is clear that S is invertible, and elementary calculations show that with this choice of S, $\bar{Q} = C^{T}S^{T}SC$, which proves our claim.

ii) Since (A,B) is controllable, there exist nonsingular transformation matrices $\bar{S}$ and $\bar{T}$ such that $\bar{S}A\bar{S}^{-1}$ and $\bar{S}B\bar{T}$ are in Luenberger's Phase canonical form. Using these transformation matrices PA can be rewritten as $\bar{S}^{-1}[(I - \bar{S}B\bar{T}(SC\bar{S}^{-1}\bar{S}B\bar{T})^{+}SC\bar{S}^{-1}\bar{S}A\bar{S}^{-1}]\bar{S}$. We will show now that there exists an invertible matrix S such that $SC\bar{S}^{-1} = \bar{T}^{T}B^{T}\bar{S}^{T}$ (i). A similar reasoning as in [7] then shows that with this choice of S, PA is nilpotent.

To show that (i) is solvable, we rewrite this equation as $SC = \bar{T}^{T}B^{T}\bar{S}^{T}\bar{S}$. From elementary matrix calculation we have that there exists an invertible matrix X such that $XC = B^{T}\bar{S}^{T}\bar{S}$ iff $\text{Im } C = \text{Im } B^{T}\bar{S}^{T}\bar{S}$. So, by taking $S = \bar{T}^{T}X$ we see that (i) is satisfied.

□

Note that in theorem 7 matrix S can be chosen arbitrarily, which is clearly not the case in theorem 8.

We conclude this paper by considering the question under which conditions it is possible to make the command error (4) zero at any time. To prove a result on this subject we need the following lemma.

Lemma 9

$\exists S$, invertible, such that $H := (I - CB(SCB)^{+}S)M = 0$ iff $\operatorname{Im} M \subseteq \operatorname{Im} CB$.

Proof

"$\Rightarrow$" Since $(I - CB(SCB)^{+}S)M = 0$, we have that $M = CB(SCB)^{+}SM$

As $\operatorname{Im}(SCB)^{+}SM \subseteq \mathbb{R}^{m}$ it is clear that $\operatorname{Im} M \subset \operatorname{Im} CB$, which yields the stated result.

"$\Leftarrow$" Let $S \in \mathbb{R}^{r\times r}$ be an arbitrary, invertible matrix. Consider now matrix H. Obviously $H = S^{-1}SH$. So H can be rewritten as $S^{-1}(I - SCB(SCB)^{+})SM$.

Since $\operatorname{Im} M \subseteq \operatorname{Im} CB$ there exists a matrix T such that $M = CBT$. So,

$H = S^{-1}(I - SCB(SCB)^{+})SCBT = S^{-1}(SCB - SCB(SCB)^{+}SCB)T = 0$, which proves the claim. □

Note that in fact we showed that, like in theorem 7, matrix S can be chosen arbitrarily in case the condition is satisfied. By taking M in this lemma I, CA and C, respectively, we obtain the next corollary. In theorem 11 and 12 we use this result to give conditions under which the MV-controller yields a closed-loop system in which the command error is zero and depends only on v(.), respectively.

Corollary 10

There exists an invertible matrix S such that

i) $I - CB(SCB)^{+}S = 0$ iff rank $CB = r$

ii) $CPA = 0$ iff rank $CA \le$ rank CB

iii) $CP = 0$ iff rank $C =$ rank CB

Theorem 11

Let S be an arbitrary invertible matrix; x(0), d(0) and y*(1) arbitrarily vectors, and u_{MV} the corresponding MV controller.

Then the command error $y(1) - y^*(1)$ equals zero iff CB is surjective.

Proof

From (4) we have that the command error equals zero for arbitrarily chosen x(0), d(0) and y*(1), iff the matrices CPA, CPG and $CB(SCB)^{+}S - I$ are zero. Since both CPA and CPG are zero whenever $CB(SCB)^{+}S - I$ is zero, it suffices to give conditions under which $CB(SCB)^{+}S - I = 0$. According to corollary 10.i) this is the case iff. CB is surjective, which completes the proof. □

Theorem 12

Let $c > 0$, S(.) a sequence of bounded invertible matrices with the additional property that all non-zero singular values of SCB(.) are at any time k greater than c, and u_{MV} the corresponding MV controller.

If rank $C(k+1)A(k) \le$ rank $C(k+1)B(k)$ $\forall$ k, then u_{MV} is successful. Moreover, the closed loop command error only depends on v(.) (see (1)).

Proof

Let $y^*(.)$ be an asymptotically admissible target path. According to theorem 2 there exist $u^*[0,.]$ and $v[0,.]$ such that $y^*(k)$ satisfies (1).

Using (4), we consequently have that the command error

$$y(k+1) - y^*(k+1) = C(k+1)P(k,Q)A(k)(x(k) - x^*(k)) +$$

$$C(k+1)B(k)(S(k+1)C(k+1)B(k))^+S(k+1) - I)v(k+1) \quad \text{(i)}$$

Due to our rank condition we have (see corollary 10.ii)) that CPA = 0.

From [10], we have that $SCB = U\Sigma V^T$, where both U and V are column unitary matrices and Σ is a diagonal matrix containing on the diagonal all non-zero-singular values of SCB. Since by assumption all these singular values are larger than c, $(SCB)^+(.)$ $(= V\Sigma^{-1}U^T(.))$ is bounded. Consequently, the last part of (i) converges to zero if k converges to infinity. □

CONCLUSION

In this paper we considered two optimal controllers for tracking asymptotically admissible target paths. The first one we considered, the LQ optimal controller, turned out to be successful under rather smooth conditions on the system parameters. The disadvantage of it is, however, that its implementation requires much information concerning the future development of the system.

The second optimal controller we considered, the minimum variance controller, has not this disadvantage. But, unfortunately, this one is not always successful. In case the weight matrix in the corresponding cost criterion is chosen such that the resulting optimal controller stabilizes the closed-loop system the successfulness is guaranteed. So, this is a selection criterion for the weight matrix. Another selection criterion is provided by the requirement that the proposed controller should be robust for small parameter perturbations.

Motivated by these two selection criteria we considered the question under which conditions the minimum variance controller becomes a deadbeat and minimum norm controller, respectively. We derived elementary results on these topics, though the problems in their most general form still remain to be solved.

REFERENCES

1. Aoki, M., Notes and Comments: On sufficient conditions for optimal stabilization policies, Rev. Econom. Studies 47 (1973), 131-138.
2. Aoki, M., On a generalization of Tinbergen's condition in the theory of policy to dynamic models; Rev. Econom. Studies 42, (1975) 293-296.
3. Aoki, M. and Canzoneri, M., Sufficient conditions for control of target variables and assignment of instruments in dynamic macroeconomic models; Internat. Econom. Rev. 20, no. 3, 605-616.
4. Brockett, R.W. and Mesarovic, M.D., The reproducibility of multivariable systems; Math. Anal. Appl. 11, (1965), 548-563.
5. Chow, G.C., Analysis and Control of Dynamic Economic Systems, John Wiley & Sons, New York, 1975.
6. Engwerda, J.C., Regulation of Linear Discrete Time-Varying Systems, Ph.D. Thesis Eindhoven Technical University, The Netherlands, 1988.
7. Engwerda, J.C. and Otter, P.W., On the choice of weighting matrices in the minimum variance controller; Automatica 25, (1989), 279-287.
8. Engwerda, J.C., Tracking strategies for asymptotically admissible target paths in economic models; Internal report Tilburg University, The Netherlands, 1989.
9. Engwerda, J.C., Admissible target paths in economic models; Internal report Tilburg University, The Netherlands, 1989.
10. Lancaster, P. and Tismenetsky, M., The Theory of Matrices, second edition, Academic Press, London, 1985.
11. Luenberger, D.G., Canonical forms for linear multivariable systems, IEEE AC 12, (1967), 290-293.
12. Maybeck, P.S., Stochastic Models, Estimation and Control vol. 141-3 in the series Mathematics in Science and Engineering, Academic Press, London, 1985.
13. Nijmeijer, H., On dynamic decoupling and dynamic path controllability in economic systems; J. Econom. Dynamics Control 13, (1989), 21-39.
14. Pindyck, R.S., Optimal Planning for Economic Stabilization, North Holland, Amsterdam, 1973.
15. Preston, A.J., A paradox in the theory of optimal stabilization; Rev. Econom. Studies 39, (1972), 423-432.
16. Preston, A.J., A dynamic realization of Tinbergen's theory of policy; Rev. Econom. Studies 41, (1974), 65-74.
17. Preston, A.J. and Pagan, A.R., The Theory of Economic Policy, Cambridge University Press, New York, 1982.
18. Tinbergen, J., On the Theory of Economic Policy, North Holland, Amsterdam, 1952.
19. Turnovsky, S.J., 1977, Applications of Control Theory to Economic Analysis, North Holland, Amsterdam, 1977.
20. Wohltmann, H.W., Target Path Controllability of linear time-varying dynamical systems; IEEE Trans. Automat. Control 30, (1985), 84-87.
21. Zeeuw, A.J. de, Difference Games and Linked Econometric Policy Models, Ph.D. Thesis Tilburg University, The Netherlands, 1984

NUMERICAL SOLUTION OF LINEAR DIFFERENTIAL GAMES

N.D.Botkin, M.A.Zarkh,V.S.Patsko
Institute of Mathematics and Mechanics
Ural Branch of the Academy of Sciences
Sverdlovsk, USSR

1. Introduction

Speaking about differential games (DGs) we shall bear in mind the aircraft landing problem in the presence of wind disturbance as an example. The landing begins from the hight of 400 m and continues 120 sec approximately.

It is necessary to control aircraft so that its deviation from the nominal trajectory would not be too much and in the moment of crossing the runway (RW) threshold parameters of motion would be permissible. We consider that wind disturbance, which interferes control process, rises from wind microburst. The microburst is caused by falling mass of air, which hits the ground surface and gives vortex. When aircraft crosses the microburst zone the value and direction of the wind change sharply along the aircraft motion.

Extension and configuration of the microburst and distribution of wind field inside could be regarded as known under too idealistic consideration only. More realistic we can hope to have information only about deviations of the wind value and its direction from some middle values.

So we meet some mathematical task, formulated as a differetial game (DG): equations of aircraft dynamics and restrictions of its control parameters are given, restrictions upon the disturbance parameters are known also. The question arises about an optimization of guaranteed result.

Nowadays the DGs theory is the developed mathematical discipline [1-8].

Essential results have been achieved, particularly, by mathematicians in Sverdlovsk: conception of positional DGs was developed, universal ways for construction of optimal strategies were suggested, and now numerical methods and algorithms are devising. These are some key-words, typical for Sverdlovsk mathematical school on DG: discrete scheme of control, maximal stable bridge, extremal strategy. Main results are stated in monographs [4,7,8,9].

Devised in Sverdlovsk numerical methods concern both linear and nonlinear problems. In this paper we are dealing with some methods, namely those, which can be applied for solving the aircraft landing problem. We shall suppose that aircraft dynamics equations are linearized respectively the nominal motion and so DG is linear. This linear DG can be considered as an auxiliary game for the original nonlinear system. Closed-loop control methods (strategies) found from solving linear DG are applied then to the original system.

2. Linear DG with fixed terminal time

The standard form of linear DG is following

$$\dot{x} = A(t)x + B(t)u + C(t)v \qquad (2.1)$$

$$x \in R^n, \quad u \in P, \quad v \in Q, \quad T, \quad \varphi(x(T)) .$$

Here x is the state vector, u is the control parameter of the first player, v is one of the second player. Compact sets P, Q restrict controls of players.

The terminal time T of DG is considered to be fixed. The quality of process is evaluated by the convex payoff function φ, which calculated at the terminal moment.

Very often one passes [7,9] from the game (2.1) to the equivalent DG of the form

$$\dot{y} = D(t)u + E(t)v$$

$$D(t) = X(T,t)B(t) , \quad E(t) = X(T,t)C(t) \qquad (2.2)$$

$$u \in P, \quad v \in Q, \quad T, \quad \varphi(y(T)) .$$

The pass is realised by transformation $y(t) = X(T,t)x(t)$, where $X(T,t)$ is the Cauchy matrix, corresponding to the matrix A of system (2.1). The advantage of DG (2.2) over DG (2.1) is that the state variable is absent in the right side, simplifying writing. More than that, in the case, when payoff function φ depends upon some m coordinates of the state vektor only, we can reduce dimension of the equivalent DG and make it equal m. For this it is necessary to use m corresponding rows of Cauchy matrix in performing pass from variable x to variable y.

3. Switch surfaces

In the Institute of Mathematics and Mechanics of the Ural Branch of the USSR Academy of Sciences effective methods and algorithms have been devised for solving linear DGs with fixed terminal moment and convex payoff function, which depends upon two, three and more coordinates of the state vector.

The numerical procedures are founded on construction in coordinates of the equivalent DG (2.2) of sections of level sets for DG value function Γ. Every section corresponds to some definite moment on the time axis. Giving number c, we find corresponding level set

$$W_c(T) = \{ y \in R^m : \varphi(y) \le c \}$$

of payoff function φ, and moving contraward in time from terminal moment T, we construct sections $W_c(t_i)$ (Fig.1) of level set of value function Γ, using chosen net for calculation.

Contraward constructions in DG theory ascend to works of R.Bellman, R.Isaacs, W.Fleming, L.S.Pontryagin, B.N.Pshenichnii. In the case of linear DGs with fixed terminal time and convex payoff function, the main difficulty to perform pass from the current section $W_c(t_i)$ (constructed for moment t_i) to the next section $W_c(t_{i+1})$ is connected with the procedure of convexing for positively-homogeneous function. Complexity of this procedure grows essentially with increasing of dimension m. There is specifical facilitation, so as before beginning the convexity procedure we have information about place of violation of local convexity. Such specifics allowes to create very fast algorithms for contraward constructions.

Dealing with level sets of value function, we can construct optimal strategies both for the first and the second players. Most clearly it appears in the case when control parameter of the first player or the second player is scalar.

Suppose, that control parameter of the first player is scalar, namely $| u(t) | \le \mu$. Then $D(t)$ is a vector with dimension m. For every t we find in space R^m the set of all points so that for every point there is a vector from vectors of the subdifferential of the value function $y \to \Gamma(t,y)$, which gives zero scalar product with vector $D(t)$. This set generates "the surface", which divides the space R^m into two parts. In the part, where the vector $D(t)$ is directed, the optimal control parameter of the first player has the value $-\mu$ in the moment t, on the other side from the surface the optimal value is $+\mu$. Just upon the switch surface one can take arbitrary values from the segment $[-\mu, \mu]$. This way of the first player control was grounded in [10,11]. It is stable in respect to errors of numerical construction of the switch surfaces. When we use the discrete scheme of control, the switch surfaces are to be constructed before upon the given net of the time moments.

The most simple constructions are carried out in the case $m = 2$. Here we have switch lines. During calculation these lines are maintained from segments (Fig.2). Numerical constructed switch lines for control law aircraft landing problem are shown in Fig.3, where $\tau = T - t$.

Analogically the optimal strategy of the second player is constructed by means of the switch surfaces in the scalar case $|v(t)| \leq \nu$. Here the vector $E(t)$ is using. But in contrast to the first player, the optimal strategy of the second player is not stable [12].

Let now the control of the first player (or the secont player) is a vector u (v) , which components u_j (v_k) are restricted by independent conditions $|u_j| \leq \mu_j$ ($|v_k| \leq \nu_k$). In this case it is possible to construct own switch surface for every component u_j (v_k) , using j (k) column of matrix $D(t)$ ($E(t)$) . So we shall have the set of switch surfaces for every time moment t . Method of closed-loop control, which uses such sets, gives an optimal result under special suggestions.

In conclusion of this section, note we can use contraward procedures for immediate constructing of value function epigraph. For this we use contraward method for definite corresponding DG in space with dimension increased by 1. In Fig.4 we show graph of value function $y \to \Gamma(t, y)$ in the model DG

$$\dot{x}_1 = x_2 + v$$
$$\dot{x}_2 = u , \qquad |u| \leqslant 1 , \ |v| \leqslant 1$$

with terminal moment T and payoff function $\varphi(x) = \max\{|x_1|, |x_2|\}$ for the moment $t = T-2$ in equivalent coordinates $y(t) = X(T,t)x(t)$.

4. Aircraft landing problem

The aircraft motion during landing is described by a differential equations system of 22-th order. The first 12 equations are correspond to trajectory and angle motion. The last 10 ones imitate the inertionality of control devices and inertial character of wind velocity along the motion. The control factors are: deviations of the elevator, the rudder, the ailerones and change of thrust force. The disturbance vector consist of three wind components. The linearization of the system with respect to the nominal motion gives linear controllable system, which desintegrates into two subsystems of vertical (longitudinal) and lateral motions.

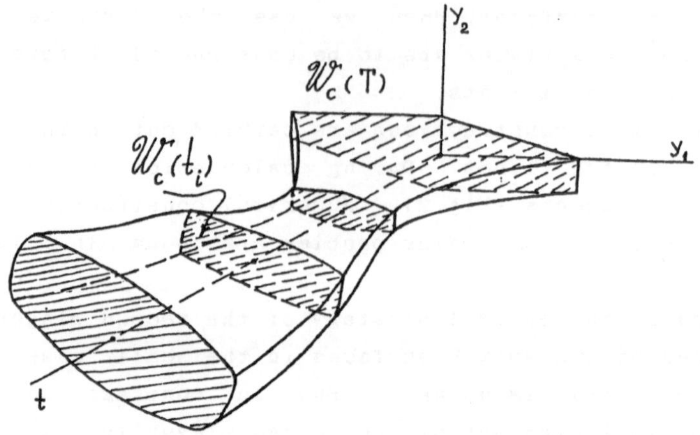

Fig.1.Maximal stable bridge

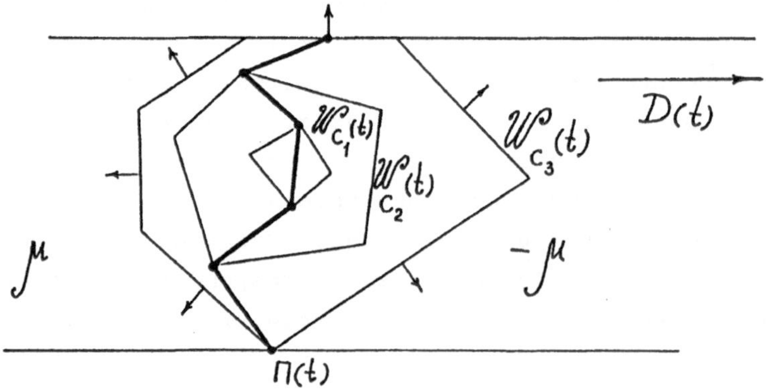

Fig.2.Switch line construction

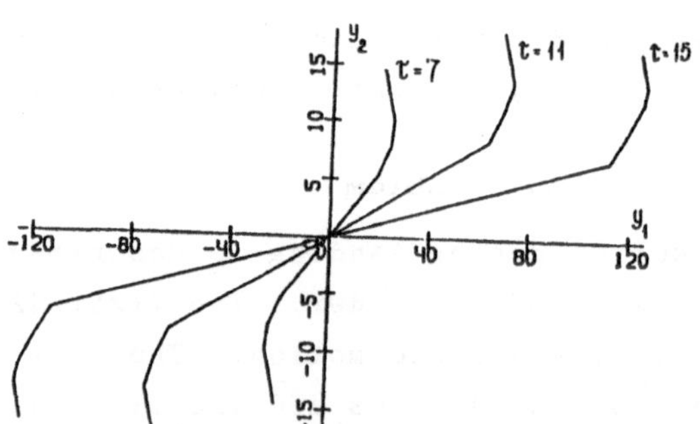

Fig.3.Switch lines

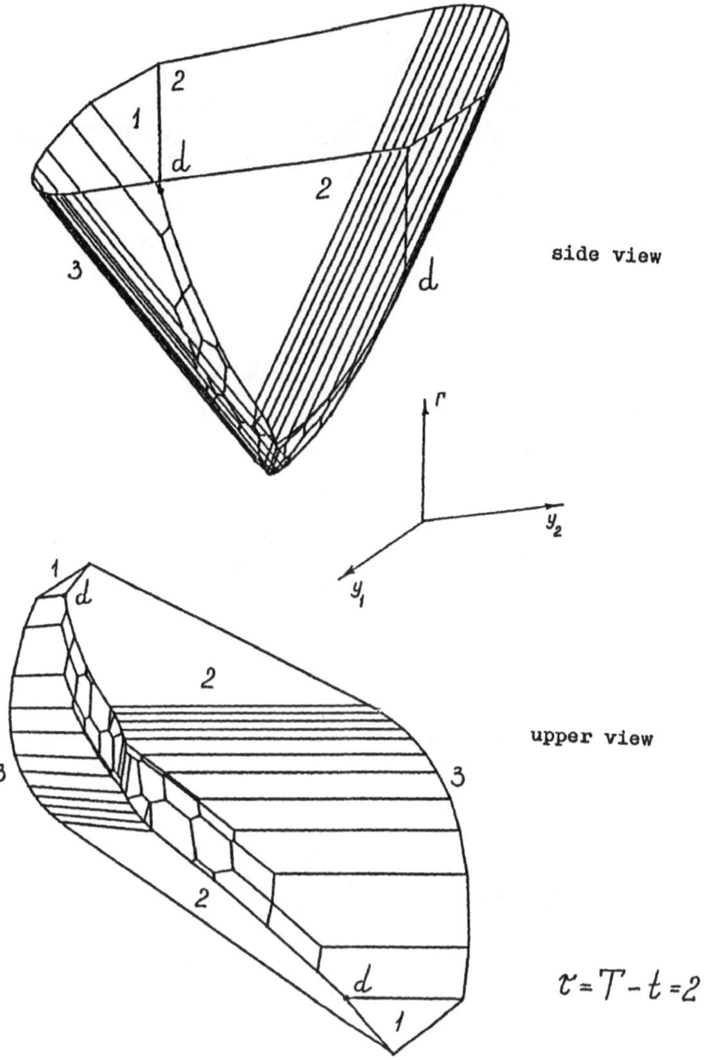

Fig. 4 .Value function

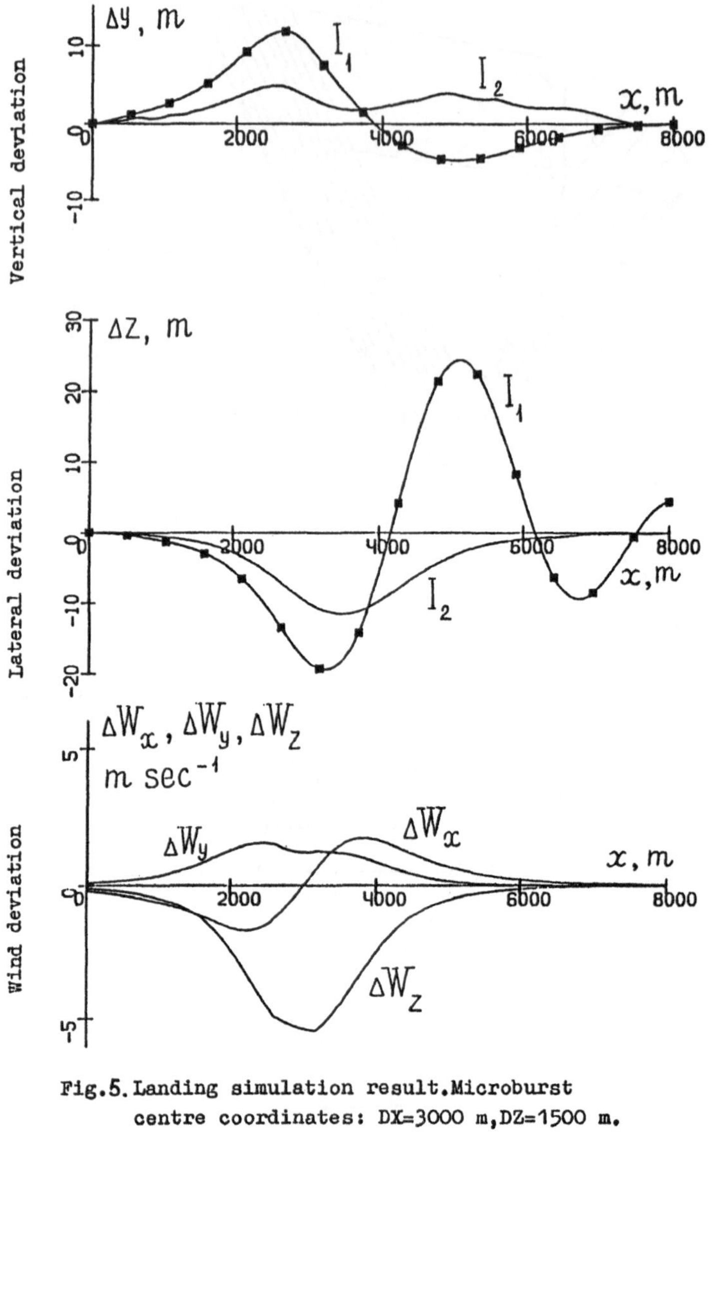

Fig.5. Landing simulation result. Microburst centre coordinates: DX=3000 m, DZ=1500 m.

For each of the subsystems we consider an auxiliary differential game with fixed terminal time T and convex payoff function depending on two state vector coordinates at the moment T . Solving the auxiliary problems on computer, we find optimal strategies for control parameters which are realized by means of switch lines.

Simulating original nonlinear system motions,we suppose that the wind disturbance is coused by the aircraft flight through the microburst zone. The microburst model we used has been taken from the paper [13].

Consider two methods of control. The method I_1 uses accepted nowadays autopilot algorithms. These algoritms are founded on the linear theory of automatic control. In the method I_2 control factors are constructed by means of switch lines obtained from the auxiliary differential games.

Simulation results for the control methods I_1, I_2 are shown in Fig 5. We give graphs of vertical Δy and lateral Δz deviations from the nominal motion and also realizations of wind velocity deviations ΔW_x , ΔW_y , ΔW_z. It can be seen that the results for the minimax method I_2 are better than for the traditional method I_1.

In conclusion we emphasize that computation of minimax control method demands neither accurate information about the disposition of external wind disturbance zone nor any information about the wind velocity field in that zone. It is enough to describe amplitude of wind velocity variation approximately . This is the principle difference of the approach, based on the DG theory, from the methods, given in [14,15], where such information is essential.

Applications of DG theory to the landing problem have been considered in [12,16-20].

References

1. Isaaks, R.(1965).Differential Games, John Wiley, New York.
2. Pontryagin, L.S. (1967). Linear differential games. 2, Soviet Math. Dokl. 8, 910-912.
3. Blaquiere, A., Gerard, F., Leitman, G. (1969). Quantitative and Qualitative Games, Academic Press, New York.
4. Krasovskii, N.N. (1970). Games Problems about Contact of Motions. Nauka, Moskow, 1970 (in Russian).
5. Pshenichnii, B.N. and Sagaidak, M.I. (1970). Differlntial games of prescribed duration, Cybernetics, 6(2), No.2, 72-83.
6. Friedman, A. (1971). Differential Games. Interscience, New York.
7. Krasovskii, N.N. (1985). Dinamic System Control, Nauka, Moskow (in Russian).
8. Krasovskii, N.N. and Subbotin, A.I. (1988). Game-Theoretical Control Problems, Springer-Verlag, New York.
9. Subbotin, A.I. and Chentsov, A.G. (1981). Optimization of Guaranteed Result in Control Problems. Nauka, Moskow (in Russian).

10. Botkin, N.D. and Patsko, V.S. (1982). Universal strategy in a differential game with fixed terminal time, Problems of Control and Information Theory, 11, No.6, 419-432.
11. Botkin,N.D. and Patsko, V.S. (1983). Positional control in a linear differential game, Engineering Cybernetics, 21, No.4, 69-75.
12. Zarkh, M.A. and Patsko, V.S. (1987). The construction of the second player control in a linear differential game on the basis of the repulsion property, The Control with the Guaranteed Result. Ural Sci. Centr. Akad. Nauk SSSR, Sverdlovsk, 37-70 (in Russian).
13. Ivan, M. (1985). A ring-vortext downburst model for real time flight simulation of severe wind shears. AIAA Flight Simulation Technologies Conf., St.Louis, Miss, 57-61.
14. Miele, A., Wang, T. and Melvin, W.W. (1986). Optimal take-off trajectories in the presence of windshear. J. Opt. Theory and Appl., 49, No.1, 1-45.
15. Miele, A., Wang, T., Wang, H. and Melvin, W.W. (1988). Optimal penetration landing trajectories in the presence of windshear, J. Opt. Theory and Appl., 57, No.1, 1-40.
16. Kein, V.M.,Parikov, A.N. and Smurov, M.Yu. (1980). On means of optimal control by the extremal aiming method. Prikl. Matem. Mekhan., 44, No.3, 434-440 (in Russian).
17. Titovskii, I.N. (1981). Game theoretical approach to the synthesis problem of aircraft control in the landing. Uchenye zapiski centr. Aerohidrodinam. Inst., XII , No.1, 85-92 (in Russian).
18. Botkin, N.D.,Kein, V.M. and Patsko, V.S. (1984). The model problem of controlling the lateral motion of an aircraft in the landing, J. Appl. Math. Mech., 48, No.4, 395-400.
19. Korneev, V.A., Melikyan, A.A. and Titivskii, I.N. (1985). Stabilization of aircraft glide path in wind disturbances in the min-max formulation, Izv. Akad. Nauk SSSR. Tekhn. Kibernet., 1985, No.3,132-139.
20. Botkin, N.D., Kein, V.M., Patsko, V.S. and Turova, V.L. (1989). Aircraft landing control in the presence of windshear, Problems of Control and Information Theory, 18, No.4, 223-235.

Robust aircraft take-off control: A comparison of aircraft performance under different windshear conditions

Veijo Kaitala*, George Leitmann**, and Sandeep Pandey**

* Systems Analysis Laboratory, Helsinki University of Technology, SF-02150 Espoo, Finland
** College of Engineering, University of California, Berkeley, CA 94720, USA

Aircraft take-off control under conditions of windshear has recently received considerable attention in the control theory literature. Severe windshear may cause difficulties during aircraft take-off; it has been a major cause of at least 30 aircraft accidents during the two last decades. We study the controlled take-off of an aircraft flying through an unavoidable windshear. The purpose of the guidance is to guarantee aircraft take-off without crash; no other performance criteria are applied. The proposed aircraft take-off control scheme consists of a memoryless state feedback control strategy for a class of continuous time aircraft models including unpredictable but bounded windshear. The design of the take-off control scheme is carried out by applying a theory of deterministic control of uncertain systems. The time or the place of encountering a windshear is not known but, according to the basic assumption, the upper bound of the uncertainty, here, the maximum rate of the change of the wind velocity, is assumed to be known; such knowledge allows one to design a deterministic controller to stabilize the relative path inclination of the aircraft.

1 Introduction

The problem of control of aircraft flying through windshear has gained considerable importance since a 1977 FAA study revealed the presence or potential of low-level windshear as a factor in many accidents involving large aircraft, [1]. This condition is hazardous for an aircraft flying at low altitudes, e.g., during take-off or landing. Over the past 20 years, some 30 aircraft accidents have been attributed to windshear [2].

Windshear is caused by several different motions of atmospheric masses. The atmospheric boundary layer in its natural state always contains some degree of windshear. Thus, during most approach and take-off operations some degree of windshear is encountered. The strength of the shear and the degree to which it becomes hazardous is dependent upon the existing combination of meteorological conditions, [5]. Generally a windshear condition involves several factors such as horizontal shear, vertical shear, wind direction change and height of shear above ground level. In [6] windshear is defined as significant changes in wind speed and/or direction up to 500 m above the ground which may adversely affect the approach, landing or take-off of an aircraft.

To study this problem we must have accurate and reliable wind profiles. However, the existing windshear models used in computer and manned flight simulator studies are not very realistic. Existing mathematical models of windshear are spatially two-dimensional and based on limited data. None of them includes time dependence. Moreover, few of these models contain small-scale microburst type windshear, [3]. The aim of present work is, however, not to improve the existing windshear models, but to develop a control strategy which is effective for flying in different kinds of situations. For modeling aspects of windshear the reader is referred to [1,3,5-7].

In one of his seminal papers, [16], Miele deduces optimal control strategies for take-off in the presence of a given windshear; that is, the controlled trajectories are optimum in the specified flow field. Several guidance strategies which achieve near optimum performance in the given windshear were studied in the pioneering investigations [9,17-19]. These strategies employ local information on windshear. Since local information is difficult to obtain on present day aircraft, several piloting strategies were studied in [4,19]. The so-called *simplified gamma guidance* strategy, which yields a quick transition to horizontal flight in a windshear, was shown to have very good survival capabilities in the prescribed windshear. This strategy is particularly suitable for flight in a severe windshear of the type considered; it is based on the philosophy of trying to avoid altitude loss while simultaneously containing velocity loss. A possible disadvantage of this kind of guidance scheme is that it is based on a particular model of windshear and may be not effective when the aircraft flies through other windshear distributions.

In [20] a control scheme is deduced by considering the so-called pseudo-energy as a measure of aircraft altitude and speed relative to the wind. A set of linearized equations about equilibrium level flight conditions together with a linear feedback scheme are employed to obtain prescribed closed-loop eigenvalues. The nonlinear guidance scheme consists of a nominal part together with the linear feedback control. There is no assurance that this feedback control will yield stable trajectories under varying windshear conditions of relatively high intensity.

The investigation reported in [21] is the first attempt to apply a theory of deterministic control of uncertain systems ([13,14]) to the problem of aircraft guidance during conditions of windshear. The resulting control scheme utilizes the full state of the aircraft in order to assure the convergence of the aircraft state to a desired end state.

In the present work, the problem of aircraft take-off under windshear conditions is considered as a problem of controlling an uncertain dynamical system: the mathematical model of the aircraft dynamics involves uncertainty due to the unknown windshear. Here, an upper bound of the uncertainty must be known to assure effective control.

The control of dynamical systems, whose mathematical models contain uncertainties, has been studied extensively, e.g., see [10-14]. These uncertainties may be due to parameters, constant or varying, which are unknown or imperfectly known, or uncertainties due to unknown or imperfectly known inputs into the system. Taking a deterministic point of view, controllers have been designed which, utilizing only information about the possible magnitudes of the uncertainties, guarantee desirable system performance such as some form of stability. The design of these control schemes is based on a constructive use of Lyapunov stability theory.

The problem of controlling an aircraft through windshear falls into a general class of control problems of nonlinear systems. In the following we use the theory developed in [10-14] to design a deterministic controller to stabilize the relative path inclination which is one of the states of the system. Then we test this controller on different windshear models of [7,17,20]. The simulations indicate that it is a promising strategy which can work in a very general windshear condition.

2 Aircraft equations of motion

Following Miele's lead, we employ equations of motion for the center of mass of the aircraft, in which the kinematic variables are relative to the ground ("inertial reference frame") while the dynamic ones are taken relative to a moving but non-rotating reference frame translating with the wind velocity at the aircraft center of mass ("wind based reference frame").

We employ the following notation:

ARL = aircraft reference line;
D = drag force, *lb*;
g = gravitational force per unit mass (= constant), *ft* sec^{-1};
h = vertical coordinate of aircraft center of mass (altitude), *ft*;
L = lift force, *lb*;
m = aircraft mass, *lb* *ft*$^{-1}$sec^2;
O = mass center of aircraft;
S = reference wing area, *ft*^{2};
t = time, sec;
T = thrust force, *lb*;
V = aircraft speed relative to wind based reference frame, *ft* sec^{-1};
V_e = aircraft speed relative to ground, *ft* sec^{-1};
W_x = horizontal component of wind velocity, *ft* sec^{-1};
W_h = vertical component of wind velocity, *ft* sec^{-1};
x = horizontal position component of mass center of aircraft, *ft* ;
α = relative angle of attack, *rad* ;
γ = relative path of inclination, *rad* ;
γ_e = path of inclination, *rad* ;
δ = thrust of inclination, *rad* ;
ρ = air density (=constant), *lb* *ft*^{2} sec^2;
Dot denotes time derivative.

The basic simplifying assumptions are as follows: 1) The rotational inertia of the aircraft is neglected; 2) The aircraft mass is constant; 3) Flight is in the vertical plane; 4) Maximum thrust is used.

In view of Assumption 1, we consider only the equations of motion of the center of mass (see Figure 1).

The kinematical equations are

$$\dot{x} = V\cos\gamma + W_x \tag{1}$$

$$\dot{h} = V\sin\gamma + W_h \tag{2}$$

and the dynamical equations are

$$m\dot{V} = T\cos(\alpha+\delta) - D - mg\sin\gamma - m(\dot{W}_x\cos\gamma + \dot{W}_h\sin\gamma) \tag{3}$$

$$mV\dot{\gamma} = T\sin(\alpha+\delta) + L - mg\cos\gamma + m(\dot{W}_x\sin\gamma - \dot{W}_h\cos\gamma). \tag{4}$$

These equations must be supplemented by specifying the thrust force $T = T(V)$, the drag $D = D(h, V, \alpha)$, the lift $L = L(h, V, \alpha)$, the horizontal windshear $W_x = W_x(x, h)$ or $W_x(t)$ and the vertical windshear $W_h = W_h(x, h)$ or $W_h(t)$. For a given value of the thrust inclination δ, the differential equation system (1)-(4) involves four state variables - the horizontal distance $x(t)$, the altitude h(t), the relative speed $V(t)$, and the relative path inclination $\gamma(t)$ - and one control variable - the angle of attack $\alpha(t)$, since maximum thrust is employed according to Assumption 4.

A. BOUNDED QUANTITIES

In order to account for the aircraft capabilities, we shall assume that there is a maximum attainable value of the relative angle of attack, α; that is,

$$\alpha \in [0, \alpha_*], \tag{5}$$

where $\alpha_* > 0$ depends on the specific aircraft and generally is taken to be the *stick shaker* angle of attack.

In view of Assumption 1, the rotational inertia of the aircraft is neglected. To account for the neglected dynamics of rotation, and of sensors and actuators, we bound the attainable magnitude of the rate of change of the relative angle of attack, $\dot{\alpha}$; that is,

$$| \dot{\alpha} | \leq C, \tag{6}$$

where $C > 0$ depends on the specific aircraft.

Furthermore, the range of practical values of the relative aircraft speed, V, is limited; that is,

$$\underline{V} \leq V \leq \bar{V} \tag{7}$$

where $\underline{V} > 0$ and $\bar{V} > \underline{V}$ depend on the specific aircraft. These bounds correspond, for instance, to the relative stall speed and the maximum attainable relative speed, respectively.

The bounds (5) and (6) on α and $| \dot{\alpha} |$ will be neglected in deducing the proposed aircraft guidance scheme; however, they will be taken into account in the numerical simulations. On the other hand, the bounds (7) on the relative speed, V, will be employed in the construction of the proposed guidance scheme.

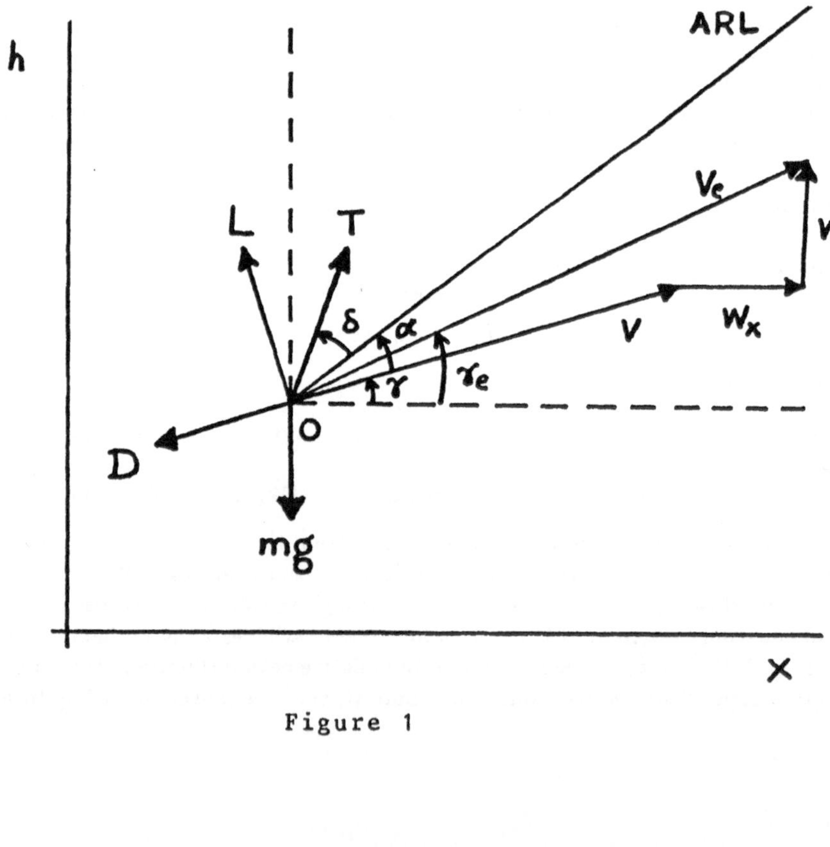

Figure 1

B. APPROXIMATIONS FOR THE FORCE TERMS

Thrust.

$$T = A_0 + A_1 V + A_2 V^2, \tag{8}$$

where the coefficients A_0, A_1, A_2 depend on the altitude of the runway, the ambient temperature and the engine power setting.

Drag.

$$D = C_D \rho S V^2/2, \tag{9}$$

where $C_D = B_0 + B_1\alpha + B_2\alpha^2$. The coefficients B_0, B_1, B_2 depend on the flap setting and the under-carriage position.

Lift.

$$L = C_L \rho S V^2/2, \tag{10}$$

where $C_L = C_0 + C_1\alpha$, if $\alpha \leq \alpha_{**} < \alpha_*$, and $C_L = C_0 + C_1\alpha + C_2(\alpha - \alpha_{**})^2$, if $\alpha_{**} \leq \alpha < \alpha_*$. The coefficients C_0, C_1, C_2 depend on the flap setting and the under-carriage position, and α_{**} depends on the particular aircraft.

3 Aircraft take-off control schemes

A. A NOMINAL CONTROL

We shall consider a *nominal angle of attack*, $\alpha = \alpha_n(V)$, which corresponds to a constant relative path inclination, $\gamma(t) \equiv \gamma_n$, in the absence of windshear. Of course, in the absence of windshear, $\gamma(t) \equiv \gamma_e(t)$.

B. STABILIZING CONTROL

The general problem of controlling an aircraft flying through windshear falls into the class of problems of stabilizing a so-called mismatched uncertain nonlinear system (see [10-14] for definitions, and [21,22] for the first applications to the windshear case). Moreover, the system does not satisfy the conditions required for the controllers of [13,14]. Hence, in the present investigation we make some simplifications which allow direct utilization of the controllers proposed in [13] and [14] and in many earlier papers referenced there.

In [21] and [22], the relative angle of attack $\alpha(t)$ was employed as control. However, in general, $\alpha(0) \neq \alpha_n(V(0))$ so that $\dot{\alpha}(t)$ usually saturates and undesirable oscillations of $\alpha(t)$ occur. To avoid this undesirable feature, we utilize

$$\Delta\alpha(t) \equiv \alpha(t) - \alpha_{n0}, \quad \alpha_{n0} \equiv \alpha_n(V(0)) \tag{11}$$

as control variable.

First of all, we consider the stabilization problem of only the relative path inclination, γ. Although it is desirable to stabilize all the states of the system (see [21]), the selection of only one state reduces the problem to a so-called matched one. Since successful stabilization of relative path inclination about a nominal relative path inclination should ensure avoidance of a crash, it was chosen as the state to stabilize.

Next we make the approximation (see [22]),

$$C_L \approx C_0 + C_1\alpha \text{ for all } \alpha \in [0, \alpha_*] \tag{12}$$

in order to satisfy a condition of the theory in [10-14] for the design of a controller. From the plot of C_L versus α, [17], we observe that this is a valid approximation. Moreover, the contribution of C_L to lift (see eq. (10)) is small compared to other terms. As mentioned earlier, we consider bounds on the relative speed when deriving the controller, that is, we impose

$$\underline{V} \leq V \leq \bar{V}$$

where $\underline{V}$ and $\bar{V}$ depend on the specific aircraft.

A major assumption in the theory of deterministic control is that uncertain terms are bounded. In our case, it suffices to assume a known bound on the magnitude of the time rate of change of the wind; that is,

$$(\dot{W}_x^2 + \dot{W}_h^2)^{1/2} \leq q, \tag{13}$$

where $q > 0$ is assumed to be known from previous flights through windshear conditions.

Now we consider the stabilization of the relative path of inclination. Let $\gamma_R(t)$ be the reference value of $\gamma(t)$ with respect to which we want to stabilize $\gamma(t)$. Let us define,

$$\Delta\gamma(t) \equiv \gamma(t) - \gamma_R(t). \tag{14}$$

Then from eqs. (4), (8), (10), (12) and (14) we have

$$\dot{\Delta\gamma} = -K\Delta\gamma + (A_0 + A_1V + A_2V^2/mV)\sin(\alpha_n + \Delta\alpha + \delta) + (C_0 + C_1(\alpha_n + \Delta\alpha)\rho SV^2/2mV$$
$$-(g/V)\cos(\Delta\gamma + \gamma_R) - (1/V)[\dot{W}_x \sin(\Delta\gamma + \gamma_R) + \dot{W}_h \cos(\Delta\gamma + \gamma_R)] - \dot{\gamma}_R + K\Delta\gamma. \tag{15}$$

The controller design, based on the theory in [10-14], is analogous to that presented in [22] and results in a feedback controller

$$\Delta\alpha = p^\epsilon = \begin{cases} -\xi^c\epsilon^{-1}\eta, & |\eta| \leq \epsilon, \\ -\xi^c \mid \eta \mid^{-1} \eta, & |\eta| > \epsilon \end{cases} \tag{16}$$

where ϵ is a design parameter and $\eta = k^c\phi$ with $\phi = \Delta\gamma/(2K)$,

$$k^c = k = \mid A_0 \mid /m\underline{V} + \mid A_1 \mid /m + \mid A_2 \mid \bar{V}/m + (C_0 + C_1\alpha_{n0})\rho S\bar{V}/2m$$
$$+(g/\underline{V}) + (1/\underline{V})2q + \bar{\dot{\gamma}}_R + K \mid \Delta\gamma \mid,$$
$$\xi^c = \xi = 2mk^c/(C_1\rho\underline{V}), \text{ and } \dot{\gamma}_R(t) \leq \bar{\dot{\gamma}}_R.$$

4 Simulations

We consider three windshear models:

Model 1. A discretized version of the model employed by Miele in [19].

Model 2. The model utilized by Bryson in [20].

Model 3. A single vortex model, [7].

The details of the windshear models can also be found in [22].

For purposes of simulations we consider aircraft data pertaining to the Boeing-727; they are given in [22], as are the initial values of the aircraft state variables and of the design parameters K and ϵ.

Figures 2-4 show the altitude, $h(t)$, and the corresponding relagive angle of attack, $\alpha(t)$, histories for flights in each of the three windshear models and under the controls of Miele [19], Bryson [20] and the one proposed here in (16), respectively. Performance deteriorates with increasing windshear intensity in all cases, i.e., for all guidance schemes in all windshear models.

The proposed guidance scheme is the most robust except in the case of Miele's windshear model for which his guidance scheme is the best relative to intensity; however, it must be noted that this guidance scheme was designed using the knowledge of the windshear structure. The proposed guidance scheme, while diminishing considerable angle of attack variations, obviates the recognition of the windshear in order to activate it; $\alpha(t) = \alpha_n(V(t))$ until onset of windshear results in $\Delta\gamma(t) \neq 0$.

More details as well as simulations for flight through a double vortex can be found in [23].

Aknowledgements

V.K. acknowledges the support by the Academy of Finland and by Finnair. This work is based on research supported in part by the NSF and AFOSR under grant ECS-8602524.

References

1. S. Zhu and B. Etkin, "Fluid Dynamic Model of a Downburst," Institute of Aerospace Studies, University of Toronto, UTIAS Report No. 271.

2. A. Miele, Wang, T., Tzeng, C.Y., Melvin, W.W., Optimization and guidance of abort landing trajectories in a windshear, Paper No. AIAA-87-2341, AIAA Guidance, Navigation and Control Conference (1987).

3. W. Frost, "Flight in a Low-Level Wind Shear," NASA CR 3678 (1983).

4. A. Miele, Wang, T., Melvin, W.W., and Bowles, R.L., "Maximum Survival Capacity of an Aircraft in a Severe Windshear," JOTA 53, No 2 (1987).

5. Frost, W., and Camp, D.W., "Wind Shear Modeling for Aircraft Hazard Definition", FAA Report No. FAA-RD-77-36, U.S. Department of Transportation, Washington, DC (1977).

6. Frost, W., and Camp, D.W., and Wang, S.T., "Wind shear modeling for aircraft hazard definition", FAA Report No. FAA-RD-78-3, U.S. Department of Transportation, Washington, DC (1978).

7. G.G. Roetcisoender, W.J. Grantham and E.K. Parks, "The DFW Microburst: Two-Dimensional Multiple Vortex Models for AAL-539," J. Aircraft (to appear).

8. F. Caracena, "The Microburst: Common Factor in Recent Aircraft Accidents," Proc. Fourth Annual Workshop on Meteorological and Environmental Inputs to Aviation Systems, 1980, NASA CP 2139 / FAA-RD-80-67 (1980).

9. A. Miele, Wang, T., Melvin, W.W., "Guidance Strategies for Near Optimum Take-Off Performance in a Windshear," JOTA 50, No 1 (1986).

10. G. Leitmann, " On the Efficacy of Nonlinear Control in Uncertain Linear Systems," Trans. ASME J. Dyn. Syst., Meas. Control 102 (1981).

11. M. Corless and G. Leitmann, "Continuous State Feedback Guaranteeing Uniform Ultimate Boundedness for Uncertain Dynamic Systems," IEEE Trans. AC, AC-26, No 5 (1981).

12. Y.C. Chen and G. Leitmann, "Robustness of Uncertain Systems in the Absense of Matching Assumptions," Int. J. Control 45 (1987).

13. M. Corless and G. Leitmann, "Deterministic Control of Uncertain Systems: A Lyapunov Theory Approach," in Deterministic Nonlinear Control of Uncertain Systems: Variable Structure and Lyapunov Control (A. Zinober, ed.), IEE Publishers (to appear).

14. G. Leitmann, " Deterministic control of uncertain systems via a contstructive use of Lyapunov stability theory," SIAM Conference "Control in the 90's", San Francisco (1989).

15. A. Miele, Wang, T., Melvin, W.W., "Optimal Flight Trajectories in the Presence of Windshear, Part I - Take-Off," Paper No. AIAA-85-1843-CP, AIAA Atmospheric Flight Mechanics Conference (1985).

16. A. Miele, Wang, T., Melvin, W.W., "Optimal Flight Trajectories in the Presence of Windshear, Parts I - IV," Rice University, Aero-Astronautics Reports Nos. 191-194 (1985).

17. A. Miele, Wang, T., Melvin, W.W., "Quasi-Steady Flight to Quasi-Steady Flight Transition in a Windshear: Trajectory Optimization and Guidance," JOTA 54, No. 2 (1987).

18. A. Miele, Wang, T., Melvin, W.W., "Optimization and Acceleration Guidance of Flight Trajectories in a Windshear," J. Guidance, Control, and Dynamics 11, No. 4 (1987).

19. A. Miele, Wang, T., Melvin, W.W. and Bowles, R.L., "Gamma Guidance Schemes for Flight in a Windshear," J. Guidance, Control, and Dynamics 11, No. 4 (1987).

20. A.E. Bryson, Jr., and Zhao, Y.," Feedback control for penetrating downburst," Paper No. AIAA-87-2343 (1987).

21. Y.H. Chen and S. Pandey, "Robust Control Strategy for Take-Off Performance in a Windshear," OCAM 10 (1989).

22. G. Leitmann and S. Pandey, " Aircraft control under condition of windshear," in Control and Dynamical Systems, (T. Leondes, ed.) Vol. XXXIV, Academic Press, N.Y. (1990).

23. V. Kaitala, G. Leitmann, and S. Pandey, "Aircraft control for take-off in windshear," Dynamics and Control, to appear.

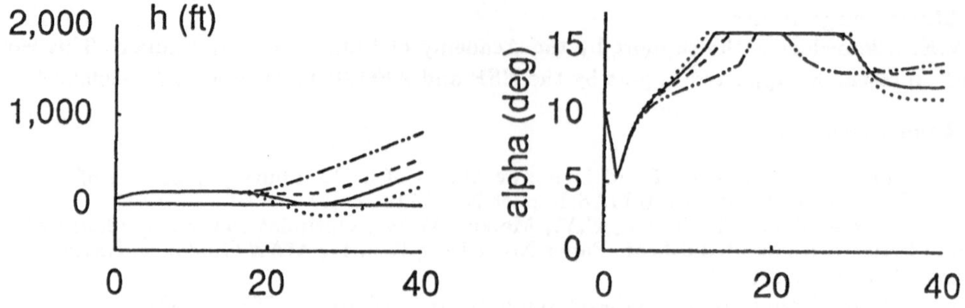

Simplified gamma-guidance, Miele [19]

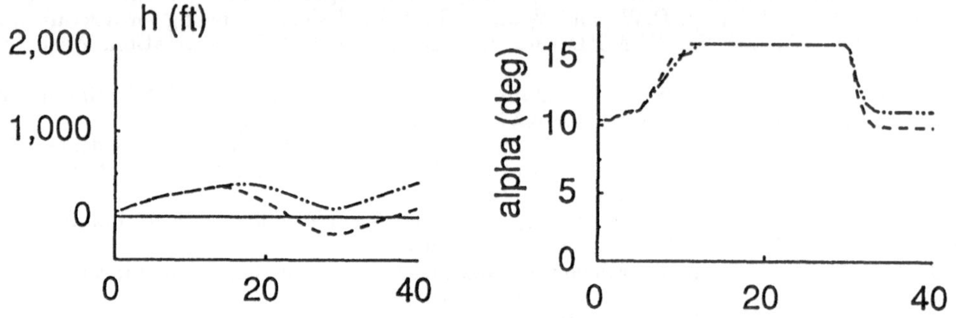

Nonlinear guidance, Bryson [20]

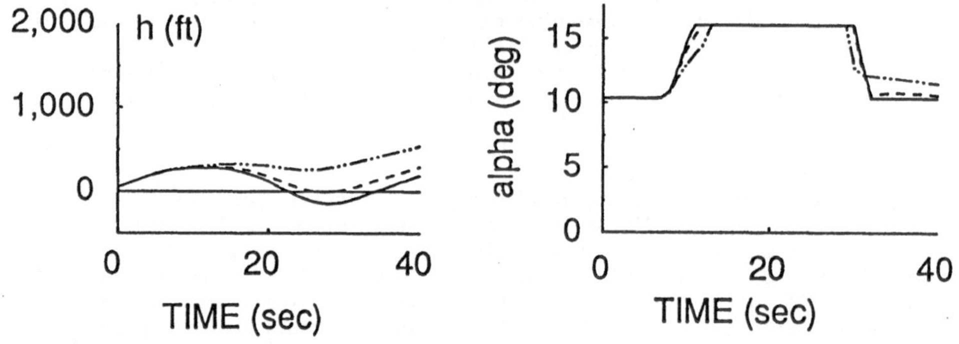

Proposed guidance (16)

Figure 2. Windshear model, Miele [20]

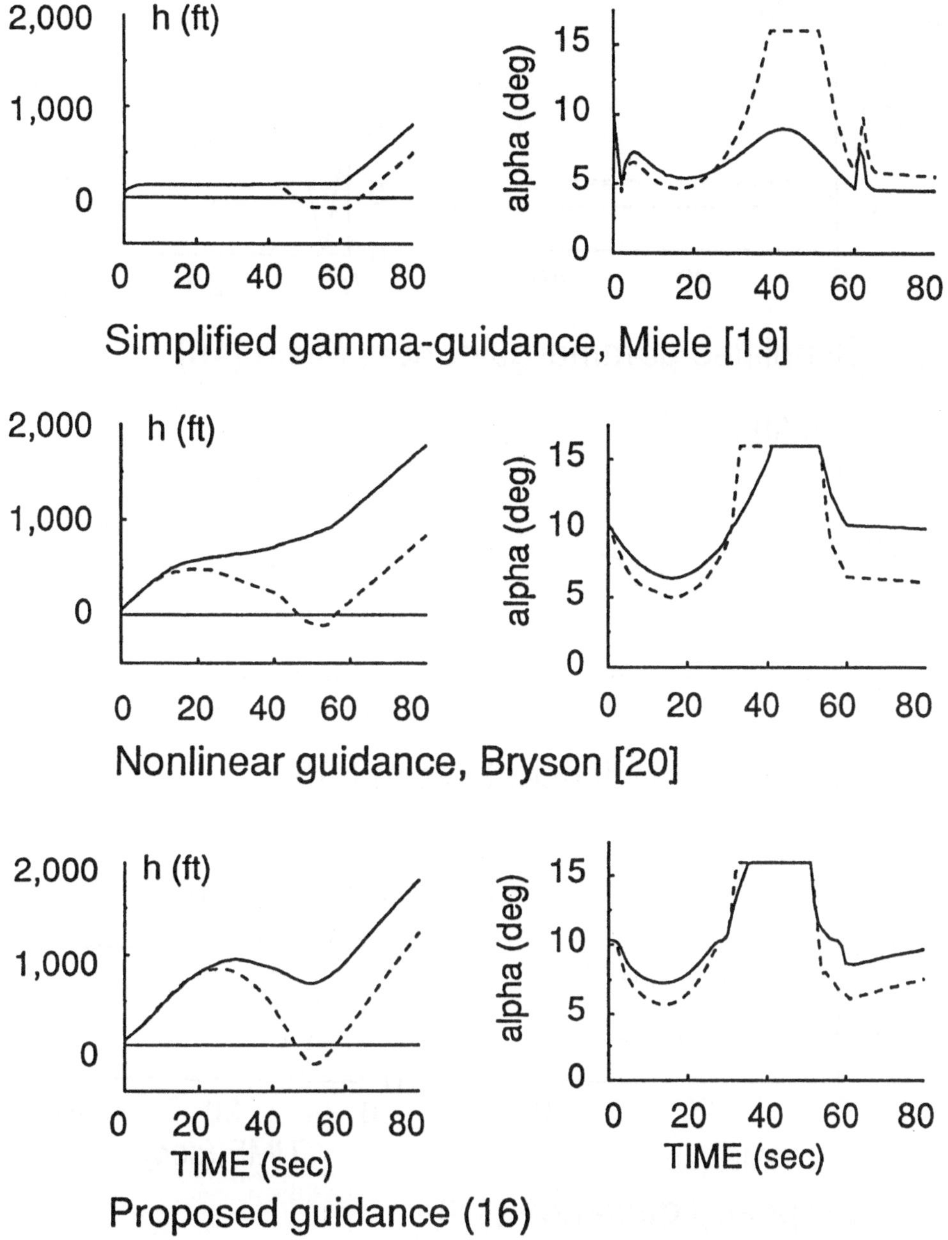

Figure 3. Windshear model, Bryson [19]

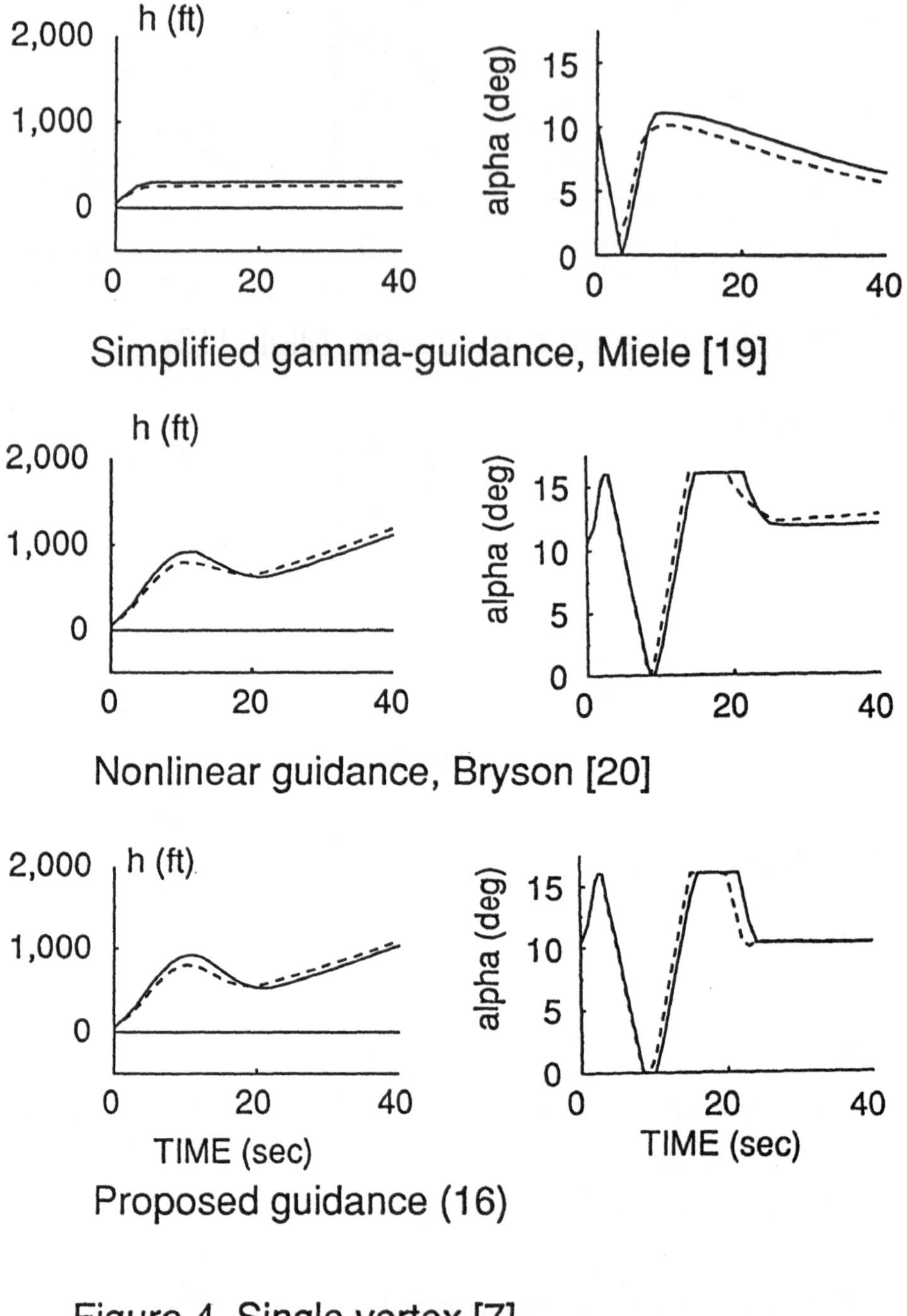

Figure 4. Single vortex [7]

ITERATIVE COMPUTATION OF NASH EQUILIBRIA IN M-PLAYER GAMES WITH PARTIAL WEAK COUPLING

Tamer Başar and R. Srikant
Coordinated Science Laboratory
University of Illinois
1101 W. Springfield Avenue
Urbana, IL 61801/USA

Abstract

We formulate two general classes of M-player deterministic and stochastic nonzero-sum games where the players can be placed into two groups such that there are strong interactions within each group and a weak interaction between the two groups. This weak interaction is characterized in terms of a small parameter ϵ which, when set equal to *zero*, leads to two independent nonzero-sum games. Under the Nash equilibrium solution concept both within and in between the groups, we study the merits of an iterative method for the construction of the equilibrium by solving simpler problems at each stage of the iteration. In this iterative scheme, the *zero*'th order solution is the Nash equilibrium of the two independent games obtained by setting $\epsilon = 0$, whereas the higher-order solutions are Nash equilibria of quadratic games, even though the original problem may have non-quadratic cost functions.

Keywords: Noncooperative nonzero-sum games, Nash equilibria, weak coupling, iterative computation.

1 Introduction and Problem Formulation

We consider, in this paper, two general classes of static nonzero-sum M-player games, where the players can be placed into two groups, say I and II, of sizes K and $M - K$, respectively. The players have relatively strong interactions within each group, but weak interactions across the two groups, with these interactions quantified solely through the cost functions. In mathematical terms, if $u_i \in \mathbb{R}^{r_i}$ denotes the decision variable of the

i'th player, $i \in \mathcal{M} := \{1, \ldots M\}$, and J_i denotes the cost function of that player, then

$$J_i(\mathrm{u}) = \begin{cases} g_i(u_1, \ldots, u_K) + \epsilon h_i(\mathrm{u}), & i \in \mathcal{M}_I \\ \\ g_i(u_{K+1}, \ldots, u_M) + \epsilon h_i(\mathrm{u}), & i \in \mathcal{M}_{II} \end{cases} \tag{1.1}$$

where $\mathrm{u} := (u_1, \ldots, u_M)$, $\mathcal{M}_I := \{1, \ldots K\}$, $\mathcal{M}_{II} := \{K+1, \ldots, M\}$, ϵ is a sufficiently small (in magnitude) scalar, and g_i, h_i are analytic functionals that are defined on appropriate dimensional Euclidean spaces. This is a deterministic game, for which a Nash equilibrium is an M-tuple u^* satisfying

$$J_i(\mathrm{u}^*) \leq J_i(\mathrm{u}^*_{-i}, u_i), \quad \forall u_i \in I\!R^{r_i}, \quad i \in \mathcal{M}, \tag{1.2a}$$

where

$$\mathrm{u}^*_{-i} := (u_1^*, \ldots, u_{i-1}^*, u_{i+1}^*, \ldots, u_M^*). \tag{1.2b}$$

One of our objectives in this paper is to study the dependence of this (possibly nonunique) Nash solution on the small parameter ϵ, and in particular to develop an iterative procedure that would provide $0(\epsilon^\ell)$ approximations to $\mathrm{u}^*(\epsilon)$, where ℓ is any positive integer and $0(\cdot)$ is the order function. This iterative procedure, which is fully discussed in the next section, starts with $\mathrm{u}^*(0)$ which is shown to constitute a Nash solution to the two independent K and $M - K$ player games obtained by letting $\epsilon = 0$. To arrive at higher order approximations, one then has to solve only linear-quadratic games, even though the kernels g_i and h_i, $i \in \mathcal{M}$, may be nonquadratic.

In Section 3, we study the stochastic counterpart of this class of games, where the cost function (1.1) is replaced by

$$J_i(\gamma) = \begin{cases} E\{g_i(u_1, \ldots, u_K; x) + \epsilon h_i(\mathrm{u}; x) \mid \mathrm{u} = \gamma(\cdot)\}, & i \in \mathcal{M}_I \\ \\ E\{g_i(u_{K+1}, \ldots, u_M; x) + \epsilon h_i(\mathrm{u}; x) \mid \mathrm{u} = \gamma(\cdot)\}, & i \in \mathcal{M}_{II} \end{cases} \tag{1.3}$$

where x is a finite-dimensional random vector with a known probability distribution P_x. We also introduce M finite-dimensional random vectors y_i, $i \in \mathcal{M}$, with joint conditional distribution $P_{\mathrm{y}|x}$, where y_i denotes the measurement vector of player i. If y_i takes values in $I\!R^{m_i}$, then γ_i, the policy of Player i, is defined as a Borel measurable mapping from $I\!R^{m_i}$ into $I\!R^{r_i}$. For future reference, let us denote the class of all such mappings for Player i by Γ_i. We assume that for each $\gamma := (\gamma_1, \ldots, \gamma_M) \in \boldsymbol{\Gamma} := \Gamma_1 \times \ldots \times \Gamma_M$, and with $u_i = \gamma_i(y_i)$, $i \in \mathcal{M}$, the functions $g_i(\cdot)$ and $h_i(\cdot)$ are well-defined random variables, so that their expected values $E\{\cdot\}$ exist.

A Nash equilibrium solution $\gamma^* \in \boldsymbol{\Gamma}$ for this stochastic game is defined as in (1.2a):

$$J_i(\gamma^*) \leq J_i(\gamma^*_{-1}, \gamma_i), \quad \forall \gamma_i \in \Gamma_i, i \in \mathcal{M}, \tag{1.4}$$

for which we again ask the same questions as in the deterministic game. The conclusions we arrive at, and the properties we derive from our analysis (in Section 3) turn out to be similar to those of Section 2. The paper ends with the concluding remarks of Section 4.

Weakly coupled systems can be viewed as a subclass of regularly perturbed systems which have been discussed extensively in the literature in the context of dynamic single decision maker (optimal control) problems[7], [12], and some additional work has been done in the context of quadratic deterministic games in [9]. Our motivation for studying weakly coupled multiple agent decision problems has been to solve (to some degree of approximation) problems which would otherwise be unsolvable, such as those with non-classical information patterns. Some of our results in this area, such as those included in [13], [5], [6], in the context of stochastic teams and stochastic differential games, indeed substantiate this objective.

2 The Deterministic Game

We first make the following two assumptions:

A1. The functions g_i and h_i, $i \in \mathcal{M}$, are infinitely many times differentiable. ◇

A2. There is an open neighborhood of $\epsilon = 0$, where the function $g_i + \epsilon h_i$ is strictly convex in u_i, for all $i \in \mathcal{M}$. ◇

Under **A2**, an M-tuple $\mathbf{u}^*(\epsilon)$ constitutes a Nash equilibrium solution if, and only if

$$\frac{\partial}{\partial u_i} g_i(u_1^*, \ldots, u_K^*) + \epsilon \frac{\partial}{\partial u_i} h_i(\mathbf{u}^*) = 0, \quad i \in \mathcal{M}_I \tag{2.1a}$$

and

$$\frac{\partial}{\partial u_i} g_j(u_{K+1}^*, \ldots, u_M^*) + \epsilon \frac{\partial}{\partial u_j} h_j(\mathbf{u}^*) = 0, \quad j \in \mathcal{M}_{II}. \tag{2.1b}$$

The following property now follows from the *Implicit Function Theorem* [10] applied to the above set of equations, at $\epsilon = 0$:

Property 2.1. *If there exists a Nash equilibrium* $\mathbf{u}^*(\epsilon)$*, then under* **A1** *it admits the expansion*

$$u_i^*(\epsilon) = \sum_{k=0}^{\infty} \epsilon^k u_i^{(k)}, \quad i \in \mathcal{M}. \tag{2.2}$$

(◇)

In view of this property, we now substitute (2.2) into (2.1a) and (2.1b), and use the differentiability of g_i*, to arrive at, for* $\epsilon = 0$,

$$\frac{\partial}{\partial u_i} g_i\left(u_1^{(0)}, \ldots, u_K^{(0)}\right) = 0, \quad i \in \mathcal{M}_I \tag{2.3a}$$

and

$$\frac{\partial}{\partial u_j} g_j\left(u_{K+1}^{(0)}, \ldots, u_M^{(0)}\right) = 0, \quad i \in \mathcal{M}_{II} \tag{2.3b}$$

which shows that the zero'th order term in the expansion (2.2),

$$\mathbf{u}^{(0)} := \left(u_1^{(0)}, \ldots, u_M^{(0)}\right) \tag{2.4}$$

provides a Nash equilibrium solution to the original game with $\epsilon = 0$, which actually involves two independent games, with K and $M-K$ players, respectively. ◇

To obtain the counterparts of (2.3a) and (2.3b) for the first-order terms in the expansion (2.2), we again substitute (2.2) into (2.1a)-(2.1b), expand the resulting equations in terms of ϵ, and equate the first powers of ϵ, to arrive at:

$$\sum_{\ell=1}^{K}\left[\frac{\partial^2}{\partial u_i \partial u_\ell} g_i\left(u_1^{(0)},\ldots,u_K^{(0)}\right)\right] u_\ell^{(1)} + \frac{\partial}{\partial u_i} h_i(\mathbf{u}^{(0)}) = 0, \quad i \in \mathcal{M}_I \tag{2.5a}$$

and

$$\sum_{\ell=K+1}^{M}\left[\frac{\partial^2}{\partial u_j u_\ell} g_j\left(u_{K+1}^{(0)},\ldots,u_M^{(0)}\right)\right] u_\ell^{(1)} + \frac{\partial}{\partial u_j} h_j(\mathbf{u}^{(0)}) = 0, \quad j \in \mathcal{M}_{II}. \tag{2.5b}$$

Note that these are two independent linear equations, from which $(u_1^{(1)},\ldots,u_K^{(1)})$ and $(u_{K+1}^{(1)},\ldots,u_M^{(1)})$ can be solved uniquely if the matrices Λ_I and Λ_{II} are nonsingular, where

$$\text{block } [\Lambda_I]_{i\ell} := \frac{\partial^2}{\partial u_i \partial u_\ell} g_i(u_1^{(0)},\ldots,u_K^{(0)}), \quad i,\ell = 1,\ldots,K \tag{2.6a}$$

$$\text{block } [\Lambda_{II}]_{i\ell} := \frac{\partial^2}{\partial u_{i+K} \partial u_{\ell+K}} g_{i+K}(u_{K+1}^{(0)},\ldots,u_M^{(0)}), \quad i,\ell = 1,\ldots,M-K. \tag{2.6b}$$

Hence, one can associate two (K and $M-K$ player) quadratic nonzero-sum games with the first-order solution, with the kernels of these games determined solely by $\mathbf{u}^{(0)}$.

To obtain the general k'th order term in the expansion (2.2), we follow the procedure described above and arrive at the following set of *linear* equations:

$$\sum_{\ell-1}^{K}\left[\frac{\partial^2}{\partial u_i \partial u_\ell} g_i(u_1^{(0)},\ldots,u_K^{(0)})\right] u_\ell^{(k)} + R_i(\mathbf{u}^{(0)},\mathbf{u}^{(1)},\ldots,\mathbf{u}^{(k-1)}) = 0, \quad i \in \mathcal{M}_I \tag{2.7a}$$

and

$$\sum_{\ell=K+1}^{M}\left[\frac{\partial^2}{\partial u_j \partial u_\ell} g_j(u_{K+1}^{(0)},\ldots,u_M^{(0)})\right] u_\ell^{(k)} + R_j(\mathbf{u}^{(0)},\mathbf{u}^{(1)},\ldots,\mathbf{u}^{(k-1)}) = 0, \quad j \in \mathcal{M}_{II}. \tag{2.7b}$$

Here R_i, $i \in \mathcal{M}$, are functions of only $\mathbf{u}^{(s)}$, $s \leq k-1$, and hence are known terms at the k'th stage of the iteration. Note that the coefficient matrices which determine the existence of a solution are exactly the same as the ones at the 1st stage, and hence again a unique solution will exist provided that Λ_I and Λ_{II} are nonsingular. This now brings us to the following Theorem.

Theorem 2.1. *Suppose that the original game admits a Nash equilibrium solution $\mathbf{u}^*(\epsilon)$, and furthermore that the zero'th order game (obtained by setting $\epsilon = 0$) admits a unique Nash equilibrium solution $\hat{\mathbf{u}}$. Then, $u^*(\epsilon)$ is expandable in ϵ, as in (2.2), in some open neighborhood of $\epsilon = 0$, such that* *i)* $\hat{u}_i = u_i^{(0)}$, $i \in \mathcal{M}$;

ii) If Λ_I *and* Λ_{II} *are nonsingular,* $u_i^{(k)}$, $i \in \mathcal{M}$, $k \geq 1$, *can be determined uniquely and iteratively, by solving for each* k *two quadratic games associated with group I and group II.* ◇

A natural question to ask at this point is the degree of approximation provided by any truncated version of (2.2) to the true Nash equilibrium solution. The following proposition answers this question even in cases when a Nash equilibrium to the original game may not exist.

Before stating the Proposition, let us first introduce the notation $u^*_{i(n)}$ to denote the n'th order truncation of (2.2), i.e.,

$$u^*_{i(n)} := \sum_{k=0}^{n} u_i^{(k)} \epsilon^k, \quad i \in \mathcal{M}, \tag{2.8a}$$

where the $u_i^{(k)}$'s are determined through the procedure outlined above. Furthermore let

$$\mathbf{u}^*_{(n)} := (u^*_{1(n)}, \ldots, u^*_{M(n)}) \tag{2.8b}$$

Proposition 2.1. *The* M*-tuple* $\mathbf{u}^*_{(n)}$ *provides an* $0(\epsilon^{2n+2})$ *Nash equilibrium, i.e., for all* $i \in \mathcal{M}$,

$$\inf_{u_i \in R^{m_i}} J_i(u^*_{1(n)}, \ldots, u^*_{i-1,(n)}, u_i, u^*_{i+1,(n)}, u^*_{M(n)}) = J_i(\mathbf{u}^*_{(n)}) + 0(\epsilon^{2n+2}). \tag{2.9}$$

Proof. Suppose that the left-hand-side of (2.9) admits a minimum, with the minimizing solution denoted by u_i^ϵ, $i \in \mathcal{M}$. Then, this solution should satisfy the first order necessary condition (here we assume, without any loss of generality, $i \in \mathcal{M}_I$):

$$\frac{\partial}{\partial u_i} g_i(u^*_{1(n)}, \ldots, u_i^\epsilon, \ldots, u^*_{K(n)}) + \epsilon \frac{\partial}{\partial u_i} h_i(u^*_{1(n)}, \ldots, u_i^\epsilon, \ldots, u^*_{M(n)}) = 0.$$

By applying the *implicit function theorem* [10] to the above, we obtain that u_i^ϵ has derivatives of all orders in ϵ, and furthermore that

$$u_i^\epsilon = u^*_{i(n)} + 0(\epsilon^{n+1}), \tag{2.10}$$

which follows from the construction of $u^*_{i(n)}$. Now, let

$$L_i(u_i) := J_i(u^*_{1(n)}, \ldots, u_i, \ldots, u^*_{M(n)}) - J_i(\mathbf{u}^*_{(n)})$$

which we expand around $u_i = u^*_{i(n)}$, and let $u_i = u_i^\epsilon$:

$$L_i(u_i^\epsilon) = \underbrace{L_i(u^*_{i(n)})}_{0} + \left[\frac{\partial}{\partial u_i} L_i(u^*_{i(n)})\right] (u_i^\epsilon - u^*_{i(n)}) + 0(| u_i^\epsilon - u^*_{i(n)} |^2). \tag{2.11a}$$

Since $\frac{\partial}{\partial u_i} L_i(u_i^\epsilon) = 0$ by the definition of u_i^ϵ, it again follows from an application of the implicit function theorem that in view of (2.10),

$$\frac{\partial}{\partial u_i} L_i(u^*_{i(n)}) = 0(\epsilon^{n+1}). \tag{2.11b}$$

Using this in (2.11a), we finally have

$$L_i(u_i^\epsilon) = 0(\epsilon^{2n+2}),$$

which shows that (2.9) holds under the assumption that there is a minimum. If there is no minimum, however, then for any fixed ϵ we can attain a value arbitrarily close to the infimum by choosing an appropriate u_i^ϵ, and hence the result stands proven for all cases.

◇

We now conclude this section by discussing the relationship between the successive approximation embodied in the procedure leading to Theorem 2.1, and the Cournot (Jacobi) and Gauss-Seidel policy iterations [4]. Let us first consider the case when we have two players and two groups (i.e., $M = 2$, $K = 1$). The Cournot iteration reads

$$\text{Cournot} \begin{cases} u^C_{1(k)} = \arg\min_{u_1} J_1(u_1, u^C_{2(k-1)}) \\ u^C_{2(k)} = \arg\min_{u_2} J_2(u^C_{1(k-1)}, u_2) \\ k \geq 1;\, u^C_{1(0)} \in I\!R^{r_1}, u^C_{2(0)} \in I\!R^{r_2} \text{ are chosen arbitrarily} \end{cases} \tag{2.12}$$

while the Gauss-Seidel (G-S) iteration is given by

$$\text{G-S} \begin{cases} u^{GS}_{1(k)} = \arg\min_{u_1} J_1(u_1, u^{GS}_{2(k-1)}) \\ u^{GS}_{2(k)} = \arg\min_{u_2} J_2(u^{GS}_{1(k)}, u_2) \\ k \geq 1; \text{ only } u^{GS}_{2(0)} \in I\!R^{r_2} \text{ is specified.} \end{cases} \tag{2.13}$$

Note that the Cournot iteration is a *parallel* policy iteration, while G-S is a sequential update mechanism. The following Theorem gives the "order" relationships between $u^C_{i(k)}$, $u^{GS}_{i(k)}$, and $u^*_{i(k)}$, where the latter was defined by (2.8a).

Theorem 2.2. *Let the sequences* $\{u^C_{i(k)}\}_{k\geq 1}$ $\{u^{GS}_{i(k)}\}_{k\geq 1}$, $i = 1, 2$ *be generated by (2.12) and (2.13), respectively. Furthermore, let* $\{u^*_{i(k)}\}_{k\geq 1}$ *be defined as in (2.8a), with* $u_i^{(k)}$, $k \geq 1$, *generated by the procedure introduced prior to the statement of Theorem 2.1. Then,*

(i) $u^C_{i(k)} = u^*_{i(k)} + 0(\epsilon^k), \quad i = 1, 2, \quad k \geq 1$

(ii) $u^{GS}_{1(k)} = u^*_{1(2k-1)} + 0(\epsilon^{2k-1}), \quad k \geq 1$

$$u^{\rm GS}_{2(k)} = u^*_{2(k)} + 0(\epsilon^{2k}), \quad k \geq 1.$$

Proof. To prove (i), first we note that the leading terms in the expansion of $u^*_{i(1)}$ and $u^C_{i(1)}(\epsilon)$ are the same, since

$$u^C_{i(1)}(0) = \arg\min_{u_i} g_i(u_i).$$

Hence, $u^C_{i(1)} = u^*_{i(1)} + 0(\epsilon)$. We now assume that the claim is true for k, and prove its validity for $k+1$. Toward this end, we write, for $i = 1$, (from the definition of Cournot iteration)

$$\begin{aligned} & u^C_{1(k+1)} = \arg\min\nolimits_{u_1} \left\{ g_1(u_1) + \epsilon h_1(u_1, u^C_{2(k)}) \right\} \\ \leftrightarrow & \\ & \frac{\partial}{\partial u_1} g_1(u^C_{1(k+1)}) + \epsilon \frac{\partial}{\partial u_1} h_1(u^C_{1(k+1)}, u^C_{2(k)}) = 0 \end{aligned} \tag{2.14}$$

Letting $G_1(u_1) := \frac{\partial}{\partial u_1} g_1(u_1)$, $H_1(u_1, u_2) := \frac{\partial}{\partial u_1} h_1(u_1, u_2)$, we expand $G_1(u_1) + \epsilon H_1(u_1, u_2)$ around $u_1 = u^C_{1(k+1)}$, $u_2 = u^C_{2(k)}$ and evaluate it at $u_1 = u^*_{1(k+1)}$, $u_2 = u^*_{2(k+1)}$:

$$\begin{aligned} G_1(u^*_{1(k+1)}) &+ \epsilon H_1(u^*_{1(k+1)}, u^*_{2(k+1)}) = G_1(u^C_{1(k+1)}) \\ &+ \epsilon H_1\left(u^C_{1(k+1)}, u^C_{2(k)}\right) + \frac{\partial}{\partial u_1} G_1(u^C_{1(k+1)})(u^*_{1(k+1)} - u^C_{1(k+1)}) \\ &+ \epsilon \frac{\partial}{\partial u_1} H_1(u^C_{1(k+1)}, u^C_{2(k)})(u^*_{1(k+1)} - u^C_{1(k+1)}) \\ &+ \epsilon \frac{\partial}{\partial u_2} H_1(u^C_{1(k+1)}, u^C_{2(k)})(u^*_{2(k+1)} - u^C_{2(k)}) \\ &+ 0(\mid u^*_{1(k+1)} - u^C_{1(k+1)} \mid^2) + 0(|u^*_{2(k+1)} - u^C_{2(k)}|^2). \end{aligned} \tag{2.15}$$

By (2.11b), the term on the left hand side is $0(\epsilon^{k+2})$, and by (2.14), the sum of the first two terms on the right hand side is zero. Furthermore, since by hypothesis, $u^*_{2(k)} - u^C_{2(k)} = 0(\epsilon^k)$, it follows that $u^*_{2(k+1)} - u^*_{2(k)} = 0(\epsilon^{k+1})$. Hence, the preceding equality implies that

$$u^*_{1(k+1)} - u^C_{1(k+1)} = 0(\epsilon^{k+1})$$

under the working assumption that the matrix $\frac{\partial}{\partial u_1} G_1(u^C_{1(k+1)})$ is nonsingular. (Otherwise there does not exists a $u^C_{1(k+1)}$ solving (2.14)). Clearly the same argument works for $i = 2$, thus proving the first part of the Theorem.

The proof of part (ii) is similar, but now we have to note that when $k = 1$, $u^{\rm GS}_{1(1)} = u^*_{1(0)} + 0(\epsilon)$, and $u^{\rm GS}_{2(1)} = u^*_{2(1)} + 0(\epsilon^2)$. This starts the inductive argument of the previous proof. From an appropriate modification of (2.15), we have

$$u^{\rm GS}_{1(k+1)} = u^*_{1(2k+1)} + \max(0(\epsilon^{2k+2}), \epsilon\, 0(|u^*_{2(2k)} - u^{\rm GS}_{2(k)}|))$$

and following a similar analysis, this time with $i = 2$,

$$u^{\rm GS}_{2(k)} = u^*_{2(2k)} + \max(0(\epsilon^{2k+1}), \epsilon\, 0(|u^*_{1(2k-1)} - u^{\rm GS}_{1(k)}|))$$

The result then follows from these two order relationships and the starting conditions. ◇

If each group has more than one player, then Theorem 2.2 equally applies provided that within each group the exact Nash solution is obtained at each stage of the iteration. If, however, a Cournot or Gauss-Seidel type iteration is adopted within each group, and actions are exchanged between the groups before internal convergence is achieved, we cannot expect to obtain an ϵ^{ℓ}-order approximation to the Nash solution since the interactions in between the players of a group are strong.

3 The Stochastic Game

We will develop here the counterparts of the results in Section 2 for the stochastic game formulated in Section 1. The presentation below will simply draw the parallels between the results for the deterministic game and the stochastic one.

Under conditions **A1** and **A2**, and the assumption that the expected values in (1.3) are well defined, we have, as the counterpart of (2.1), the following necessary and sufficient conditions for an M-tuple $\gamma^* \in \Gamma$ to constitute a Nash equilibrium:

$$E\left\{\frac{\partial}{\partial u_i} g_i\left(\gamma_1^*(y_1), \ldots, \gamma_K^*(y_K); x\right) + \epsilon \frac{\partial}{\partial u_i} h_i(\gamma^*(\mathrm{y}); x) \mid y_i\right\} = 0; \quad i \in \mathcal{M}_I \qquad (3.1a)$$

$$E\left\{\frac{\partial}{\partial u_j} g_i\left(\gamma_{K+1}^*(y_{K+1}), \ldots, \gamma_M^*(y_M); x\right) + \epsilon \frac{\partial}{\partial u_j} h_j(\gamma^*(\mathrm{y}); x) \mid y_j\right\} = 0; \quad j \in \mathcal{M}_{II} \qquad (3.1b)$$

where we have used the compact notation

$$\gamma^*(\mathrm{y}) := \{\gamma_1^*(y_1), \ldots, \gamma_M^*(y_M)\},$$

and have assumed the usual conditions for the interchange of differentiation and expectation operation in (3.1a) and (3.1b) (see, for example, [11]).

Now, consider the above set of equations written for $\epsilon = 0$:

$$E\left\{\frac{\partial}{\partial u_i} g_i\left(\gamma_i^{(0)}(y_1), \ldots, \gamma_K^{(0)}(y_K); x\right) \mid y_i\right\} = 0; \quad i \in \mathcal{M}_I \qquad (3.2a)$$

and

$$E\left\{\frac{\partial}{\partial u_j} g_j\left(\gamma_{K+1}^{(0)}(y_{K+1}), \ldots, \gamma_M^{(0)}(y_M); x\right) \mid y_j\right\} = 0; \quad j \in \mathcal{M}_{II} \qquad (3.2b)$$

and let us assume that they admit a solution denoted $\gamma^{(0)} \in \Gamma$. Then, this is the Nash equilibrium solution of the *zero*'th order game (obtained by setting $\epsilon = 0$), and the following counterpart of Property 2.1 follows again from the Implicit Function theorem in function spaces:

Property 3.1. *If (3.2a)-(3.2b) admit a solution $\gamma^{(0)} \in \Gamma$, then the solution $\underline{\gamma}^*(\epsilon)$ to (3.1a)-(3.1b) and hence the Nash solution to the original game admits an expansion*

$$\gamma_i^*(y_i; \epsilon) = \gamma_i^{(0)}(y_i) + \sum_{k=1}^{\infty} \epsilon^k \gamma_i^{(k)}(y_i; \epsilon), \quad i \in \mathcal{M} \qquad (3.3)$$

in an open neighborhood of $\epsilon = 0$. Furthermore, if $\gamma^{(0)}$ is unique, so is $\underline{\gamma}^*(\epsilon)$ in an open-neighborhood of $\epsilon = 0$. ◇

Now to obtain the first-order terms in the expansion (3.3), we substitute this form into (3.1a) and (3.1b) and expand the resulting equation in terms of ϵ, retaining only the coefficient of ϵ. The result is the set of equations:

$$\begin{aligned}\sum_{\substack{\ell=1\\ \ell\neq i}}^{K} E\left\{\left[\frac{\partial^2}{\partial u_i \partial u_\ell} g_i\left(\gamma_1^{(0)}(y_1),\ldots,\gamma_K^{(0)}(y_k)\right)\right]\gamma_\ell^{(1)}(y_\ell) \mid y_i\right\} \\ +E\left\{\frac{\partial^2}{\partial u_i^2} g_i\left(\gamma_1^{(0)}(y_1),\ldots,\gamma_K^{(0)}(y_k)\right) \mid y_i\right\}\gamma_i^{(1)}(y_i) \\ +E\left\{\frac{\partial}{\partial u_i} h_i\left(\gamma^{(0)}(\mathrm{y})\right) \mid y_i\right\} = 0, \quad i \in \mathcal{M}_I\end{aligned} \tag{3.4a}$$

and

$$\begin{aligned}\sum_{\substack{\ell=K+1\\ \ell\neq j}}^{K} E\left\{\left[\frac{\partial^2}{\partial u_j \partial u_\ell} g_j\left(\gamma_{K+1}^{(0)}(y_{K+1}),\ldots,\gamma_M^{(0)}(y_M)\right)\right]\gamma_\ell^{(1)}(y_\ell) \mid y_j\right\} \\ +E\left\{\frac{\partial^2}{\partial u_j^2} g_j\left(\gamma_{K+1}^{(0)}(y_{K+1}),\ldots,\gamma_M^{(0)}(y_M)\right) \mid y_j\right\}\gamma_j^{(1)}(y_j) \\ +E\left\{\frac{\partial}{\partial u_j} h_j\left(\gamma^{(0)}(\mathrm{y})\right) \mid y_j\right\} = 0, \quad j \in \mathcal{M}_{II}.\end{aligned} \tag{3.4b}$$

Note that, as in (2.5a)-(2.5b), these are again linear equations from which the first-order Nash policies $\gamma^{(1)}$ can be obtained. It is possible to obtain here the precise conditions under which this set of linear equations will admit a unique solution (see [1], [2], [3], but the actual derivation will in general involve the inversion of some infinite-dimensional linear operators, or an infinite recursion (unless the coefficient terms whose conditional expectations are taken are independent of $y_1, \ldots, y_M$). Hence, we will not give the precise conditions here. What we note, however, is that (using some abstract notation) the solution to (3.4) above, and hence the first-order terms in the expansion (3.3),can be written as

$$\gamma^{(1)}(\mathrm{y}) = \mathcal{L}^{-1}_{\gamma^{(0)}}\left(m^{(1)}_{\gamma^{(0)}}\right) \tag{3.5}$$

where $\mathcal{L}_{\gamma^0}$ is a linear operator mapping Γ into itself, and $m^{(1)}_{\gamma^{(0)}} \in \Gamma$; both depend on $\gamma^{(0)}$ which solves (3.2a)-(3.2b). Now, following the reasoning that led to (2.7a)-(2.7b), it is not conceptually difficult (though tedious in terms of the manipulations involved) to see that the k'th term in the expansion (3.3) will be given by

$$\gamma^{(k)}(\mathrm{y}) = \mathcal{L}^{-1}_{\gamma^{(0)}}\left(m^{(k)}_{\gamma^{(0)},\ldots,\gamma^{(k-1)}}\right), \quad k \geq 1 \tag{3.6}$$

where the linear operator $\mathcal{L}_{\gamma^{(0)}}$ is precisely the one used in (3.5), while the vector on which it acts is different for different k. We now summarize these conclusions in the following Theorem:

Theorem 3.1. *Let the stochastic game admit a Nash equilibrium solution $\gamma^*(y;\epsilon)$, and furthermore that the zero'th order stochastic game (obtained by setting $\epsilon = 0$) admit a unique Nash equilibrium solution $\hat{\gamma}$. Then, there exists an open neighborhood $\mathcal{N}$ of $\epsilon = 0$ such that the following are true:*

i) The expansion (3.3) holds for all $\epsilon \in \mathcal{N}$.

ii) $\hat{\gamma}_i = \gamma_i^{(0)}$, $i \in \mathcal{M}$.

iii) If the linear operator $\mathcal{L}_{\gamma^{(0)}}$ introduced in (3.5) (in abstract terms) is strongly nonsingular, then $\gamma_i^{(k)}$, $i \in \mathcal{M}$, $k \geq 1$, in the expansion (3.3), are uniquely determined iteratively (on k) using (3.6). ◇

Again parallel to (2.12) and (2.13), one can introduce the *Cournot* and *Gauss-Seidel* policy iterations (on the normal form of the game), leading to essentially the result of Theorem 2.2, with u's replaced by γ's. Instead of reproducing this result, we conclude this section with a scalar example that will serve to illustrate the Gauss-Seidel iteration in stochastic games.

Example 3.1. In the framework of the stochastic game formulation of this section, let $M = 2$, $K = 1$,

$$J_1 = E\left\{\frac{1}{2}u_1^2 + u_1 x + \epsilon u_1 u_2 \mid u_i = \gamma_i(y_i), \quad i = 1,2\right\} \tag{3.7a}$$

$$J_2 = E\left\{\frac{1}{2}u_2^2 + 2u_2 x + \epsilon u_1 u_2 \mid u_i = \gamma_i(y_i), \quad i = 1,2\right\}. \tag{3.7b}$$

Let the measurement vectors be given by

$$y_1 = x + v_1; \quad y_2 = x + v_2; \tag{3.8}$$

where x, v_1, v_2 are independent Gaussian random variables with mean zero and equal variance 1.

The complete solution to this quadratic game can be obtained using the theory of [1]. The Nash equilibrium solution (as a function of ϵ, and for $\epsilon^2 < 4$) is *unique*, and is given by

$$u_1 = \gamma_1^*(y_1;\epsilon) = \left(\frac{2\epsilon - 2}{4 - \epsilon^2}\right) y_1; \quad u_2 = \gamma_2^*(y_2;\epsilon) = -\left(\frac{4 - \epsilon}{4 - \epsilon^2}\right) y_2. \tag{3.9}$$

We evaluate these for two different values of ϵ, which will serve to gauge the degree of approximation provided by the Gauss-Seidel iteration:

$$\epsilon = 0.1 : \gamma_1^*(y_1;0.1) = -0.451128 y_1; \quad \gamma_2^*(y_2;0.1) = -0.977444 y_2 \tag{3.10a}$$

$$\epsilon = 0.01 : \gamma_1^*(y_1;0.01) = -0.495012 y_1; \quad \gamma_2^*(y_2;0.01) = -0.997525 y_2 \tag{3.10b}$$

Let us now apply the G-S policy iteration on this problem. Starting with $\gamma_2^{(0)}(y_2) \equiv 0$, the first step yields

$$\min_{u_1}\left\{\frac{1}{2}u_1^2 + u_1 E\left[x \mid y_1\right]\right\} \Rightarrow \gamma_{1(1)}^{\mathrm{GS}}(y_1) = -\frac{1}{2}y_1.$$

At the next step:

$$\min_{u_2}\left\{\frac{1}{2}u_2^2+2u_2E\left[x\mid y_2\right]-\frac{\epsilon}{2}u_2E\left[y_1\mid y_2\right]\right\}\Rightarrow \gamma_{2(1)}^{\mathrm{GS}}(y_2;\epsilon)=-\frac{4-\epsilon}{4}y_2$$

and with one more step:

$$\min_{u_1}\left\{\frac{1}{2}u_1^2+u_1E\left[x\mid y_1\right]-\frac{\epsilon(4-\epsilon)}{4}u_1E\left[x\mid y_1\right]\right\}\Rightarrow \gamma_{1(2)}^{\mathrm{GS}}(y_1;\epsilon)=-\frac{1}{2}\left(1-\frac{\epsilon(4-\epsilon)}{4}\right)y_1.$$

Now, numerical evaluation yields

$$\gamma_{1(2)}^{\mathrm{GS}}(y_1;0.1)=-0.45125y_1;\quad \gamma_{2(1)}^{\mathrm{GS}}(y_2;0.1)=-0.975y_2$$

$$\gamma_{1(2)}^{\mathrm{GS}}(y_1;0.01)=-0.4950125y_1;\quad \gamma_{2(1)}^{\mathrm{GS}}(y_2;0.01)=-0.9975y_2$$

which, when compared with (3.10a) and (3.10b), do indeed corroborate the order relationships given in Theorem 2.1.

4 Conclusion

The results presented in this paper can be extended to static games where there are more than two groups of players, with weak interactions among all the groups. Further extensions would be to dynamic game models, which have in fact been studied in the two-player case in [14], [6] for deterministic and stochastic systems. Other related references on the optimization of weakly coupled systems are [7], [12], [13], [5], [8] and [9].

Acknowledgement. This work was supported in part by the U.S. Department of Energy under Grant DE-FG-02-88-ER-13939.

References

[1] T. Başar. Equilibrium solutions in two-person quadratic decision problems with static information structures. *IEEE Transactions on Automatic Control*, AC-20(3):320–328, June 1975.

[2] T. Başar. Equilibrium solutions in static decision problems with random coefficients in the quadratic cost. *IEEE Transactions on Automatic Control*, AC-23(5):960–962, October 1978.

[3] T. Başar. Two-criteria LQG decision problems with one-step delay observation sharing pattern. *Information and Control*, 38(1):21–50, July 1978.

[4] T. Başar and G. J. Olsder. *Dynamic Noncooperative Game Theory.* Academic Press, London/New York, 1982.

[5] T. Başar and R. Srikant. Approximation schemes for stochastic teams with weakly coupled agents. *Proceedings of 11th IFAC World Congress, Tallinn, USSR*, August 1990.

[6] T. Başar and R. Srikant. Stochastic differential games with weak spatial and strong informational coupling. *Proceedings of the 9th International Conference on Analysis and Optimization of Systems*, June 1990, Antibes, France.

[7] A. Bensoussan. *Perturbation Methods in Optimal Control.* John Wiley, Gauthier-Villars, Chichester, England, 1988.

[8] Ü. Özgüner and W. R. Perkins. A series solution to the Nash strategy for large scale interconnected systems. *Automatica*, 13(3):313–315, May 1977.

[9] Z. Gajic, D. Petkovski, and X. Shen. *Singularly Perturbed and Weakly Coupled Linear Control Systems: A Recursive Approach.* Lecture Notes in Control and Information Sciences, vol.140, Springer-Verlag, Berlin, 1990.

[10] L. A. Liusternik and V. J. Sobolev. *Elements of Functional Analysis.* Ungar, New York, 1965.

[11] M. Lóeve. *Probability Theory.* Van Nostrand, Princeton, NJ, 3rd edition, 1963.

[12] A. A. Pervozvanskii and V. G. Gaitsgori. *Theory of Suboptimal Decisions.* Kluwer Academic Press, Amsterdam, The Netherlands, 1988.

[13] R. Srikant and T. Başar. Optimal solutions in weakly coupled multiple decision maker Markov chains with nonclassical information. In *Proceedings of the 29th IEEE Conference on Decision and Control*, pp. 168-173, Tampa, FL, December 1989.

[14] R. Srikant and T. Başar. Iterative computation of noncooperative equilbria in nonzero-sum differential games with weakly coupled players. *Journal of Optimization Theory & Applications*, to appear, 1991.

Approximate Solutions to Continuous Stochastic Games[1]

Michèle Breton
GERAD and École des Hautes Études Commerciales

Pierre L'Écuyer
Dept. d'Informatique et de Recherche Opérationnelle
Université de Montréal

Abstract

In this paper, we suggest an approximation procedure for the solution of two-player zero-sum stochastic games with continuous state and action spaces similar to finite elements and modified policy iteration approaches used for the solution of Markov Decision Problems.

1 Introduction

Zero-sum, two-player discounted stochastic games were introduced by L. S. Shapley [13] who gave a (constructive) existence proof of saddle points in stochastic games with finite state and action sets which provided a first iterative algorithm for the computation of the value of such games. Since then, other iterative algorithms have been proposed (Pollatscheck and Avi-Itzhak [11], Filar and Tolwinsky [6], Tolwinsky [14]); all these algorithms can be related to methods used for the solution of Markov Decision Problems, i.e. value iteration, policy iteration and modified policy iteration.

We are interested in devising an approximation procedure to solve stochastic games with continuous (or very large) state or action spaces. There already exists a large literature on discretization and approximation methods in dynamic programming (see [9] and the references cited there). Approximation methods are used in order to define "smaller" problems which can then be solved using any available algorithm for discrete problems. For dynamic games, theoretical considerations pertaining to their approximation have been presented by [15]; we will address here a general method for the computation of an equilibrium point in a zero-sum game.

[1]Research supported by NSERC-Canada, Grants #OGPIN020 and #A5463 and FCAR-Québec Grants #90-NC-0252 and #EQ2831.

In this paper, we consider a two-player zero-sum stochastic game model with continuous state and action spaces as studied in [10] (the extension of the approximation procedure to more general models, e.g. locally contracting renewal games (see [3]) is straightforward). We describe a finite element computational approach to deal with continuous or very large state spaces. The algorithm used with the finite element approach can be viewed as an extension of the "modified policy iteration" algorithm studied in [12] for Markov Decision Problems.

The outline of the paper is as follows: In section 2, we state the basic stochastic game model and its associated dynamic programming operators. In section 3, we briefly recall some existing algorithms for the solution of discrete stochastic games. Finally, in section 4, we describe a finite element computational approach, using an approximate policy iteration algorithm.

2 Zero-sum Two-Player Stochastic Game model

Consider the following two-player game model with Borel state space S and separable metric action spaces A and B. For each state s in S, let $A(s) \subset A$ and $B(s) \subset B$ be the non empty compact set of admissible actions to player 1 and player 2 respectively when the system is in state s. To allow for randomized strategies, we assume that each action in $A(s)$ and $B(s)$ is in fact a mixed action, i.e. a probability measure over an underlying set of pure actions. At each of an infinite sequence of stages (decision times), the players observe the state s of the system and independently select actions $a \in A(s)$ and $b \in B(s)$. For the current stage, the expected return to player 1, paid by player 2, is $r(s, a, b)$ and the system moves to a new state s' according to a probability measure $q(\cdot|s, a, b)$ over S. A new action pair is then selected by the players, and so on. The expected one-stage return function of player 1 $r(s, a, b)$ is a bounded Borel-measurable real valued function of $s \in S$, $a \in A(s)$ and $b \in B(s)$ and the law of motion is given by the family of probability measures $\{q(\cdot|s, a, b) : s \in S, a \in A(s), b \in B(s)\}$ which form a Borel-measurable stochastic kernel on S given $s \in S$, $a \in A(s)$ and $b \in B(s)$.

A policy δ for player 1 is a Borel-measurable function from $s \in S$ into his admissible action set $A(s)$ under which player 1 takes the mixed action $\delta(s)$ whenever the system is in state s. In the same way, a policy γ for player 2 is a Borel-measurable function $\gamma : s \in S \to \gamma(s) \in B(s)$. Let Δ and Γ denote the set of policies for player 1 and 2 respectively. A stationary strategy pair for players 1 and 2, denoted $[\delta, \gamma]$, consists in using respectively the policies δ and γ at each stage of the game. In this paper, we consider only stationary strategies.

Let $v_{[\delta,\gamma]}(s)$ denote the expected discounted sum of the rewards of player 1 when the initial state of the system is s and the players use the stationary strategy pair $[\delta, \gamma]$ with discount factor ρ, $0 < \rho < 1$. Player 1 wishes to maximize the expected sum of his discounted rewards as player 2 wishes to minimize the same. In zero-sum games, an equilibrium point is called a saddle point; If it exists, a saddle point in stationary strategies $[\delta^*, \gamma^*]$ is a strategy pair such that, for any strategy pair $[\delta, \gamma]$ and for all $s \in S$,

$$v_{[\delta,\gamma^*]}(s) \le v_{[\delta^*,\gamma^*]}(s) = v^*(s) \le v_{[\delta^*,\gamma]}(s) \tag{1}$$

and the function v^* is called the value of the game. Sufficient conditions for the existence of saddle points in continuous zero-sum games are given in [10].

Let V represent the Banach space of all Borel-measurable bounded functions $v : S \to \mathbb{R}$, endowed with the supremum norm. In order to use a dynamic programming operators formalism, we define the local return function h by

$$h(s,a,b,v) = r(s,a,b) + \rho \int_S v(s')q(ds'|s,a,b) \tag{2}$$

for $v \in V, s \in S, a \in A(s)$ and $b \in B(s)$. It represents the expected return to player 1 for a fictive auxiliary game starting in state s, if the players use the action pair (a,b) and if the expected returns to player 1 from the next stage on are described by the function v. For every policy pair $[\delta, \gamma]$, the associated return operator $H_{[\delta,\gamma]} : V \to V$ is defined by:

$$H_{[\delta,\gamma]}(v)(s) = h(s, \delta(s), \gamma(s), v). \tag{3}$$

Finally, we define the operator $F : V \to V$ by

$$F(v)(s) = \sup_{a \in A(s)} \left(\inf_{b \in B(s)} h(s,a,b,v) \right). \tag{4}$$

$H_{[\delta,\gamma]}$ and F are both monotone contracting operators on V with modulus ρ.

3 Value Iteration and Policy Iteration

Value iteration and policy iteration are two general methods for solving dynamic programs. They operate as follows.

Value iteration.
Select initial v_0 in V;
For $n := 1$ to $\bar{n}$ do

$$v_n := F(v_{n-1}); \tag{5}$$

Retain $[\delta^*, \gamma^*]$ such that $H_{[\delta^*,\gamma^*]}(v_n) = F(v_n)$;
End.

Policy iteration.

Select initial policy pair $[\delta_0, \gamma_0]$;
For $n := 1$ to $\bar{n}$ do
Policy evaluation: find v_n such that

$$H_{[\delta_{n-1},\gamma_{n-1}]}(v_n) = v_n; \tag{6}$$

Policy update: find $[\delta_n, \gamma_n]$ such that

$$H_{[\delta_n,\gamma_n]}(v_n) = F(v_n); \tag{7}$$

Retain $[\delta_{\bar{n}}, \gamma_{\bar{n}}]$;
End.

In both cases, the value of $\bar{n}$ may be chosen in advance or depend on some stopping criterion.

The algorithm proposed by Shapley [13] corresponds to value iteration. Each step requires the solution of $|S|$ matrix games in equation (5). It converges to v^* from any starting v_0.

The algorithm proposed by Pollatschek and Avi-Ithzak [11] corresponds to policy iteration. Each step requires the solution of the system of $|S|$ linear equations (6) and $|S|$ matrix games (7). This algorithm does not converge in general for stochastic games.

It is well known that value iteration converges linearly (sometimes very slowly) while policy iteration (when it works) is equivalent to applying Newton's method to the equation $F(v) - v = 0$ (see [11]). When v is not too far from v^*, it typically has quadratic convergence. Empirical evidence presented in [2] and [1] suggests that policy iteration is the fastest method for solving stochastic games in cases when it converges. Motivated by this fact, Filar and Tolwinsky [6] have recently proposed a modified Newton's method (MNM) which is guaranteed to converge and has the same rate of convergence as policy iteration when the latter converges.

In the context of MDPs having large state spaces, Puterman and Shin [12] proposed an adaptation of policy iteration, the so-called "modified policy iteration" method, where at each iteration, (6) is solved approximately by applying only a few iterations of the value iteration method with a fixed policy $[\delta_{n-1}, \gamma_{n-1}]$, starting from the previous v. Tolwinsky [14] proposed a modified iteration algorithm combining the MNM scheme with the ideas of [12]. In numerical experimentation, the modified policy iteration seemed to perform better than the MNM in cases where the number of states was large relative to the number of actions.

Modified policy iteration.
Select initial v_0 in V;
For $n := 1$ to $\bar{n}$ do
Policy update: find $[\delta_n, \gamma_n]$ such that

$$H_{[\delta_n,\gamma_n]}(v_n) = F(v_n); \tag{8}$$

Set $d := v_n$;
Search direction: select k and repeat k times

$$d := H_{[\delta_n,\gamma_n]}(d); \tag{9}$$

Choose a step size α ensuring descent and set $v_{n+1} := v_n + \alpha(d - v_n)$;
Retain $[\delta_n, \gamma_n]$;
End.

For a descent criterion in the modified Newton method, in order to choose the step size α according to Armijo's rule, see [6].

Obviously, for continuous (or very large) state spaces, these algorithms cannot be applied exactly in general. Some form of approximation must be used. From an approximate solution to the functional equation (5), one can obtain bounds on v^* and on the suboptimality of a given policy pair (see [4]).

4 A Finite Element Approach

We now introduce an approximate policy iteration algorithm, with finite element approximation of the value function. For more details on the finite element method, see e.g. [8]. Generally speaking, we assume that an expected "value-to-go" function v associated with a fixed policy can be approximated reasonably well by a linear combination of a small set of (simple) base functions $w_1, \ldots, w_J$:

$$v(s) = \sum_{j=1}^{J} d_j w_j(s). \tag{10}$$

One particular finite element approach [9] is to select a finite number of points $\sigma_1, \ldots, \sigma_J$ in S and to express directly $v(s)$ as a convex combination of the values of v at the J evaluation points:

$$v(s) = \sum_{j=1}^{J} v(\sigma_j) w_j(s) \tag{11}$$

where, for all $s \in S$,

$$0 \le w_j(s) \le 1, \tag{12}$$

$$w_j(\sigma_i) = \delta_{ij} \tag{13}$$

(the Kronecker's delta), and

$$\sum_{j=1}^{J} w_j(s) = 1. \tag{14}$$

The σ_j's are in fact the *nodes* of the finite elements. The interesting point in this particular scheme is that it permits the evaluation of v easily at any point in S, and thus on any set of nodes. In this setting, an analog to (9) is to apply pre-Jacobi iterations to the linear system

$$d = c + Md, \tag{15}$$

where d and c are the column vectors $(d_1, \ldots, d_J)'$ and $(c_1, \ldots, c_J)'$ respectively and M is the $J \times J$ matrix (m_{ij}), with, for a given policy pair $[\delta, \gamma]$,

$$c_j = r(\sigma_j, \delta(\sigma_j), \gamma(\sigma_j)), \tag{16}$$

and

$$m_{ij} = \rho \int_S w_j(s') q(ds' | \sigma_i, \delta(\sigma_i), \gamma(\sigma_i)). \tag{17}$$

In general, policies must also be approximated: it is usually not possible to find $[\delta, \gamma]$ such that (7) is satisfied exactly when the action space is very large or continuous. As we did for the state space, we can define a finite dimensional subspace of the action spaces A and B and consider only the actions that belong to that subspace. Since the detailed way to do that is rather problem-dependent, we will content ourselves with the following description. For any v in V, $\epsilon \geq 0$ and $\sigma \in S^J$, define

$$\phi_\epsilon(v) = \left\{ [\delta, \gamma] \in \Delta \times \Gamma : |H_{[\delta,\gamma]}(v) - F(v)| \leq \epsilon \right\} \tag{18}$$

and

$$\phi_\epsilon(v, \sigma) = \left\{ [\delta, \gamma] \in \Delta \times \Gamma : |H_{[\delta,\gamma]}(v)(\sigma_i) - F(v)(\sigma_i)| \leq \epsilon \right\}. \tag{19}$$

At every "policy update" step of the algorithm (equation (8)), we will in fact seek a new policy in $\phi_\epsilon(v)$ for some given value of ϵ. Often, in practice, we will first find a policy pair $[\delta, \gamma]$ and then estimate the smallest ϵ for which $[\delta, \gamma] \in \phi_\epsilon(v)$. Under this setting, the approximation algorithm is given by:

Approximate policy iteration

Select $\epsilon > 0$, initial v_0 in V and initial policy pair $[\delta_0, \gamma_0]$;
Outer loop: For $n := 1$ to $\bar{n}$ do
Step 1: Select $J_n, \sigma = (\sigma_1, \ldots, \sigma_J)' \in S^J$ and $\{w_1, \ldots, w_J\} \subset V$ such that (12– 14) are satisfied;
For the policy pair $[\delta_{n-1}, \gamma_{n-1}]$, compute c and M by (16–17) and set $d := (v(\sigma_1), \ldots, v(\sigma_J))'$;
Inner loop (search direction): select k and repeat k times: $d := Md + c$;
Set $\alpha = 1$;
Step 2: set $v_{n+1}(\sigma_i) := v_n(\sigma_i) + \alpha(d_i - v_n(\sigma_i))$;
Define v by (11);
Find a new policy pair $[\delta_n, \gamma_n]$ in $\phi_0(v, \sigma)$;
If descent is not verified (see [6]), set $\alpha = \mu\alpha, \mu \in (0, 1)$, and go to step 2.
If desired, perform a stopping test: compute or estimate a bound $\bar{\epsilon}$ on $\|v^*, v_{[\delta_n, \gamma_n]}\|$. If $\bar{\epsilon} \leq \epsilon$, or other stopping criteria is satisfied, stop;
Endloop
End.

Obviously, as it stands, the approximation algorithm is not completely defined. For instance, the stopping criteria, the way of choosing $\epsilon, \bar{\epsilon}, \alpha, J$ and the base functions w_j, the method used to update the policy pair and to compute or estimate $\bar{\epsilon}$ are all left open. These are usually problem dependent. In practice, they may vary from iteration to iteration.

The stopping test can be costly and should not be performed at each iteration. The bound $\bar{\epsilon}$ may have to be estimated heuristically, for instance as in [7]: Recompute $F(v)$ and $H_{[\delta,\gamma]}(v)$ at a large number of new points, compute the approximation error at there points and take the largest and smallest to estimate the bounds.

Notice that for $k = 1$, the modified policy iteration method becomes the value iteration algorithm, as for $k = \infty$, it becomes policy iteration. A good choice for k is probably problem dependent. It could be chosen adaptively, based on the previous iterations; intuitively, the more costly it is to compute c and M, the larger the value of k should be. But the inner loop should also stop when progress gets too slow, i.e. when d is not changing significantly enough anymore of if d does not appear to converge geometrically.

The choice of σ determines a grid over the state space S. A coarser grid should be chosen at the early stages of the algorithm and the grid should be refined only when progress is stalling. Multigrid techniques [5] or various other techniques for the iterative solution of linear systems can also be used.

We have described a finite element approach to solve stochastic game models with continuous or very large state spaces. It can deal with most reasonably smooth value functions, provided that the state space is bounded and has few (continuous) dimensions. Numerical experiments with this approach are presently in progress.

References

[1] Breton, M., Filar, J. A., Haurie, A. and Schultz, T. "On the Computation of Equilibria in Discounted Stochastic Games", in *Dynamic Games and Applications in Economics*, T. Başar Ed., Springer-Verlag, Berlin, (1986), 64–87.

[2] Breton, M. "Équilibres pour des jeux séquentiels", Ph. D. Thesis, Université de Montréal (1987).

[3] Breton, M. and L'Écuyer, P. "Noncooperative Stochastic Games Under a N-Stage Local Contraction Assumption", *Stochastics*, **26**, (1989), 227–245.

[4] Breton, M. "Algorithms for Stochastic Games", in *Stochastic Games and Related Topics - Shapley Honor Volume*, T. E. S. Raghavan, T. S. Ferguson, T. Parthasarathy and O. J. Vrieze Eds., Kluer, The Netherlands (to appear).

[5] Briggs, W. L. *A Multigrid Tutorial*, SIAM, Philadelphia, 1987.

[6] Filar, J. A. and Tolwinsky, B. "On the Algorithm of Pollatschek and Avi-Itzhak", in *Stochastic Games and Related Topics - Shapley Honor Volume*, T. E. S. Raghavan, T. S. Ferguson, T. Parthasarathy and O. J. Vrieze Eds., Kluer, The Netherlands (to appear).

[7] Haurie, A. and L'Écuyer, P. "Approximation and Bounds in Discrete Event Dynamic Programming", *IEEE Transactions on Automatic Control*, **AC-31**, 3 (1986), 227–235.

[8] Hugues, T. J. R. *The Finite Element Method: Linear Static and Dynamic Finite Element Analysis*, Prentice-Hall, Englewood, New Jersey, 1987.

[9] L'Écuyer, P. "Computing Approximate Solutions to Markov Renewal Programs with Continuous State Spaces", Technical Report DIUL-RR-8912, Université Laval, Québec, 1989.

[10] Nowak, A. S. "On Zero-Sum Stochastic Games with General State Space I", *Probability and Mathematical Statistics*, **4**, (1984), 13–32.

[11] Pollatschek, M. and B. Avi-Itzhak "Algorithms for Stochastic Games with Geometrical Interpretation", *Management Science*, **15**, (1969), 399–415.

[12] Puterman, M. L. and Shin, M. C. "Modified Policy Iteration Algorithms for Discounted Markov Decision Problems", *Management Science*, **24**, 11 (1978), 1127–1137.

[13] Shapley, L. S. "Stochastic Games", *Proceedings of the National Academy of Sciences of USA*, **39**, (1953), 1095–1100.

[14] Tolwinsky, B. "Newton-Type Methods for Stochastic Games", in *Differential Games and Applications*, T. S. Başar and P. Bernhard Eds., Springer-Verlag, Berlin (1989), 128–143.

[15] Whitt, W. "Representation and Approximation of Noncooperative Sequential Games", *SIAM Journal on Control*, **18**, 1 (1980), 33–48.

Solving Dynamic Games via Markov Game Approximations

Boleslaw Tolwinski
Department of Mathematics, Colorado School of Mines
Golden, Colorado 80401

1 Introduction

In the task of solving a differential game, the ultimate goal is to obtain optimal strategies of the feedback type. In most cases this leads to a dynamic programming problem that, except for some special classes of games, can be solved only numerically if at all. The 'curse of dimensionality' and the resulting tremendous computational complexity of dynamic programming is well known, so it is not surprising that few attempts have been made to solve differential games of any practical significance via dynamic programming algorithms. Since other numerical techniques used in the context of differential games can as a rule produce only open-loop solutions, unsatisfactory for most applications, the inability to solve the dynamic programming equations arising in the theory of differential games may have been the single most important factor preventing that theory from finding wider applications to practical problems.

A key step in the numerical solution of a differential game, either deterministic, stochastic, or piecewise deterministic, is the discretization of an appropriate Hamilton-Jacobi-Bellman equation. This can be achieved by the method due to H.J. Kushner [5] and orginally developed to deal with stochastic control problems. In the case of dynamic games, Kushner's method produces a sequence of finite-state Markov games characterized by sparse transition matrices. Relatively large problems of that type can be solved by recently developed algorithms based on the policy iteration method [4,8,9].

This paper has two goals. The first goal is to reexamine, in view of the recent advances in numerical techniques for finite-state Markov games, the usefulness of dynamic programming as a tool for solving non-trivial differential games. The second goal is to present a new approach to the solution of Markov games, which seems especially well adapted to problems resulting from the discretization of certain differential games. The new method is related to the learning algorithms considered in the neural network literature [2] and, like dynamic programming, it is based on the computation of the Bellman-Isaacs value function. Unlike classical dynamic programming that treats all possible states of the game in the same manner and calculates the value function for all of them, the new method, given a set of initial states, produces the most accurate approximations to the value function only for the states that are most likely to occur during the play, with less accurate approximations for the states less likely to occur, and ignoring entirely the states whose probability of occuring is negligible. The early experiments indicate that such an approach can dramatically reduce the severity of the 'curse of dimensionality' by decreasing the number of states for which the value function has to be found to a small fraction of the total.

The next section contains the definition of the piecewise deterministic differential game, a model that generalizes the classical differential game and is relevant to many applications; the corresponding dynamic programming equation, and its discretized version. Section 3 describes dynamic programming algorithms of the modified policy iteration type applicable to large Markov games with sparse transition matrices, and discusses their performance and limitatations. Finally, Section 4 presents an alternative, neural net type algorithm for dynamic programming.

2 Formulation of the Problem

The piecewise deterministic differential games (PDDG) extend the classical dynamic model of conflict to problems evolving through several phases, with transitions from one phase to another taking place at random times and with system dynamics differently defined in each phase. A duel between two aircrafts armed with missiles [6,7] is a good example of such a situation. We chose to discuss numerical methods in the context of PDDG rather than in the context of classical differential games to emphasize the flexibility of these methods, and also to further popularize PDDG as models with great potential for applications.

The definition of piecewise deterministic differential game to be considered consists of the following elements.

- The extended state space $S = X \times E$, where $X \subset R^n$ and E is a finite set enumerating possible phases or modes of the play. The set E contains a nonempty subset ∂E of terminal states.

- The decision (control) sets U_i^k for $i = 1, 2$ and $k \in E$.

- The state dynamics

$$\dot{x} = f^k(x, u_1, u_2), \text{ where } x \in X, u_i \in U_i^k, \text{and } k \in E \tag{1}$$

- The transition rate functions $\lambda_{kl}(x)$, where

$$\lambda_{kl}(x) = \lim_{dt \to 0} \frac{1}{dt} \Pr[\xi(t + dt) = l \mid \xi(t) = k, x(t) = x]; \; k, l \in E \tag{2}$$

- The payoff function $g^k(x)$, defined for $x \in X$ and $k \in \partial E$.

The play starts at some initial state $s_0 \in S$. At each moment of time, the players observe the current extended state $s(t)$ of the system and choose their controls u_i. The state $x(t)$ evolves according to the state dynamics and transitions within E occur at random stopping times, the probability distributions of which are determined by the transition rate functions. When the play enters a terminal state $k \in \partial E$ player 2 pays player 1 the amount $g^k(x)$, which in general may depend also on the value of the state variable x, and the play terminates. Thus, player 1 and player 2 seek respectively to maximize (with respect to u_1) and minimize (with respect to u_2) the expected (discounted) payoff

$$R(s_0; u_1, u_2) = E\{e^{-\rho T} g^{k(T)}[x(T)]\} \tag{3}$$

where $k(T)$ is a terminal state reached at a random stopping time T, and $\rho \geq 0$ denotes a discount factor. The presence of the discount factor in eq.3 is equivalent to the assumption that the play may terminate at a random stopping time τ, the probability distribution of which is given by

$$Pr(\tau \in [t, t+dt] \mid T \geq t) = \rho dt + o(dt)$$

If ρ is strictly positive, the play will terminate with probability one.

Let $V^k(x)$ be the Bellman value function corresponding to $k \in E$ and $x \in X$, i.e.,

$$V^k(x) = \sup_{u_1} \inf_{u_2} R[(x,k); u_1, u_2] \tag{4}$$

Then, the Hamilton-Jacobi-Bellman equation for the problem at hand can be written as

$$\begin{aligned} \rho V^k(x) &= \sup_{u_1 \in U_1^k} \inf_{u_2 \in U_2^k} \{\frac{\partial}{\partial x} V^k(x) f(x, u_1, u_2) \\ &+ \sum_{l \in E, l \neq k} \lambda_{kl}(x)[V^l(x) - V^k(x)]\} \end{aligned} \tag{5}$$

for $k \in E \setminus \partial E$, and

$$V^k(x) = g^k(x) \tag{6}$$

for $k \in \partial E$.

Let h denote the finite difference interval, in each coordinate direction, and e_j the unit vector in the jth coordinate direction. Let R_h^n be the finite difference grid on R^n and define $X_h = X \cap R_h^n$. Following [5] we shall use the approximations

$$\begin{aligned} \frac{\partial V^k(x)}{\partial x_j} &\longrightarrow [V^k(x + e_j h) - V^k(x)]/h \text{ if } f_j^k(x, \underline{u}) \geq 0 \\ &\longrightarrow [V^k(x) - V^k(x - e_j h)]/h \text{ if } f_j^k(x, \underline{u}) < 0 \end{aligned} \tag{7}$$

Define for all $x \in X_h$ and $\underline{u} \in U_1^k \times U_2^k$

$$F^k(x,\underline{u}) = \sum_{j=1}^{n} | f_j^k(x,\underline{u}) | \tag{8}$$

$$p^k(x + e_j h \mid x, \underline{u}) = \max[0, f_j^k(x,\underline{u})]/F^k(x,\underline{u}) \tag{9}$$

$$p^k(x - e_j h \mid x, \underline{u}) = \max[0, -f_j^k(x,\underline{u})]/F^k(x,\underline{u}) \tag{10}$$

$$p^k(z \mid x, \underline{u}) = 0 \text{ if } z \neq x + e_j h \text{ and } z \neq x - e_j h \tag{11}$$

$$p^x(l \mid k, \underline{u}) = \frac{\lambda_{kl}(x)h}{F^k(x,\underline{u}) + \lambda_{kk}(x)h} \tag{12}$$

where $\lambda_{kk} = \sum_{l\in E, l\neq k} \lambda_{kl}$. The dynamic programming equations can be now written in the form

$$\begin{aligned} V^h(x;k) &= \max_{u_1} \min_{u_2} \{ \\ &\quad \frac{1}{1 + \frac{\rho}{F^k(x,\underline{u})/h + \lambda_{kk}(x)}} [\sum_{z\in X_h} p^k(z \mid x, \underline{u}) V^h(x;k)] \\ &+ \sum_{l\in E, l\neq k} p^x(l \mid k, \underline{u}) V^h(x;l) \} \end{aligned} \tag{13}$$

for $k \in E \setminus \partial E$, and

$$V^h(x;k) = g^k(x) \tag{14}$$

for $k \in \partial E$.

Solving the above equation for a fixed discretization step h amounts to the solution of a zero-sum finite-state Markov game. The next section will take a closer look at this problem.

3 Modified Policy Iteration Methods

To simplify the description of the Markov game at hand, we introduce the following notation. Let $S_h = X_h \times E$, and for $s = (x,k) \in S_h$, $s' = (y,l) \in S_h$ define

$$p(s' \mid s, \underline{u}) = \begin{cases} p^k(y \mid x, \underline{u}) & \text{if } k = l \\ p^x(l \mid k, \underline{u}) & \text{if } x = y \\ 0 & \text{otherwise} \end{cases} \tag{15}$$

$$r(s, s', \underline{u}) = \begin{cases} \frac{1}{1 + \frac{\rho}{F^k(x,\underline{u})/h + \lambda_{kk}(x)}} & \text{if } k = l \\ 1 & \text{otherwise} \end{cases} \tag{16}$$

Notice that $0 \leq r(s, s', \underline{u}) \leq 1$. Now, the equations 13 and 14 can be rewritten in the form

$$V^h(s) = \max_{u_1} \min_{u_2} \sum_{s' \in S_h} r(s, s', \underline{u}) p(s' \mid s, \underline{u}) V^h(s') \tag{17}$$

for $k \in E \setminus \partial E$, and

$$V^h(s) = g^k(x) \tag{18}$$

for $k \in \partial E$.

The policy iteration algorithm for the Markov game defined by equations 17 and 18 can be formulated as follows.

1. Select an initial approximation, say $V^{(0)}(\cdot)$, to the value function $V^h(\cdot)$. Set m to zero.

2. Given $V^{(m)}(\cdot)$, compute a pair of strategies $\underline{\mu}^{(m)}(\cdot) = (\mu_1^{(m)}(\cdot), \mu_2^{(m)}(\cdot))$ such that for every $s \in S_h$

$$\begin{aligned} &\sum_{s' \in S_h} r(s, s', \underline{\mu}^{(m)}(s)) p(s' \mid s, \underline{\mu}^{(m)}(s)) V^{(m)}(s') \\ &= \max_{u_1} \min_{u_2} \sum_{s' \in S_h} r(s, s', \underline{u}) p(s' \mid s, \underline{u}) V^{(m)}(s') \end{aligned} \tag{19}$$

3. Solve the system of linear equations

$$V(s) = \sum_{s' \in S_h} r(s, s', \underline{\mu}^{(m)}(s)) p(s' \mid s, \underline{\mu}^{(m)}(s)) V(s') \tag{20}$$

   for $V(\cdot)$, where $s = (x, k) \in S_h$, $k \in E \setminus \partial E$;

$$V(s) = g^k(x) \tag{21}$$

   where $s = (x, k) \in S_h$, $k \in \partial E$. Denote the solution by $V^{(m+1)}$ (this is the expected payoff corresponding to strategies $\underline{\mu}^{(m)}(\cdot)$), set m to $m + 1$, and go back to step 2.

Remark 1. It is usually required that strategies $\underline{\mu}$ obtained in step 2 be in fact equilibrium strategies, or in other words, that the order of operators max and min in eq.19 be interchangeable. If the decision sets U_i^k are finite (the original decision spaces may be discretized), the problem of computing strategies $\underline{\mu}$ is equivalent to solving a matrix game for every $s \in S_h$.

Remark 2. The policy iteration method in its pure form presented above, as well as its various modifications discussed below, is not guaranteed to converge to the solution of the Markov game defined by eqs. 17 and 18. For the case of a constant discount parameter r, the convergence can be assured by additional modifications [4,8]. The case of state and control dependent discount factor appears to be as yet unresolved. Fortunatly, the computational experience indicates that instances when policy iteration fails to converge are extremly rare.

The policy iteration algorithm as defined above is very effective for Markov games with a moderate number of states. If N denotes the number of elements of S_h less the number of states $s = (x, k)$ for which $k \in \partial E$, then step 3 of the algorithm requires solving a system of N equations by Gaussian elimination. For larger values of N this approach quickly ceases to be effective. The natural alternative is to solve the system defined by eqs.20 and 21 by an iterative method, e.g., Jacobi, Gauss-Seidel, multigrid, or conjugate-gradient technique. The methods that use iterative techniques instead of Gaussian elimination to compute $V^{(m)}(\cdot)$ are called *modified policy iteration algorithms.*

The computational experience concerning modified policy iteration methods is as yet rather limited. In [9] we describe an experiment with the piecewise deterministic differential game model of a duel between two vehicles moving in a two dimensional plane. In that problem the state space X was a subset of R^3 and the set E contained three elements. This led to approximating Markov games with up to about 50,000 states. These games were solved by modified policy iteration using Jacobi, Gauss-Seidel, and multigrid techniques. The most effective approach appeared to be to use the Gauss-Seidel method in combination with 'one-way multigrid' that consists in obtaining initial approximations to V^h on coarser grids (relatively large h) and then to proceed to finer grids corresponding to smaller values of h. The results obtained with the full multigrid method [3,1] that moves repeatedly up and down between different grids did not as yet provide evidence that in the case of Markov games such an approach might be more effective than the much simpler one-way multigrid technique.

One reason why the modified policy iteration methods are much superior to the traditional policy iteration algorithm when applied to the system given by eqs.20, and 21 is that the matrix of that system tends to be very large and sparse. Another reason is that to achieve convergence to the value function $V^h(\cdot)$ in the policy iteration algorithm described above, the function $V^{(m)}(\cdot)$ in step 3 of that algorithm need not to be computed exactly. In all our experiments the total computation time was minimized when for each m only five to twenty Gauss-Seidel iterations were performed (out of hundreds necessary to obtain an almost exact value of $V^{(m)}(\cdot)$).

There is now enough evidence to confirm the thesis that deterministic, piecewise deterministic or stochastic differential games of moderate size can be solved via Markov games approximations. By games of moderate size we mean here problems where the dimension of the state space X does not exceed three or perhaps four, leading to the approximating Markov games with state space cardinality of order 10^4 or even 10^5. For larger problems the modified policy iteration becomes impractical and methods other than traditional dynamic programming must be sought. One such a method will be discussed in the next section.

4 Neural Net Approach to Dynamic Programming

Solving a control problem or sequential game by the traditional dynamic programming methods requires one to compute the value function of the problem for every possible state of the system. In many applications, however, the set of possible initial states, say S^0, constitutes only a small subset of the state space, and not all of the remaining states are equally likely to be visited by a trajectory that originates in S^0. In the air combat example [6,7], the duel always starts when the adversaries are at the certain distance from one another, defined by the range of their radars, and then only a very small fraction of the total number of all possible states has positive probability of being visited, at least if one assumes intelligent and reasonably well trained pilots. This leads to the conclusion that given a set of initial states, one may need to know the value function and the optimal strategies only for a, possibly quite small, subset of the state space. Moreover, it can be argued that if the probability that some state will be visited is small, then the value function for this state need not be computed with a very high accuracy. The traditional dynamic programming methods like policy iteration or successive approximations are ill-suited to take advantage of that property since they compute the value function on the entire state space, with a similar accuracy for every state.

An approach that promises to be better adapted to exploit the special structure of Markov decision problems arising in the approximations of certain stochastic control problems and differential games is suggested by an idea coming from the study of neural networks, and more specifically from the paper by Barto, Sutton, and Anderson [2] concerning a model of learning by natural or artificial intelligence. The key element of that model is the so-called *adaptive critic element* (ACE), which bears close resemblance to the value function of dynamic programming. The learning process consists of a sequence of simulation runs. Each simulation run generates a trajectory, that is, a sequence of states. In each state a decision is to be made that affects the probability of transition to a next state. The trajectory terminates in an absorbing state, which has a reward associated with it. The objective of learning is to discover decisions that maximize the expected reward. During each run, a decision is made based on the current values of ACE's for the neighboring states. The ACE's are updated after each choice of a decision at a currently visited state.

The learning algorithm described in [2] has been designed to model the behavior of animal intelligence. The algorithm presented below is based on the same general idea, but, since its purpose is quite different, most of the important details have been changed. The algorithm is defined in terms of two functions (or subroutines), called *basic iteration* and *simulation run*, that manipulate data on three arrays: *stack* that stores the states generated in the process of simulation, *value* containing the current approximations to the value function, and *control* storing the corresponding optimal strategies. The two functions are defined as follows.

$$[s_{new}, t_{new}, er] = basic\ iteration(s_{old}, t_{old})$$

Comment: Given input variables $s_{old} \in S_h$ and t_{old} that points to the top of the data stacks, this procedure adds s_{old} to *stack*, if s_{old} is not already there; updates the value function at s_{old}; and generates a new state s_{new}. The output variables t_{new} and er represent, respectively, the new top of the stacks, and the difference between the new and old values of the value function.

1. Verify whether s_{old} is on the *stack*. If not, assign $t_{new} \leftarrow t_{old} + 1$ and $stack(t_{new}) \leftarrow s_{old}$. Determine j such that $s_{old} = stack(j)$.

2. Compute the transition probability functions $p(s \mid s_{old}, \underline{u})$ for $s = (x, k) \in S_h$ and $\underline{u} \in U_1^k \times U_2^k$.

3. For all s such that $p(s \mid s_{old}, \underline{u}) > 0$ for some $\underline{u} \in U_1^k \times U_2^k$, define $v(s)$ as follows. If $k \in \partial E$, set $v(s) = g^k(x)$; else if $s = stack(j_1)$ for some j_1 satisfying $1 \leq j_1 \leq t_{new}$, set $v(s) = value(j_1)$; else set $v(s)$ to a random number from an appropriately chosen probability distribution.

4. Compute v_{new} and $\underline{\mu}$, where

$$\begin{aligned} v_{new} &= \sum_{s \in S_h} r(s_{old}, s, \underline{\mu}) p(s \mid s_{old}, \underline{\mu}) v(s) \\ &= \max_{u_1} \min_{u_2} \sum_{s \in S_h} r(s_{old}, s, \underline{u}) p(s \mid s_{old}, \underline{u}) v(s) \end{aligned} \tag{22}$$

5. Assign $v_{old} \leftarrow value(j)$ and update the value function according to

$$value(j) = v_{old} + \beta(v_{new} - v_{old}) \tag{23}$$

   where β is a parameter satisfying $0 < \beta \leq 1$.

6. Assign $er \leftarrow \mid v_{new} - v_{old} \mid$ and $control(j) = \underline{\mu}$.

7. Draw s_{new} from the probability distribution defined by $p(\cdot \mid s_{old}, \underline{\mu})$.

8. Return.

$$[\Delta, ter] = simulation\ run(s, t)$$

Comment: Given an initial state s and the pointer t to the top of the data stacks, this procedure generates a trajectory terminating in an absorbing state, and returns the cumulative error ter as well as the total number Δ of new states added to *stack* during the simulation run.

1. Set Δ and ter to zero.

2. Set $[s, t_{new}, er] \leftarrow$ *basic iteration*(s, t).

3. Assign $ter \leftarrow ter + er$; $\Delta \leftarrow \Delta + (t_{new} - t)$; and $t \leftarrow t_{new}$.

4. If $k \in \partial E$ return; else go to step 2.

The algorithm consists of repeated simulation runs performed for a subset of initial states. It terminates when, for a number of consecutive runs, Δ remains zero (no new states are added to the stack), and the total error ter is less than some given tolerance parameter epsilon.

The process of numerical testing of the above described algorithm is still in very preliminary stages. The early results are, however, quite encouraging. Experiments with a pursuit-evasion problem with about 100,000 states, produced solutions in a fraction of time needed by the modified policy iteration procedures. The value function in that example had to be computed in only one to two percent of the total number of states, and this percentage actually decreased when the total number of states increased.

References

[1] Akian, M. *Resolution numerique d'equations d'Hamilton-Jacobi-Bellman au moyen d'algorithmes multigrille et d'iterations sur les politiques*, Lecture Notes in Control and Inf. Sc., No. 111, Springer-Verlag, 1988.

[2] Barto, A.G., Sutton, R.S., and Anderson C.W. *Neuronlike adaptive elements that can solve difficult learning control problems*, IEEE Trans. on Systems, Man, and Cybernetics, SMC-13,pp. 834-846, 1983.

[3] Hackbush, W. and Trottenberg, U. (Eds.) *Multigrid Methods*, Lecture Notes n Mathematics, No. 960, Springer-Verlag, 1981.

[4] Filar, J.A. and Tolwinski B. *On the algorithm of Pollatschek and Avi-Itzhak*, in T.E.S. Raghavan, T.S. Ferguson, and O.J. Vrieze (Eds.), Stochastic Games and Related Topics, Kluwer Academic Publishers, to appear.

[5] Kushner, H. *Probability Methods for Approximation in Stochastic Control and for Elliptic Equations*, Academic Press 1977.

[6] Moritz, K., Polis R., and Well, K.H. *Pursuit-evasion in medium-range air-combat scenarios*, Comp. Math., Appl., 13, pp. 167-180, 1987.

[7] Shinar J., Siegel, A.W., and Gold Y.I. *On the analysis of a complex differential game using artificial intelligence techniques*, Proceedings of the 27th Conference on Decision and Control, Austin, Texas, pp. 1436-1441, 1988.

[8] Tolwinski, B. *Newton-type methods for stochastic games*, in T. Basar and P. Bernhard (Eds.) "Differential Games and Applications", pp. 128-144, Springer-Verlag, Heidelberg 1989.

[9] Tolwinski, B., *Numerical solutions to differential games based on approximations by Markov games*, in Proc. 28th IEEE Conference on Decision and Control, 1, pp. 174-179, 1989.

A GAME PROGRAMMING APPROACH TO EFFICIENT MANAGEMENT OF INTERCONNECTED POWER NETWORKS*

A. Haurie**, G. Zaccour***

GERAD
5255, avenue Decelles, Montréal QC Canada H3T 1V6

Abstract: This paper presents a stochastic dynamic cooperative game model of power exchange between interconnected utilities. The model is used for the qualitative analysis of an efficient management of the interconnected networks. It also permits an analysis of the different pricing schemes through which the utilities effectuate the side-payments associated with the bargaining solutions. The case of two utilities is approached via the Nash-Harsanyi bargaining solution. Different approaches concerning the design of acceptable pricing schemes are proposed and discussed.

1. Introduction

The aim of this paper is to propose a stochastic dynamic game formalism permitting the analysis of the economics of long term power exchanges between interconnected utilities. Power exchanges between interconnected utilities have existed for a long time. They were motivated by security, load curve diversity and differences in production costs. Recently, long term power exchanges between neighboring countries have considerably developed. This is particularly the case between Canada and USA. In Eastern Canada, Provinces like Quebec or New Brunswick are proposing American utilities in New England and New York State to engage in *firm power* exchange contracts. This means that the Canadian utilities will commit themselves to a long term power export, possibly through specific investments exploiting relatively cheap hydro or nuclear sources. The model proposed in this paper is the simplest structure which retains the most important basic features of this economic problem: (a) one deals with a capital accumulation process; (b) the exchanges involve a small number of participating utilities and can thus be viewed as a cooperative game; (c) the pricing schemes proposed in power exchange contracts are designed to effectuate the side-payments necessary to the definition of a fair cooperative solution; (d) uncertainty is inherent in the planning process which should be adapted

* Research supported by FCAR-Actions spontanées, Grant #87-AS-2434, NSERC, Grants #A4952, #OPG0037525, MESS-Action structurante 6.1/7.4 (28)

** Département d'économie commerciale et industrielle, Université de Genève, Suisse.

*** Dept. of Marketing, École des HEC, Montréal, Canada.

to conjonctural changes. Although it may look simplistic, this model is structurally similar to the the very detailed models commonly used in practice for the modeling of long term energy choices. We refer the reader to [1] for such a detailed model, based on a linear programming formalism and used to analyse the power exchange perspectives between Quebec and New York state. However, in detailed models, a game theoretic analysis has not yet been fully developed, particularly in a stochastic context. The present paper is thus a preliminary qualitative study which could orient a more precise one based on real data and using a large-scale mathematical programming model. The paper is organized as follows: In section 2 the production and investment model is formulated. In contrast with a previous work dealing with a deterministic differential game model [2], we use a discrete-time stochastic programming formalism as in [3]. In section 3 we characterize an efficient exchange program. In section 4 we discuss different pricing schemes realizing the optimal side-payments. In conclusion we discuss the possibility to extend the analysis to a full-fledged power exchange model.

2. A Stochastic Production and Investment Model

Consider two power utilities deserving two regions, $i = 1, 2$. Let T be the (discrete) time horizon. At period $t = 0, \ldots, T$ let $s^t \in S^t$ be the sample value of a random event affecting the electricity demand as well as the power generation costs. The set S^t is supposed to be finite and the sequence of random events is decribed as an event tree. Let $\theta(s^t)$ be the a priori probability of being at event-node s^t at period t. We shall denote by $D(s^t)$ the set of immediate successors of s^t in the event tree.

We represent the electricity demand associated with node s^t in region i by a single number $d_i(s^t)$ (in a detailed model the demand would be better described through a stepwise approximation of the load curve as in [1]). In order to produce electricity, utilities must install some equipment (capacity). Again we assume that the installed capacity in region i at node s^t is represented by a single value $K_i(s^t)$ (in a detailed model the installed capacity would be better described through a vector corresponding to a list of technologies as in [1]). We call $I_i(s^t)$ the additional capacity (investment) installed in region i at node s^t. We assume, as in general capital accumulation models, that the capacity depreciates at a constant rate μ_i. Electricity production at node s^t is denoted by $q_i(s^t)$. This quantity has to meet the demand plus the exports to (or minus the imports from) the other region. Call $e_{ij}(s^t)$ the energy flowing from region i to region j at node s^t; a positive value means an export from i to j, while a negative value corresponds to an import. In order to exchange power the utilities have to install specific equipment (e.g. transmission lines, AC-DC converters, etc). We assume that the installed exchange capacity is represented by a single value $k_i(s^t)$ which depreciates at a rate ν_i and we call $u_i(s^t)$ the additional exchange capacity (investment) installed in region i at node s^t.

The cost at node s^t for utility i has three components

- A production cost $C_P^i(K_i(s^t), q_i(s^t))$ which is a convex function increasing in q_i, decreasing in K_i.

- A production capacity investment cost function $C_I^i(K_i(s^t), I_i(s^t))$ which is also assumed to be convex but increasing in its two arguments.

- An exchange capacity investment cost function $C_u^i(k_i(s^t), u_i(s^t))$ which is also assumed to be convex and increasing in its two arguments. We call C_i^T the expected total cost over the horizon of T periods, discounted at a rate $r = \frac{1-\rho}{\rho}$, $0 < \rho < 1$, which is supposed to be the same for the two regions. The model is specified by the following equations

$$
\begin{aligned}
C_i^T &= \sum_{t=0}^{T} \rho^t \sum_{s^t \in S^t} \theta(s^t)[C_P^i(K_i(s^t), q_i(s^t)) \\
&\qquad + C_I^i(K_i(s^t), I_i(s^t)) \quad + C_u^i(k_i(s^t), u_i(s^t))] && (1) \\
K_i(s^{t+1}) &= (1-\mu_i)K_i(s^t) + I_i(s^t) \quad \forall s^{t+1} \in D(s^t) && (2) \\
K_i(s^0) &= K_i^o \quad \text{given} && (3) \\
k_i(s^{t+1}) &= (1-\nu_i)k_i(s^t) + u_i(s^t) \quad \forall s^{t+1} \in D(s^t) && (4) \\
k_i(s^0) &= k_i^o \quad \text{given} && (5) \\
q_i(s^t) &= d_i(s^t) + e_{ij}(s^t) && (6) \\
q_i(s^t) &\leq K_i(s^t) && (7) \\
|e_{ij}(s^t)| &\leq \min\{k_1(s^t), k_2(s^t)\} && (8) \\
q_i(s^t), &I_i(s^t), \ u_i(s^t) \geq 0 \\
&t = 0, \ldots, T, \quad s^t \in S^t, \quad i = 1, 2
\end{aligned}
$$

where (1) is the cost accumulation equation, (2)-(3) represent the production capacity accumulation process, (4)-(5) represent the exchange capacity accumulation process, (6) is the demand satisfaction constraint, (7) is the capacity constraint, (8) is the exchange capacity constraint.

3. Efficient Exchange Program

3.1 An efficient management program

We assume that the two utilities have the same objective, namely *satisfaction of demand at a minimal cost.* We assume that the time horizon is long enough to eliminate the need to define a bequest function at period $T+1$. An efficient management program of both power systems is then obtained as the solution of the following *stochastic programming problem*:

$$
\begin{aligned}
\min \sum_{i=1}^{2} [\sum_{t=0}^{T} \rho^t \sum_{s^t \in S^t} \theta(s^t)[C_P^i(K_i(s^t), q_i(s^t)) \\
+ C_I^i(K_i(s^t), I_i(s^t)) + C_u^i(k_i(s^t), u_i(s^t))] && (1') \\
\text{s.t.}(1) - (8).
\end{aligned}
$$

Introduce the Lagrangean

$$\mathcal{L} = \sum_{t=0}^{T} \rho^t \{ \sum_{i=1}^{2} [\sum_{s^t \in S^t} \theta(s^t)[(C_P^i(K_i(s^t), q_i(s^t)) + C_I^i(K_i(s^t), I_i(s^t))$$
$$+ C_u^i(k_i(s^t), u_i(s^t))]$$
$$+ \lambda_K^i(s^0)(K_i(s^0) - K_i^0)$$
$$+ \lambda_k^i(s^0)(k_i(s^0) - k_i^0)$$
$$+ \sum_{s^{t+1} \in D(s^t)} [\lambda_K^i(s^{t+1})(K_i(s^{t+1}) - (1-\mu_i)K_i(s^t) - I_i(s^t))$$
$$+ \lambda_k^i(s^{t+1})(k_i(s^{t+1}) - (1-\nu_i)k_i(s^t) - u_i(s^t))]$$
$$- \gamma_i(s^t)(q_i(s^t) - d_i(s^t) - e_{ij}(s^t))$$
$$- \delta_i(s^t)(q_i(s^t) - K_i(s^t))$$
$$- \epsilon_{ij}^1(s^t)(e_{ij}(s^t) - k_1(s^t))$$
$$- \epsilon_{ij}^2(s^t)(e_{ij}(s^t) - k_2(s^t))$$
$$- \omega_i(s^t)q_i(s^t) - \phi_i(s^t)I_i(s^t) - \varphi_i(s^t)u_i(s^t)] \} \tag{9}$$

The sufficient optimality conditions (this is a convex programming problem)are given by

$$\frac{\partial \mathcal{L}}{\partial K_i(s^t)} = \rho^t \{ \theta(s^t)[\frac{\partial C_p^i}{\partial K_i(s^t)} + \frac{\partial C_I^i}{\partial K_i(s^t)}]$$
$$+ \frac{1}{\rho}\lambda_K^i(s^t) - (1-\mu_i) \sum_{s^{t+1} \in D(s^t)} \lambda_K^i(s^{t+1}) + \delta_i(s^t)\} = 0 \tag{10}$$

$$\frac{\partial \mathcal{L}}{\partial k_i(s^t)} = \rho^t \{ \theta(s^t)\frac{\partial C_u^i}{\partial k_i(s^t)} + \frac{1}{\rho}\lambda_k^i(s^t)$$
$$- (1-\nu_i) \sum_{s^{t+1} \in D(s^t)} \lambda_k^i(s^{t+1}) + \epsilon_{ij}^i(s^t)\} = 0 \tag{11}$$

$$\frac{\partial \mathcal{L}}{\partial I_i(s^t)} = \rho^t \{ \theta(s^t)\frac{\partial C_I^i}{\partial I_i(s^t)} - \sum_{s^{t+1} \in D(s^t)} \lambda_K^i(s^{t+1}) - \phi_i(s^t)\} = 0 \tag{12}$$

$$\frac{\partial \mathcal{L}}{\partial u_i(s^t)} = \rho^t \{ \theta(s^t)\frac{\partial C_u^i}{\partial u_i(s^t)} - \sum_{s^{t+1} \in D(s^t)} \lambda_k^i(s^{t+1}) - \varphi_i(s^t)\} = 0 \tag{13}$$

$$\frac{\partial \mathcal{L}}{\partial q_i(s^t)}\} = \rho^t \{ \theta(s^t)\frac{\partial C_P^i}{\partial q_i(s^t)} - \delta_i(s^t)\} = \rho^t \{ \theta(s^t)\frac{\partial C_P^j}{\partial q_j(s^t)} - \delta_j(s^t)\} = 0 \tag{14}$$

$$\frac{\partial \mathcal{L}}{\partial e_{ij}(s^t)} = \rho^t \{ \gamma_i(s^t) - \epsilon_{ij}^1(s^t) - \epsilon_{ij}^2(s^t)\} = 0 \tag{15}$$

$$\forall\, s^t \in S^t, \quad t = 0, \ldots, T \qquad i = 1, 2$$

with the complementary slackness conditions and the transversality conditions

$$\lambda_K^i(s^{T+1}) = \lambda_k^i(s^{T+1}) = 0, \quad i = 1,2 \tag{16}$$

for the marginal values of both capital goods.

3.2 Value decomposition over time

Let us introduce, at each node s^t, the *extended Hamiltonian*

$$\begin{aligned}
\mathcal{H}(s^t) = \sum_{i=1}^{2} \rho^t \{ &\theta(s^t)[C_P^i(K_i(s^t), q_i(s^t)) + C_I^i(K_i(s^t), I_i(s^t)) \\
&+ C_u^i(k_i(s^t), u_i(s^t))] \\
&+ \lambda_K^i(s^t) K_i(s^t) + \lambda_k^i(s^t) k_i(s^t) \\
&- \sum_{s^{t+1} \in D(s^t)} [\lambda_K^i(s^{t+1})((1-\mu_i)K_i(s^t) + I_i(s^t)) + \lambda_k^i(s^{t+1})((1-\nu_i)k_i(s^t) + u_i(s^t))] \\
&- \gamma_i(s^t)(q_i(s^t) - d_i(s^t) - e_{ij}(s^t)) \\
&- \delta_i(s^t)(q_i(s^t) - K_i(s^t)) \\
&- \epsilon_{ij}^1(s^t)(e_{ij}(s^t) - k_1(s^t)) \\
&- \epsilon_{ij}^2(s^t)(e_{ij}(s^t) - k_2(s^t)) \\
&- \omega_i(s^t) q_i(s^t) - \phi_i I_i(s^t) - \varphi_i u_i(s^t) \}
\end{aligned} \tag{17}$$

$$\forall s^t \in S^t, \quad t = 0, \ldots, T.$$

The optimality conditions show that, at each node s^t, the extended Hamiltonian is minimized w.r.t. $K_i(s^t)$, $k_i(s^t)$, $q_i(s^t)$, $I_i(s^t)$, $u_i(s^t)$, $e_{ij}(s^t)$, $\quad i = 1,2, \quad j \neq i$. Denote by $\Gamma_i(s^t)$ the cost incurred at s^t by utility i,

$$\Gamma_i(s^t) = C_P^i(K_i(s^t), q_i(s^t)) + C_I^i(K_i(s^t), I_i(s^t)) + C_u^i(k_i(s^t), u_i(s^t))$$

With the complementary slackness conditions, the minimized extended Hamiltonian is equal to

$$\begin{aligned}
\mathcal{H}(s^t) = \sum_{i=1}^{2} \rho^t \{ \theta(s^t)\Gamma_i(s^t) - &\sum_{s^{t+1} \in D(s^t)} \lambda_K^i(s^{t+1}) K_i(s^{t+1}) + \lambda_K^i(s^t) K_i(s^t) \\
- &\sum_{s^{t+1} \in D(s^t)} \lambda_k^i(s^{t+1}) k_i(s^{t+1}) + \lambda_k^i(s^t) k_i(s^t) \}] \\
&s^t \in S^t, \quad t = 0, \ldots, T.
\end{aligned}$$

This represents the total variation in value, at node s^t, for both utilities (i.e. a net discounted cost taking into account the production and investment costs for the period and also the increased capital value due to current investment). At each node s^t this net cost is minimized with respect to both *control*($q_i(s^t)$, $I_i(s^t)$, $u_i(s^t)$) and *state* ($K_i(s^t)$, $k_i(s^t)$) variables. These conditions show that the efficient exchange policy is obtained by operating the two utilities as if they were a single production system, the interconnecting capacity being treated as just another kind of production capital.

4. Sharing the Gains from Cooperation

4.1 The bargaining solution for two utilities

When they exchange power, the two utilities are in a game situation. Usually the selling utility has no other opportunity offered for its excess power, and the buying utility has no other potential supplier. The problem thus reduces to the definition of a fair sharing of the *dividends of cooperation*. This is a classical cooperative game problem. Since we assumed transferable utility functions (negative of the total cost), the definition of a fair outcome is relatively simple. If one represents on two axes the utility function of each player, there are two points called respectively SQ for *status quo* and ES for *efficient solution* (see Fig.1). The coordinates of these points correspond to the negative of the expected costs accrued to each player over the whole planning horizon ($t = 0, \ldots, T$). The SQ point corresponds to the situation where the two players do not cooperate. In the present case, no cooperation means no exchange. Since a player can always compensate the other one through a *side-payment*, the set of efficient outcomes is represented by a line of slope -1 passing through the point ES. Now, if one draws two lines, parallel to the axes and passing though the SQ point, they intersect the set of efficient outcomes at two points denoted $P1$ and $P2$ respectively. At point $P1$ all the dividends of cooperation go to player 1 and player 2 has the same total cost as in SQ. Conversely, at point $P2$ player 2 reaps all the dividends of cooperation. Any point on the segment ($P1$-$P2$) is *acceptable* by both players. Of course a fair division of the dividends of cooperation corresponds to the middle point P^* of this segment. This fair outcome is reached by having the two power systems managed in a jointly efficient maner and by divising an appropriate side-payment to move the outcome from ES to P^*.

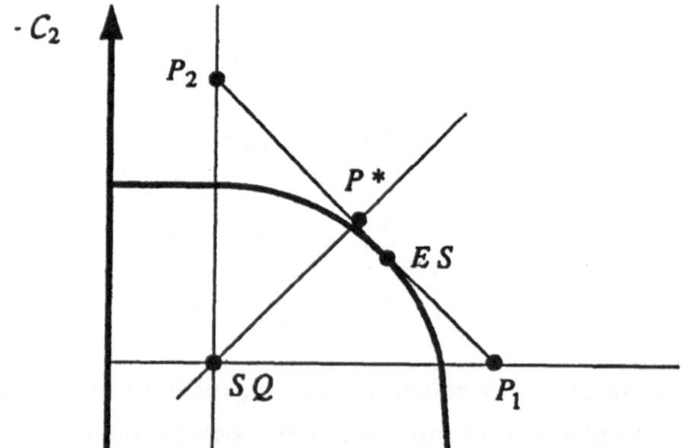

Figure 1: Nash-Harsanyi bargaining solution

4.2 Pricing schemes to implement a side-payment

To implement this side-payment the utilities define a pricing scheme for the exchanged power. Since the side-payment is effectuated on a planning horizon $[0,T]$ there is a lot of degrees of freedom for its design. To be more precise let us define the cost sequences

$$\Gamma_i^*(s^t) = C_P^i(K_i^*(s^t), q_i^*(s^t)) + C_I^i(K_i^*(s^t), I_i^*(s^t)) + C_u^i(k_i^*(s^t), u_i^*(s^t)) \tag{18}$$

$$\Gamma_i^o(s^t) = C_P^i(K_i^o(s^t), q_i^o(s^t)) + C_I^i(K_i^o(s^t), I_i^o(s^t)) \tag{19}$$

where * refers to the efficient solution (ES) and ^{o} to the status quo, no exchange, solution (SQ). Notice that, in the SQ solution, there is no cost associated with investment in exchange capacities. We can then define the total *dividends of cooperation* $\mathcal{D}^*$ as the difference between the joint costs without and with efficient exchanges

$$\mathcal{D}^* = \sum_{t=0}^{T} \rho^t \sum_{s^t \in S^t} \theta(s^t)\{\Gamma_i^o(s^t) - \Gamma_i^*(s^t) + \Gamma_j^o(s^t) - \Gamma_j^*(s^t)\}. \tag{20}$$

If the price set for traded power at node s^t is $p(s^t)$, then the expected side-payment of player j to player i is defined as

$$\sum_{t=0}^{T} \rho^t \sum_{s^t \in S^t} \theta(s^t) p(s^t) e_{ij}^*(s^t). \tag{21}$$

In order for the side-payment to realize the fair outcome P^* it must be associated with a pricing scheme $p^*(s^t)$, $t = 0, \ldots T$ that satisfies the following equality

$$\sum_{t=0}^{T} \rho^t \sum_{s^t \in S^t} \theta(s^t) p^*(s^t) e_{ij}^*(s^t) = \frac{1}{2} \sum_{t=0}^{T} \rho^t \sum_{s^t \in S^t} \theta(s^t)\{\Gamma_i^*(s^t) - \Gamma_i^o(s^t) + \Gamma_j^o(s^t) - \Gamma_j^*(s^t)\}. \tag{22}$$

It means that the side-payment is equal to half the expected discounted cost increase for player i plus half the expected discounted cost decrease for player j.

4.2.1 A fixed price for the energy exchanged

A *constant price* achieving this side-payment is thus defined by

$$p^* = \frac{\frac{1}{2} \sum_{t=0}^{T} \rho^t \sum_{s^t \in S^t} \theta(s^t)\{\Gamma_i^*(s^t) - \Gamma_i^o(s^t) + \Gamma_j^o(s^t) - \Gamma_j^*(s^t)\}}{\sum_{t=0}^{T} \rho^t \sum_{s^t \in S^t} \theta(s^t) e_{ij}^*(s^t)}. \tag{23}$$

This price is the ratio of the required expected side-payment over the expected discounted sum of traded energy.

4.2.2 A decomposition over time of the side-payment

A possible, node dependent, pricing scheme would be defined by

$$p^*(s^t) = \frac{\frac{1}{2}\{\Gamma_i^*(s^t) - \Gamma_i^o(s^t) + \Gamma_j^o(s^t) - \Gamma_j^*(s^t)\}}{e_{ij}^*(s^t)}, \quad s^t \in S^t \quad t = 0, \ldots, T, \tag{24}$$

provided that $e^*_{ij}(s^t) \neq 0$ for all s^t. This price schedule decomposes the expected side-payments into a sequence such that, at each node, the splitting of cost increases and decreases is performed. Although this schedule achieves the required expected side-payment over the complete planning horizon, it might induce negative dividend

$$\mathcal{D}^*(s^t) = \Gamma_i^o(s^t) - \Gamma_i^*(s^t) + \Gamma_j^o(s^t) - \Gamma_j^*(s^t), \tag{25}$$

particularly for the early periods t. This could induce a very erratic price schedule.

4.2.3 A decomposition based on the extended Hamiltonians

Another possibility would be to base the export pricing at period t on the net value for each player, as represented by the extended Hamiltonian. For this we introduce the extended costs

$$\begin{aligned}\mathcal{X}_i^*(s^t) = \Gamma_i^*(s^t) + \frac{1}{\theta(s^t)}\{- &\sum_{s^{t+1}\in D(s^t)} \lambda_K^i(s^{t+1})K_i^*(s^{t+1}) + \lambda_K^i(s^t)K_i^*(s^t) \\ - &\sum_{s^{t+1}\in D(s^t)} \lambda_k^i(s^{t+1})k_i^*(s^{t+1}) + \lambda_k^i(s^t)k_i^*(s^t)\}\rho^{-t}\end{aligned} \tag{26}$$

$$\begin{aligned}\mathcal{X}_i^o(s^t) = \Gamma_i^o(s^t) + \frac{1}{\theta(s^t)}\{- &\sum_{s^{t+1}\in D(s^t)} \lambda_K^i(s^{t+1})K_i^o(s^{t+1}) + \lambda_K^i(s^t)K_i^o(s^t) \\ - &\sum_{s^{t+1}\in D(s^t)} \lambda_k^i(s^{t+1})k_i^o(s^{t+1}) - \lambda_k^i(s^t)k_i^o(s^t)\}\rho^{-t}\end{aligned} \tag{27}$$

$$s^t \in S^t \quad t = 0,\dots,T,$$

where the capital accumulation paths are either indexed by ^{o} or * if they correspond to the status quo or the efficient solution. In this expression the adjoint variables $\lambda_K^i(s^t)$ and $\lambda_k^i(s^t)$ are those defined by the optimality conditions (9)–(16). Due to the maximality property of the extended Hamiltonian, at each node s^t, one has positive dividends $\mathcal{G}^*(s^t)$, expressed in net values, and given by

$$\mathcal{G}^*(s^t) = \mathcal{X}_i^o(s^t) - \mathcal{X}_i^*(s^t) + \mathcal{X}_j^o(s^t) - \mathcal{X}_j^*(s^t) \geq 0, \tag{28}$$

while, due to the initial conditions (3) and (5) and transversality conditions (16) one still has

$$\mathcal{D}^* = \sum_{t=0}^{T} \rho^t \sum_{s^t\in S^t} \theta(s^t)\mathcal{G}^*(s^t). \tag{29}$$

Therefore one can define a price schedule which, in association with the amounts of energy traded, will realize the desired side-payment

$$\pi^*(s^t) = \frac{\frac{1}{2}\{\mathcal{X}_i^*(s^t) - \mathcal{X}_i^o(s^t) + \mathcal{X}_j^o(s^t) - \mathcal{X}_j^*(s^t)\}}{e^*_{ij}(s^t)}, \quad s^t \in S^t \quad t = 0,\dots,T. \tag{30}$$

In terms of these extended costs, the dividends to share at any period are always positive. This does not guarantee however that the price $\pi^*(s^t)$ will always remain positive. A negative price means that the exporting utility is paying the importing one.

4.2.4 A general class of power exchange pricing schemes

When we base a power exchange pricing decomposition on the extended Hamiltonians we exploit two properties: (a) the fact that the fixed initial conditions for the state variables and the zero terminal values of the shadow prices $\lambda^i_K(\cdot)$ and $\lambda^i_k(\cdot)$ imply that (22) holds; (b) since the extended Hamiltonian is minimized w.r.t. state and control for the cooperative solution the dividend is positive. The positiveness of the dividend is a good property, however, as it can be easily checked on numerical examples, this does not prevent the occurence of negative prices at some periods. Instead of using the shadow prices provided by the Lagrange multipliers associated with the efficient management solution we could use any weighting of the capital stock variables in a definition of extended costs as in (26), (27), provided that these implicit shadow prices be zero at period $T+1$. For example we could consider the pricing scheme where, in (26) and (27), the shadow prices $\lambda^i_K(\cdot)$ would be set at a zero value while the shadow prices $\lambda^i_k(\cdot)$ would correspond to the optimal Lagrange multiplier values. It can be shown that this scheme leads also to a positive dividend to share at each period but with a different exchange pricing decomposition. The positive dividend results again from the minimization of the extended Hamiltonian since the state variables K and k are completely decoupled.

4.2.5 Piecewise constant exchange prices

Finally one may define *piecewise constant* exchange prices by partitionning the whole time horizon $[0,T]$ into L subintervals $[T^0,T^1]$, $[T^1,T^2]$, ..., $[T^{L-1},T^L]$, where $T^0=0$, $T^L=T$, and by defining on the ℓ-th interval the constant exchange price

$$p^*_\ell = \frac{\frac{1}{2}\sum_{t=T^{\ell-1}}^{T^\ell-1}\rho^t\sum_{s^t\in S^t}\theta(s^t)\{\Gamma^*_i(s^t)-\Gamma^o_i(s^t)+\Gamma^o_j(s^t)-\Gamma^*_j(s^t)\}}{\sum_{t=T^{\ell-1}}^{T^\ell-1}\rho^t\sum_{s^t\in S^t}\theta(s^t)e^*_{ij}(s^t)}, \tag{31}$$

or alternatively

$$\pi^*_\ell = \frac{\frac{1}{2}\sum_{t=T^{\ell-1}}^{T^\ell-1}\rho^t\sum_{s^t\in S^t}\theta(s^t)\{\mathcal{X}^*_i(s^t)-\mathcal{X}^o_i(s^t)+\mathcal{X}^o_j(s^t)-\mathcal{X}^*_j(s^t)\}}{\sum_{t=T^{\ell-1}}^{T^\ell-1}\rho^t\sum_{s^t\in S^t}\theta(s^t)e^*_{ij}(s^t)}. \tag{32}$$

The length of the intervals can be defined in such a manner that one has always positive prices.

5. Conclusion

This qualitative analysis of power exchange between interconnected utilities could be extended in different directions. Firstly one has to make some numerical simulations to study the behavior of the different pricing schemes proposed. Secondly, the game considered in this paper can be easily extended to more than two utilities by defining a game in characteristic function form and using any classical solution (Shapley value, Core, etc.) in order to share the dividends of cooperation among the interconnected utilities. (Notice, however that the task of designing a price mechanism would be more complicated than in the two-utility case and thus further investigation is needed.) Finally, and more importantly, the model could be implemented by linking large scale models (e.g. as in [1]) used to describe the energy systems in different regions. In such models, one has a full description of the technologies used, or to be introduced at a later date, to satisfy the required expected demand.

References

[1] Berger, C., Dubois, R., Haurie, A., Lessard, E., Loulou, R. "Assessing the Dividends of Power Exchange Between Quebec and New York State: A Systems Analysis Approach", *International Journal of Energy Research*, 14, (1990), 253–273.

[2] Haurie, A., Zaccour, G. "A Differential Game Model of Power Exchange Between Interconnected Utilities", *Proceedings of the 25th IEEE Conference on Decision and Control*, 1, Athens, (1986).

[3] Haurie, A., Zaccour, G., Smeers, Y. "A Stochastic Programming Equilibrium Approach for Dynamic Oligopolistic Markets", *JOTA*, 66(2), August (1990), 243–253.

Computation of cooperative and Stackelberg solutions when players are described by linear programs

R. Loulou, G. Savard, E. Lessard
GERAD, Montreal, Canada

1. Introduction

The main objective of this article is to devise computationally tractable methods for the composition of several process models. Model composition may mean different things in different contexts: for instance, one may want to optimize a single global objective common to all the models (global optimization), or alternatively to simulate a well-defined game between two or more players, each player being represented by one process model. For clarity, the methodology will be developed in the case of two models (players) only; generalization to several models would proceed in similar fashion. Each process model is a linear programming representation of a region or a sector of a global market. Although this work is presented in the context of energy markets, the same approach is entirely transposable to other economic activities with only minor changes in the vocabulary. The two models are linked together via exchange variables representing the trading of energy forms between the two players, at prices which may either be marginal costs or not, depending on the game to be simulated. Throughout this paper, it is implicit that each player's model is sufficiently large so as to make impractical the running of the two models as a single large model encompassing the two players, even if this made theoretical sense.

In the case of the cooperative optimization of a multi-component model, our work amounts to the decomposition of a large scale mathematical program (equivalently, the computation of a cooperative equilibrium), whereas in the case of non cooperative behaviour by the players, the method computes an equilibrium. In the latter case, we shall simulate a leader-follower Stackelberg equilibrium to be described later.

These two types of equilibria have received much attention in the literature. Cooperative equilibria correspond to optimization, of a joint objective. When the joint program has large scale, decomposition methods may be required; a large literature on this subject is available, see [1] [2] for decomposition methods in energy modelling. Static Stackelberg equilibria are directly related to the resolution of bilevel programming problems, a relatively recent field of investigation, see [3], [4], [5] for the basic formulations and some proposed solution methods, and [6] for an energy application of bilevel programming. There are no standard algorithms for bilevel programs of large dimensions, even in the linear case.

All approaches discussed in this paper proceed from the same general idea, viz. the sequential resolution of the individual players programs adequately modified to converge to the desired type of equilibrium. In the case of a cooperative equilibrium between suppliers and consumers, our approach can be looked at as an improvement of the simple Cob-Web algorithm (CWA), or alternatively as a generalization of the PIES method [7] to the case when consumers are not described by an explicit demand function, but rather by an implicit one. In the case of the Leader-Follower Stackelberg equilibrium, our algorithm seems to have no direct ancestor.

In section 2, we formally present the structure of the Process Models (Linear Programs) and of their inter-relationships, and we define the two types of player interactions considered in this research. In section 3 we describe and justifiy the Improved Cob-Web Algorithm (ICWA) to treat the cooperative case. In section 4 we do the same for the case of a Stackelberg equilibrium. Preliminary results are briefly discussed in section 5, which concludes the article.

2. Statement of the problems

For notational convenience, we assume that player 1 is the seller, and player 2 the buyer of all energy forms exchanged. In reality, each player may sell and/or buy some energy forms, and the special case presented here is easily extended to the general situation (at least in the cooperative case), at the expense of complicating the notation.

The following two L.P.'s will have meaning only after the mechanism for choosing the values of s and p is specified.

Seller's L.P.

$$\begin{aligned} &\min c_1 x - ps \\ \text{s.t.}\quad & Ax - s = b_1 \\ & x \in L_1 \\ & s \geq 0 \end{aligned} \tag{P1}$$

Buyer's L.P.

$$\begin{aligned} &\min c_2 y + ps \\ \text{s.t.}\quad & By + s = b_2 \\ & y \in L_2 \\ & s \geq 0 \end{aligned} \tag{P2}$$

where:

x (resp.y) is the vector of activity variables of the seller (resp. the buyer), including possibly investment, maintenance, and operations activities.

s is the vector of energy forms traded between the two players.

p is the vector of prices paid by the buyer to the seller.

A (resp. B) is the technological matrix of the seller (resp. the buyer) that links the trade variables s with the other activities of the player.

E is a diagonal matrix of coefficients ($0 \leq e_{ij} \leq 1$) representing the transmission losses for the various energy forms traded.

b_1, b_2 are the demand vectors for players 1 and player 2.

L_1, L_2 are the polyhedrons defined by other linear constraints to be satisfied by x and y, respectively.

Cooperative Equilibrium (Global Optimization)

If the players cooperate, they simply optimize their combined cost, in which case the price vector p no longer appears in the global L.P. below

$$\begin{aligned} \min_{x,y,s}\ & c_1x + c_2y \\ \text{s.t.}\quad & Ax - s = b_1 \\ & By + Es = b_2 \\ & x \in L_1 \\ & y \in L_2 \\ & s \geq 0 \end{aligned} \qquad (PC)$$

It is a well-known result of LP duality theory that if (x^*, y^*, s^*) is an optimal solution of problem (PC), and if p^* is the vector of the optimal dual variables attached to constraints $Ax - s = b_1$, then (x^*, s^*) is optimal for problem $(P1)$ with p constrained to be equal to p^*, and (y^*, s^*) is optimal for problem $(P2)$ with p constrained to be equal to p^*. This result has suggested the intuitively appealing simple cob-web algorithm (see e.g.[2]):

Cob-Web algorithm CW

0. Choose an initial price vector p^0, set $k = 0$.
1. Solve $(P2)$ with $p = p^k$. The solution is denoted y^k, s^k.
2. Solve $(P1)$ without the component ps in the objective, and with $s = s^k$. Let x^k be the optimal solution, and p^k=optimal dual variables of constraints $Ax - s = b_1$. If (x^{k+1}, y^{k+1}) is sufficiently close to (x^k, y^k), STOP; else GO TO 1.

Algorithm CW amounts to a succession of LP resolutions. It is however frequent to observe non-convergence, with a tendency for the successive x^k and y^k vectors to oscillate. The cause of such oscillations is the discontinuity of the shadow prices in Linear Programming, which are constant on intervals and have discontinuities at either ends of the intervals.

In the improved cob-web described in section 3, additional information is passed from step 1 to step 2; a local approximation of the demand function around the current value of the vector s^k is computed at step 1, and transfered to the seller. In this fashion, the seller will no longer be induced to "over-react" to the quantities s^k demanded by the buyer. Such a scheme is called a local equilibrium relaxation-decomposition method by Hogan and Weyant [1]. As observed earlier, this case is easily extended to the one where both players are sellers and buyers.

Stackelberg Equilibrium

If the seller acts as a leader, it sets a price vector which is not necessarily on its marginal cost curve, whereas the buyer, acting as a follower, reacts to this price vector by purchasing quantities that are optimal for him given the price selected by the seller. Of course, the leader should anticipate the follower's reaction when selecting a price vector. The formulation of such a situation is in the form of a bi-level optimization program (PS), as follows:

$$\begin{aligned}
&\min_{x,p} c_1x - ps \\
&\text{s.t.} \quad Ax - s = b_1 \\
&\qquad x \in L_1 \\
&\qquad \text{and s solution of} \\
&\qquad \min_{y,s} c_2y + ps \\
&\qquad\quad \text{s.t.} \quad By - Es = b_2 \\
&\qquad\qquad y \in L_2 \\
&\qquad\qquad s \geq 0
\end{aligned} \tag{PS}$$

Figure 1 describes schematically the cooperative equilibrium (point C), and the Stackelberg equilibrium (point S), when there is only one commodity exchanged by the two players. The ascending curve represents the seller's marginal production cost as a function of s, whereas the descending curve is the buyer's marginal cost as a function of s. The jointly optimal level of exchange is located at the intersection of the two marginal cost curves, whereas the Stackelberg solution maximizes the dashed area (representing the seller's profit, i.e. revenue minus cost). Note that in reality the two curves are not smooth but rather step functions, as always in linear programming.

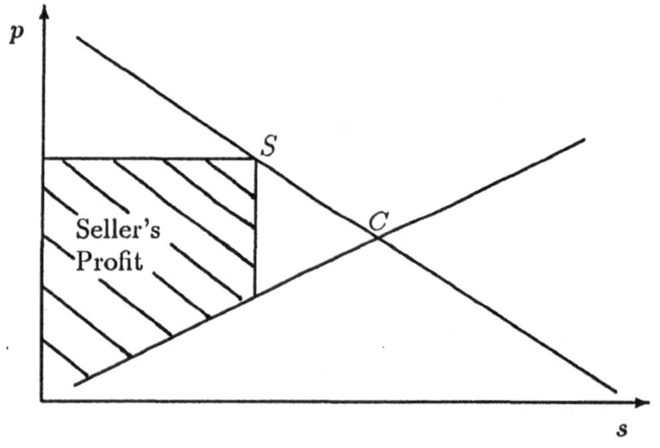

Figure 1: Cooperative and Stackelberg equilibria

3. Decomposition via the Improved Cob-Web algorithm

In this section, we describe an algorithm for the global optimization of (PC) via a succession of resolutions of modified versions of $(P1)$ and $(P2)$. The modification amounts to sending more information than just the quantity s^k, from step 1 to step 2 of the cob-web. The additional information consists of an approximation of the buyer's implicit inverse demand function $p = p(s)$.

An exact computation of $p(s)$ is very cumbersome. Instead, we will approximate $p(s)$ by a separable function $p'(s) = \sum p_i(s_i), i = 1, ..., q$. Figure 2 indicates the typical shape of $p_i(s_i)$ (dashed line). Each function $p_i(s_i)$ is defined as the optimal price of commodity i purchased, as a function of its demanded quantity, assuming all other quantities remain constant. The computation of such an approximation involves the resolution of q separate parametric linear programs based on $(P2)$, where the constraint on exchange is successively replaced by $By = (b_2 - s^k + l_i e_i), i = 1, ..., q$, where e_i is the vector with only the ith coordinate equal to 1, and all others equal to zero, and l_i sweeps the range $(-\infty, +\infty)$.

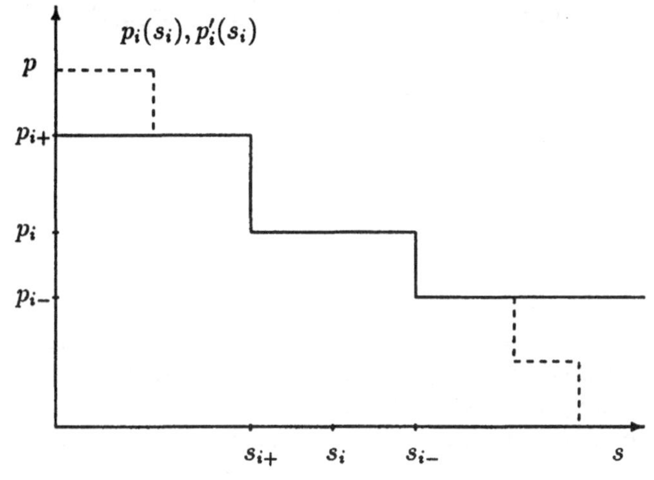

Figure 2: Approximation of the inverse demand function

The computation of $p'(s)$ is still too costly, and probably a waste of effort. Instead, we will be content to further approximate each $p_i(s_i)$ by a three-step function that coincides with $p_i(s_i)$ locally. This new approximation is denoted $p_i'(s_i)$, and is easily computed by sensitivity analysis; it is pictured in figure 2 (bold line). Note that $p_i'(s_i)$ and $p_i(s_i)$ coincide at least on the interval (s_{i+}^k, s_{i-}^k). It is known from the theory of L.P. duality that this last approximation is exact in some neighbourhood of s^k, as long as s^k lies in the interior of the interval, i.e. if program $(P2)$ is not degenerate.

The proposed approximation amounts to neglecting all cross elasticities of the demand function $p(s)$, and to retain only the own price elasticities; such an approach is similar to the one used in the PIES algorithm in the case of continuous demand functions, see [7]. Therefore, the present approach can be considered a generalization of PIES to the case where the demand function is (implicitly) defined by a Linear Program rather than analytically.

Improved Cob-Web Algorithm (ICW)

0. $s = s^0; k = 0;$

1a. Solve $(P2)$ with $s = s^k$ (note that we now omit the term ps in the objective); let y^k be an optimal solution.

1b. Using a local parametric analysis of the constraint on exchange of $(P2)$, compute for each $i = 1,..q$, the range of s_i^k such that the current basis remains optimal for $(P2)$. Let (s_{i+}^k, s_{i-}^k) be this range, with $s_{i+}^k \leq s_i^k \leq s_{i-}^k$. Furthermore the same parametric analysis provides the value p_{i+}^k and p_{i-}^k, where:

   p_{i+}^k is the optimal price p_i when s_i is just below s_{i+}^k

   p_{i-}^k is the optimal price p_i when s_i is just above s_{i-}^k.

2. Compute an equilibrium between the complete seller's program and the approximation of the buyer's demand function described at step 1b. To do so, solve the following linear program:

$$\min c_1 x - \sum_i (p_{i+}^k u_{1i} + p_i^k u_{2i} + p_{i-}^k u_{3i})$$

$$\begin{aligned} \text{s.t.} \quad & Ax - (u_1 + u_2 + u_3) = b_1 \\ & 0 \leq u_1 \leq s_+^k \\ & 0 \leq u_2 \leq s_-^k - s_+^k \\ & 0 \leq u_3 \\ & x \in L_1 \end{aligned} \qquad (LEP)$$

Program (LEP) (for local equilibrium program) computes the local equilibrium by minimizing the seller's cost minus the buyer's surplus. It has well been established in the literature that such a scheme indeed computes an equilibrium (see e.g. [7]). Figure 3 shows graphically why the maximization of the objective function of (LEP) corresponds to the intersection of the supply curve and the approximate demand curves.

3. Stopping test: let s^{k+1} be the local equilibrium computed at step 2. If $s^{k+1} = s^k$, then STOP, else set $k = k + 1$, go to step 1a.

Note: At step 1b, if $(P2)$ is degenerate at the current point s_i^k, we may have $s_{i+}^k = s_i^k$ or $s_i^k = s_{i-}^k$, in which case, the approximate function $p_i'(s_i)$ will be a two-step function. The description of the algorithm remains valid.

4. Computation of a static Stackelberg Equilibrium

Algorithm ICW above needs very little modification to solve the Stackelberg case. The only change required is in the seller's program at step 2, which must now maximize the seller's profit rather than the net social profit. The algorithm is as follows:

Stackelberg Algorithm (SA)

0. $s = s^0; k = 0;$

1a. Solve $(P2)$ with $s = s^k$; let y^k be an optimal solution.

1b. Using a local parametric analysis of the constraint on exchange of $(P2)$, compute for each $i = 1,..q$, the range of s_i^k such that the current basis remains optimal for $(P2)$. Let (s_{i+}^k, s_{i-}^k) be this range, with $s_{i+}^k \leq s_i^k \leq s_{i-}^k$. Furthermore the same parametric analysis provides the value p_{i+}^k and p_{i-}^k, where:

   p_{i+}^k is the optimal price p_i when s_i is just below s_{i+}^k

   p_{i-}^k is the optimal price p_i when s_i is just above s_{i-}^k.

2. Compute the seller's net profit assuming the approximation of the buyer's demand function described at step 1b. To do so, solve the following mixed-integer linear program:

$$\begin{aligned}
\max \ & \sum_i (p_{i+}^k w_{1i} + p_i^k w_{2i} + p_{i-}^k w_{3i}) - c_1 x \\
\text{s.t.} \ & Ax - (w_1 + w_2 + w_3) = b_1 \\
& z_{1i} + z_{2i} + z_{3i} = 1, \quad \forall i \\
& 0 \le w_{1i} \le z_{1i} s_{i+}^k \quad \forall i \\
& 0 \le w_{2i} \le z_{2i} s_i^k \quad \forall i \\
& 0 \le w_{3i} \le z_{3i} s_{i-}^k \quad \forall i \\
& z_{ji} \in \{0,1\}, \quad \forall i,j \\
& x \in L_1
\end{aligned} \qquad (LSEP)$$

Program ($LSEP$) (for local Stackelberg equilibrium program) computes the local equilibrium by maximizing the seller's profit.

3. Stopping test: let s^{k+1} be the local equilibrium computed at step 2. If $s^{k+1} = s^k$, then STOP, else set $k = k + 1$, go to step 1a.

Note: Program (LSEP) contains $3q$ binary variables, where q is the number of exchange variables. For many practical situations, this number remains reasonably small, even though the sizes of matrices A and B may be large. It should thus be possible to apply the algorithm to realistic instances.

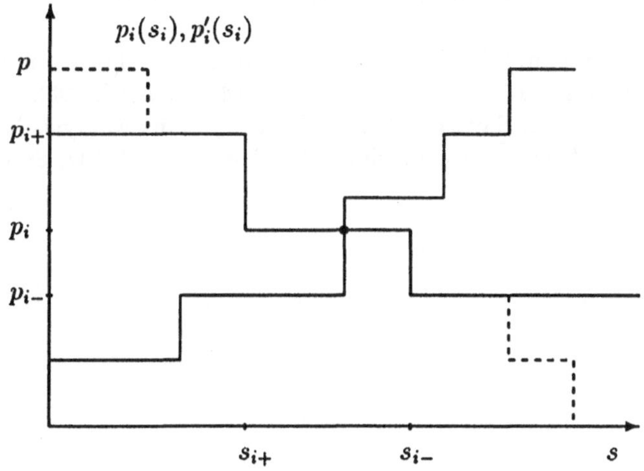

Figure 3: An approximate equilibrium

5. Preliminary results and conclusions

We have not yet conducted extensive experiments on our algorithms on realistic problems. So far, algorithm ICW has been used on a model of electricity exchange in New England, where the seller is the pulp and paper industry (selling its cogenerated power surplus), and the buyer is the electric utility. Reference [8] describes the two systems in detail. Matrices A and B have size 100x100 approximately, and the five exchange variables represent electricity at different time divisions: Summer-Day, Summer-Night, Winter-Day, Winter-Night, and Peak electricity.

Starting with almost any initial exchange vector s^0, the correct cooperative solution is reached in 3 or 4 iterations of algorithm ICW. This is a very encouraging result which should be attempted on larger problems in the future.

6. References

[1] W.W. Hogan and J.P. Weyant, "Methods and algorithms for energy model composition: Optimization in a network of process models", in: Lev (ed.), Energy Models and Studies, North-Holland, Amsterdam, 1983.

[2] F.H. Murphy, "Equation Partitioning Techniques for Solving Partial Equilibrium Models", European Journal of Operationnal Research, vol. 32, pp. 380-392, 1987.

[3] E. Aiyoshi and K. Shimizu, "A solution method for the static constrained Stackelberg problem via penalty method", IEEE Transactions on Automatic Control, vol. AC-29, no. 12, pp. 1111-1114, December 1984.

[4] J.F. Bard and J.E. Falk, "An Explicit Solution to the Multi-Level Programming Problem", Computers and Operations Research, vol. 9, no. 1, pp. 77-100, 1982.

[5] P. Hansen, B. Jaumard and G. Savard, "New Branching and Bounding Rules for Linear Bilevel Programming", Cahiers du GERAD, March 1989.

[6] P.B. Luh, Y.-C. Ho and R. Muralidharan, "Load adaptive pricing: an emerging tool for electric utilities", IEEE Transactions on Automatic Control, vol. AC-27, no. 2, pp. 320-329, April 1982.

[7] B.H. Ahn and W.W. Hogan, "On Convergence of the PIES Algorithm for Computing Equilibria", Operations Research, vol. 30, no. 2, March-April 1982.

[8] A. Haurie, R. Loulou and G. Savard, "A Two-Level Systems Analysis Model of Power Cogeneration Under Asymmetric Pricing", Cahiers du GERAD, September 1989.

Lecture Notes in Control and Information Sciences

Edited by M. Thoma and A. Wyner

Lecture Notes in Control and Information Sciences

Edited by M. Thoma and A. Wyner

Vol. 137: S. L. Shah, G. Dumont (Eds.)
Adaptive Control Strategies for Industrial Use
Proceedings of a Workshop
Kananaskis, Canada, 1988
VI, 360 pages. 1989

Vol. 138: D. C. McFarlane, K. Glover
Robust Controller Design Using Normalized Coprime Factor Plant Descriptions
X, 206 pages. 1990

Vol. 139: V. Hayward, O. Khatib (Eds.)
Experimental Robotics I
The First International Symposium
Montreal, June 19-21, 1989
XVII, 613 pages. 1990

Vol. 140: Z. Gajic, D. Petkovski, X. Shen
Singularly Perturbed and Weakly Coupled Linear Control Systems
A Recursive Approach
VII, 202 pages. 1990

Vol. 141: S. Gutman
Root Clustering in Parameter Space
VIII, 153 pages. 1990

Vol. 142: A. N. Gündeş, C. A. Desoer
Algebraic Theory of Linear Feedback Systems with Full and Decentralized Compensators
V, 176 pages. 1990

Vol. 143: H.-J. Sebastian, K. Tammer (Eds.)
System Modelling and Optimization
Proceedings of the 14th IFIP Conference
Leipzig, GDR, July 3-7, 1989
X, 960 pages. 1990

Vol. 144: A. Bensoussan, J. L. Lions (Eds.)
Analysis and Optimization of Systems
Proceedings of the 9th International Conference, Antibes, June 12-15, 1990
XII, 992 pages. 1990

Vol. 145: M. B. Subrahmanyam
Optimal Control with a Worst-Case Performance Criterion and Applications
X, 133 pages. 1990

Vol. 146: D. Mustafa, K. Glover
Minimum Entropy H_∞ Control
IX, 144 pages. 1990

Vol. 147: J. P. Zolésio (Ed.)
Stabilization of Flexible Structures
Third Working Conference, Montpellier, France, January 1989
V, 327 pages, 1990

Vol. 148: In preparation

Vol. 149: K.H. Hoffmann, W. Krabs (Eds.)
Optimal Control of Partial Differential Equations
Proceedings of the IFIP WG 7.2 International Conference, Irsee, April 9-12, 1990
VI, 245 pages. 1991

Vol. 150: L. C. G. J. M. Habets
Robust Stabilization in the Gap-topology
IX, 126 pages. 1991

Vol. 151: J. M. Skowronski, H. Flashner, R. S. Guttalu (Eds.)
Mechanics and Control
Proceedings of the 3rd Workshop on Control Mechanics in Honor of the 65th Birthday of George Leitmann
January 22-24, 1990, University of Southern California
IV, 497 pages. 1991

Vol. 152: J. D. Aplevich
Implicit Linear Systems
XI, 176 pages. 1991

Vol. 153: O. Hájek
Control Theory in the Plane
X, 272 pages. 1991

Vol. 154: A. Kurzhanski, I. Lasiecka (Eds.)
Modelling and Inverse Problems of Control for Distributed Parameter Systems
Proceedings of IFIP (W.G. 7.2)-IIASA Conference
Laxenburg, Austria, July 24-28, 1989
VII, 179 pages. 1991

Vol. 155: M. Bouvet, G. Bienvenu (Eds.)
High-Resolution Methods in Underwater Acoustics
V, 249 pages. 1991

Lecture Notes in Control and Information Sciences

Edited by M. Thoma and A. Wyner

Vol. 156: R. P. Hämäläinen, H. K. Ehtamo (Eds.)
Differential Games –
Developments in Modelling and Computation
Proceedings of the Fourth International Symposium
on Differential Games and Applications
August 9-10, 1990, Helsinki University of Technology,
Finland
XIII, 292 pages. 1991